Seung S. Park
Nov 27, 1979

POWER SYSTEM ANALYSIS

POWER SYSTEM ANALYSIS

CHARLES A. GROSS
Auburn University

JOHN WILEY & SONS
New York • Santa Barbara • Chichester • Brisbane • Toronto

Library of Congress Cataloging in Publication Data:

Gross, Charles A
    Power system analysis.

    Includes index.
    1. Electric power systems.   2. System analysis.
I. Title.
TK1005.G76   621.31   78-8631
ISBN 0-471-01899-6

Printed in the United States of America

10 9 8 7 6 5 4 3 2 1

To David, Dodie, Gramp, Larry,
Michael, Mom, Nanny, and Robert

PREFACE

This book is intended to serve as a textbook for an introductory course in power system analysis. It is estimated that the material presented can be easily covered in six semester credit hours. Prerequisite topics are sinusoidal steady state circuit theory, basic matrix notations and operations, and basic computer programming; a course in machines is desirable but not essential. The student should have at least a junior standing.

It is assumed that the book will be used at some schools in courses required of all electrical engineering students and so the topics included were selected because of their "double-valued quality"—their importance not only to power specialty students, but their possession of general academic merit. This is particularly true of the first half of the book. Thus, the study of symmetrical components is important not only because of its application to power systems, but also because of its contribution to an understanding of degrees of freedom and linear transformations. This does not mean the material is "watered down", but that topics of central importance are developed with an attempt to avoid being sidetracked with necessary, but minor, details.

The introductory chapter was written to demonstrate the basic importance of electrical energy conversion and delivery systems in a technological society. A short history of the electrical power industry is provided. The SI unit system is presented and electrical power transmission is discussed. Consideration is given to available sources of energy.

The remainder of the book is divided into five basic parts. Chapter 2 presents circuit theory as applied to power systems. This material is expected to be part review, and part new material for most students, and features an early emphasis on symmetrical components. The student is encouraged to think of the balanced case as a special case of the general unbalanced situation. Matrix methods are used to keep the equations organized, but are developed in such a way that physical understanding is not compromised. Chapter 3 explains the basics of power system representation and the per unit system. Chapters 4, 5, and 6 develop traditional mathematical models for the transmission line, the transformer, and the synchronous machine. The approach used compromises completeness with the need to develop simplified system models. The book is not a machines text, but harmonizes with rigorous and thorough work in machines.

Chapter 7 presents a discussion of the power flow problem. The objective is to give students a clear understanding of the problem, its solution, its application, and the use of the computer. Students should finish with the confidence that they could write their own working power flow programs.

Chapters 8 through 11 deal with fault analysis and system protection. The four basic fault types are discussed, with symmetrical components used as the analytical approach. The fault analysis problem is formulated for computer solution. Considerable emphasis and detail is given to the system protection problem. Breakers, relays, and instrument transformers are covered. Distance relaying is explained.

Chapter 12 of the book deals with the general problem of power system stability. Transient, and steady state stability concepts are explained. The single machine problem is discussed in some detail, with governor and exciter effects included. The equal area method is applied to the problem. The discussion is extended to the multimachine siutation and the problem solved by Runge–Kutta methods adapted to computer formulation.

The extent to which computer methods are used to solve power system problems is of necessity a compromise; there was not available the time, the space, or the inclination to provide much detail on how to optimally program power system problems. More advanced courses in the area would be taken by power specialists, and other textbooks would be more appropriate.

The units used throughout the book are those of the SI system because of its simplicity and its increasing acceptance. Also it was decided to work mainly in the per unit system because the use of per unit in some situations simplifies the system circuit models, and all indications are that the power industry will continue to operate in this system. The advantage of numerical simplicity produced by per unit is offset somewhat by some differences between the system equations before and after scaling. Experience has indicated that per unit can cause students some difficulties and appropriate coverage has been given to this topic.

Finally, this book was written for the student, not the instructor. The role of the instructor is to clarify difficult points, to "color" the material by adding anecdotes from his or her own experience, and to generally inspire the student in this area. This book is intended to help both partners in the learning experience. Also, it should satisfy the dual needs of the nonpower student studying the material as a required subject, and the future power specialist who needs instruction in the fundamentals of power system analysis.

Charles A. Gross

ACKNOWLEDGMENTS

The production of a textbook requires the cooperation and effort of a great many people and institutions. My students over the years have unwittingly contributed to this work through their questions and quest for understanding. More recently they have been most helpful in their criticisms of the manuscript and in ferreting out errors. Colleagues have been helpful in their criticisms and their encouragement. Auburn University, Georgia Power, Westinghouse, General Electric, Alabama Power, and Southern Co. Services have all provided assistance. My family deserves a special acknowledgment for their understanding and sacrifices required by the demand this project has made on my time. It has been a pleasure to work with Ms. Tujunnah Jones, who typed most of the manuscript. Bill Feaster has been most helpful in providing moral support for this project.

C. A. G.

CONTENTS

Contents

POWER SYSTEM ANALYSIS

1

INTRODUCTION

"Go placidly amid the noise and the haste, and remember
what peace there may be in silence."
Max Ehrmann, DESIDERATA

Introduction

Human civilized progress has historically been in proportion to the human ability to control energy. Humanity can feed itself through the efforts of a small minority of the population only because we know how to channel sufficient energy into these activities. In the past we could use energy in essentially unlimited quantities (per person) with little regard for its impact on the environment—there were relatively few of us and nature by comparison was of "infinite" extent. Today the inexorable geometric progression of population growth has caught up with us, making us acutely aware that our planet resources are indeed finite and the simple ways of producing controllable energy are no longer reasonable.

The forms energy assumes in nature are manifold, including:

- Radiant—the most obvious example is sunlight.
- Thermal—an example is the thermal energy stored in the earth's interior.
- Chemical—the energy content of fuels such as wood, coal, and oil is stored in chemical form.
- Kinetic—moving bodies, such as planets revolving in their orbits, represent energy.
- Potential—any system with forces varying with position has potential energy.
- Nuclear—forces that bind atom parts together relate to energy.
- Electrical—an example of "natural" electricity is lightning.

Like all living things, we use natural energy directly (e.g., sunlight for light) or make use of natural conversion processes for our purposes (e.g., sunlight for growing crops). Human intelligence allows us to go beyond these primitive methods and contrive ways of storing, controlling, and converting energy into forms suitable for use when and where we decide. The problem is not that of having insufficient energy; it is incredibly abundant. What is needed are technologies that can channel energy into beneficial applications.

What are these beneficial applications? We require light sources, cooling and heating capability, transportation systems, communication systems, industrial and manufacturing processes, construction applications, and agricultural production. All are broad areas of energy use; a detailed list would be voluminous. In each area, energy converted into the electrical form can be used extensively. Some of the inherent advantages of electrical energy include the following.

- It is amenable to sophisticated control. Consider the incredibly complicated control exerted on an electron beam to produce a TV picture.

- It can be transmitted at the speed of light.

- It can be transmitted and converted to other forms at typically high efficiencies.

- It is inherently pollution free. Conversion into the electrical form does, of course, involve many important environmental problems.

- Conversion to other forms is direct.

Electrical energy generation and delivery systems are and will continue to be of fundamental importance to a technological society, and therefore to engineering. We shall discuss the basic structure of such systems, focusing our study on electrical considerations. Bear in mind that electrical power systems are extremely large by virtually any measure: capital invested, physical size, amount of energy delivered, and so on. It is not practical to design a totally new system "from the ground up"; we must always take into account the existing system. The system will be, is being, and has been continually modified to take advantage of technological advances. To appreciate the existing system it is useful to review its evolution from a historical perspective.

1-1 A Brief History of the Power Industry

Prior to 1800 the study of electrical and magnetic phenomena was of interest to only a few scientists. William Gilbert, C. A. de Coulomb, Luigi Galvani, Otto von Guericke, Benjamin Franklin, Alessandro Volta, and a few others had made significant contributions to a meager store of piecemeal knowledge about electricity, but at that time no applications were known, and studies were motivated only by intellectual curiosity. People illuminated their homes with candles, whale oil lamps, and kerosene lamps, and motive power was supplied mostly by people and draft animals.

From about 1800 to 1810 commercial illuminating gas companies were formed, first in Europe and shortly thereafter in the United States. The tallow candle and kerosene interests, sensing vigorous competition from this young industry, actively opposed gas lighting, describing it as a health menace and emphasizing its explosive potential. However, the basic advantage of more light at cheaper cost could not be suppressed indefinitely, and steady growth in the industry occurred throughout the nineteenth century, with the industry at its zenith in about 1885.

Introduction

Exciting advances in understanding electrical and magnetic phenomena occurred during this same period. Humphrey Davy, Andre Ampere, Georg Ohm and Karl Gauss had made significant discoveries, but the discovery that was to become basic to elevating electricity from its status as an interesting scientific phenomena to a major technology with far reaching social implications was made by two independent workers, Michael Faraday and Joseph Henry. Ampere, and others, had already observed that magnetic fields were created by electric currents; yet no one had discovered how electrical currents could be produced from magnetic fields. Faraday worked on such problems from 1821 to 1831, finally succeeding in formulating the great law that bears his name. He subsequently built a machine that generated a voltage based on magnetic induction principles. Workers now had an electrical source that rivalled—and ultimately far exceeded—the capacities of voltaic piles and Leyden jars. Independently, Joseph Henry also discovered electromagnetic induction at about the same time, and went on to apply his discoveries to many areas, including electromagnets and the telegraph.

Several workers, including Charles Wheatstone, Alfred Varley, Werner and Carl Siemens, and Z. T. Gramme applied the induction principle to the construction of primitive electrical generators in the period from about 1840 into the 1870s. About the same time a phenomenon discovered some years earlier began to receive serious attention as a practical light source. It was observed that when two current-carrying carbon electrodes were drawn apart an electric arc of intense brilliance was formed.

Commercialization of arc lighting was achieved in the 1870s, with the first uses in lighthouse illumination; additional applications were street lighting and other outdoor installations. Predictably, arc lighting provided the stimulus to develop better and more efficient generators. An American engineer, C. F. Brush, made notable contributions in this area with his series arc lighting system and associated generator. The system was practical and grew into a successful business with little opposition from gas illuminating companies, since they did not directly compete for the same applications. The principle objection to arc lighting was its high intensity, making it unsuitable for most indoor applications. For those uses, gas lighting was still the best choice.

Observers had noted as early as 1809 that current-carrying materials could heat to the point of incandescence. The idea of use as a light source was obvious, and a great many workers tried to produce such a device. The main problem was that the incandescent material quickly consumed itself. In an effort to retard or prevent this destruction, the material was encased in a globe filled with inert gas or a vacuum. The problem of placing a material with a high melting point, proper conductance, and good illuminating properties into a globe with a proper atmosphere proved too much for the technology of the time. Some small improvement was noted from time to time, but until the 1870s the electric lamp was far from a practical reality. The struggle never quite ended, however, chiefly because of continued improvements in electrical generators. It became clear that if and when an incandescent electric light was developed, an electrical source would be available.

A 29-year-old inventor named Thomas Edison came to Menlo Park, New Jersey, in 1875 to establish an electrical laboratory to work on a number of projects, including the development of an incandescent electric lamp. In October 1879, after innumerable unsuccessful trials and experiments, an enclosed evacuated bulb containing a carbonized cotton thread filament was energized. The lamp glowed for about 44 hours until it finally burned out. There now was no doubt that a practical incandescent lamp could be developed. Edison subsequently improved the lamp, and also proposed a new generator design that proved to have an unbelievable efficiency of almost 90%. Some three years later, in 1882, the first system installed to sell electrical energy for incandescent lighting in the United States began operation from Pearl Street Station in New York City. The system was dc, three wire, 220/110 volts, and supplied a load of Edison lamps with a total power requirement of 30 kilowatts. This, and other early systems, were the beginnings of what would develop into one of the world's largest industries.

The early electrical companies referred to themselves as "illuminating companies" since lighting was their only service. However, very soon a technical problem was encountered that persists today: a company's load would build, starting at dark, hold roughly constant throughout the early evening and drop precipitously at about 11 P.M. to about half or less. It was obvious that here was an elaborate system that lay idle, or at least underutilized, for most of the time. Could other applications be found that could take up the slack? The electric motor was already known, and the existence of an electrical supply was a ready-made incentive to its refinement and commercial acceptance. Use of electrical motive power quickly became popular, and was employed for many applications. In recognition of their broader role, electric companies began to name themselves "power and light companies."

Another technical problem was encountered. Increasing loads meant increasing currents, which caused unacceptable voltage drops if generating stations were located any appreciable distance from the loads. The requirement of keeping generation in close proximity to loads became increasingly difficult because acceptable generation sites were frequently unavailable. It was known that electrical power was proportional to the product of voltage and current. Clearly, less current would be needed at higher voltage. Unfortunately higher voltage was not desirable from either the viewpoint of present technology or customer safety. What was needed was to transmit power at higher voltage over long distances and then to change it to lower values at the load point. The key was to design a device that could transform voltage and current levels efficiently and reliably.

In the 1890s the newly formed Westinghouse Company had experimented with a new form of electricity, christened "alternating current" (ac), inspired by the fact that current alternately reversed its flow direction in synchronism with generator rotation. This approach had many inherent advantages; for example, the commutation problems associated with dc generators were eliminated. A lively controversy between Edison of the young General Electric Company and Westinghouse developed as to whether the industry should standardize on dc or ac. The ac form finally won out for the following reasons:

- The ac transformer could perform the much needed capability to easily change voltage and current levels.

- ac generators were inherently simpler.

- ac motors, although not as versatile, were simpler and cheaper.

Having standardized on ac, the central station concept became firmly established, and remote loads posed no problems. The soft yellow glow of the Edison lamp was more convenient, cleaner, and quickly becoming cheaper than its gas counterpart. More and more customers were added to the power companies' rolls; since most of this load growth could be handled without increased capital investment, the cost per unit energy dropped, attracting even more customers.

Local electric utilities expanded until they shared boundaries. An operating advantage was apparent; since the loads in neighbouring systems did not necessarily peak at the same time, why not interconnect the systems and meet the peak load conditions with the combined generation? The advantages of interconnecting different generating sites and loads were already well known; this step would be a logical extension of the principle and better utilize everyone's equipment. A technical problem was immediately apparent; many different frequencies were in use at the time, including dc, 25, 50, 60, 125, and 133 Hz (circa 1900). Since interconnected systems must operate at the same frequency, expensive frequency conversion equipment would be required. The incentive to standardize on frequency was obvious. At the time generating units at Niagara Falls and at other hydroelectric installations were 25 Hz, since hydroturbines can be designed to operate somewhat more efficiently at corresponding mechanical speeds; hence there was strong support for using that frequency. The problem with 25 Hz was that it caused a noticeable flicker in incandescent lamps. A higher frequency, 60 Hz, was eventually adopted in this country as standard because it has acceptable electrical characteristics and because steam turbines performed satisfactorily at the corresponding mechanical speeds of 3600 and 1800 rev/min.

Technological advancement in the design of power apparatus continued: when a utility expanded its system, the new generators and transformers it purchased were inevitably of larger capacity and higher efficiency. Better electric lamps were developed, giving the consumer more light per unit energy. With the steady drop in the cost of electrical energy, selection of electric motors as mechanical drives for all sorts of applications became popular. As an example of this progress, examine the improvement of electric light sources as presented in Table 1-1.

Table 1-1 Light output for selected sources

LIGHT SOURCE	EFFICACY (LUMENS/WATT)	YEAR
Candle	0.1	—
Edison original incandescent	1.4	1879
Bamboo filament incandescent	1.7	1881
Tungsten filament incandescent	7.9	1904
Tungsten filament incandescent (100 W)	10.0	1910
Tungsten filament incandescent (100 W)	12.6	1920
Tungsten filament incandescent (100 W)	14.1	1930
Tungsten filament incandescent (100 W)	16.2	1940
Tungsten filament incandescent (100 W)	17.5	1970
400 W mercury vapor	56.2	1970
40 W cool white fluorescent	78.8	1970
400 W sodium vapor	115.0	1970

Increasing demand for power created an incentive to transmit at progressively higher voltages. Table 1-2 lists the maximum transmission line voltage in service in the United States by year.

To avoid the proliferation of an unlimited number of operating voltages, the power industry has selected certain levels as standard. The most common of these voltages are presented in Table 1-3; all are 60 Hz ac three phase, unless otherwise noted.

The next transmission voltages under consideration are 1000, 1200, and 1500 kV. It is interesting to note that Edison's initial voltage of 110 volts,

Table 1-2 Highest power transmission voltage in service in the United States.

YEAR	VOLTAGE (KILOVOLTS)
1890	3.3
1900	40
1910	120
1920	150
1930	244
1940	287
1950	287
1960	345
1970	765

Table 1-3 U.S. standard power operating voltages.

VOLTAGE CLASS	NOMINAL LINE VOLTAGE
Low	120/240 V (Single Phase)
	208 V
	240 V
	480 V
	600 V
Medium	2.4 kV
	4.16 kV
	4.8 kV
	6.9 kV
	12.47 kV
	13.2 kV
	13.8 kV
	23.0 kV
	24.94 kV
	34.5 kV
	46.0 kV
	69.0 kV
High	115 kV
	138 kV
	161 kV
	230 kV
Extra high	345 kV
	500 kV
	765 kV

with successive revisions to 115 V and 120 V, has remained the American service level standard.

The American electrical power system is unique when compared with other large national systems—although it is interconnected, it is not centrally owned. It is composed of hundreds of individual investor and government owned companies. In combination with related companies that manufacture power apparatus, lighting equipment, cables, and appliances, the electrical power industry is one of the world's largest. The fundamental criterion for the creation of such a system might be stated as follows. Design, build, and operate an electrical energy delivery system that has favorable aspects of:

- Safety

- Reliability

- Availability

- Impact on the environment

- Economics

Although occasionally shortsighted, self-serving, disorderly, and opposed to progress, the historical development of the industry records remarkable achievement in all of these areas. The future will require continued engineering excellence to pioneer in new technologies and to maintain the development of older technologies so as to satisfy the electrical energy needs of society.

1-2 The SI System of Units

The nature of engineering requires quantitative measurements and calculations. It is not sufficient to simply understand that y changes when x changes, but we must take the next step and predict "how much" and "in what direction" changes occur. This need requires a system of measure. Most units of measure were created by early pioneers in a given field and were usually defined arbitrarily or as a matter of measuring convenience. For example, the German physicist Fahrenheit selected the coldest temperature he could achieve with a salt-ice mixture to define his "zero" and the normal body temperature to define an upper value, which he set at 96 degrees. These units were quite satisfactory for Fahrenheit's purposes, but have unfortunate consequences when applied to a broader range of problems within the field of thermodynamics.

9

Table 1-4 SI units for physical quantities.

SECTION	QUANTITY	SI UNIT	SYMBOL
I	length	metre	m
	mass	kilogram	kg
	time	second	s
	current	ampere	A
	angle	radian	rad
II	area	square metre	m^2
	volume	cubic metre	m^3
	linear velocity	metre per second	m/s
	linear acceleration	metre per second squared	m/s^2
	angular velocity	radian per second	rad/s
	angular acceleration	radian per second squared	rad/s^2
III	frequency	hertz (1/s)	Hz
	force	newton (kg·m/s^2)	N
	energy	joule (N·m)	J
	power	watt (J/s)	W
	charge	coulomb (A·s)	C
	voltage	volt (W/A)	V
	electric flux	coulomb (A·s)	C
	capacitance	farad (C/V)	F
	resistance	ohm (V/A)	Ω
	conductance	siemen (A/V)	S
	magnetic flux	weber (V·s)	Wb
	magnetic flux density	tesla (Wb/m^2)	T
	inductance	henry (Wb/A)	H
IV	electric field intensity	volt/metre	V/m
	electric flux density	coulomb/square metre	C/m^2
	magnetic field intensity	ampere/metre	A/m
	torque	Newton metre	N·m
V	reactance	ohm (V/A)	Ω
	impedance	ohm (V/A)	Ω
	susceptance	siemen (A/V)	S
	admittance	siemen (A/V)	S
	reactive power	voltampere reactive (V·A)	var
	apparent power	voltampere (V·A)	VA

Section I: Base Units
Section II: Derived Units without Special Names
Section III: Derived Units with Special Names
Section IV: Compound Derived Units without Special Names
Section V: Special Quantities

To bring the problem into sharper focus consider Ohm's law in the following form:

$$V = kIR$$

The conversion factor k is necessary to make the units of current and resistance produce voltage in its desired units. Commonly k is one, *but only because the ohm is defined as one volt per ampere!*

An inconsistent system of units not only unnecessarily complicates equations, but tends to obscure physical truths. Some students (and even engineers) are uncertain that foot-pounds, calories, joules, ergs, newton-metres, British-Thermal-Units, and Kilowatt-hours are measures of the same entity (they are — energy).

The SI system (which stands for Le Systéme International d'Unités, or International System of Units) was adopted by the General Conference on Weights and Measures, with the National Bureau of Standards acting as the United States representative, as the official international system of measurement in 1960. The system possesses the following advantages:

- One and only one unit is used for each physical quantity.
- The system has international acceptance.
- The system, like older metric systems, has decimal relations between multiple and submultiple units.
- The system is coherent; no conversion factors are required.

Table 1-5 Prefixes.

PREFIX	SYMBOL	MEANING
exa	E	10^{18}
peta	P	10^{15}
tera	T	10^{12}
giga	G	10^{9}
mega	M	10^{6}
kilo	k	10^{3}
hecto	h	10^{2}
deka	da	10^{1}
deci	d	10^{-1}
centi	c	10^{-2}
milli	m	10^{-3}
micro	μ	10^{-6}
nano	n	10^{-9}
pico	p	10^{-12}
femto	f	10^{-15}
atto	a	10^{-18}

This system of measure is used throughout this book. The SI units for most of the physical quantities we encounter are summarized in Table 1-4. Note particularly the symbols since we will use these freely. Multiples of ten of SI units are created by using the appropriate prefix. Table 1-5 defines these prefixes, along with their symbol.

We will consistently use SI units except in a few cases where an alternate unit has virtually universal acceptance. The per unit system, explained in Chapter 3, is used to scale many of our equations to dimensionless quantities.

1-3 Electrical Energy Transmission

Consider the situation shown in Figure 1-1. The rate of electrical energy flow (power) from network A to network B is

$$p = vi \tag{1-1}$$

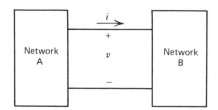

Figure 1-1 Power transmission between two networks.

Lowercase letters are used to indicate instantaneous values, that is, that p, v, and i may vary with time. High power levels require high voltage and current values. For a given value of current, higher power flows may be obtained by increasing the voltage, and vice versa. Unfortunately, the existing technology sets practical upper limits on allowable currents and voltages.

What are the limiting factors on current? We fabricate power conductors using materials with high conductivity, appropriate mechanical characteristics, and that are economical: aluminum is the most common choice, with copper used for some applications. The current-carrying capacity of a conductor is related to its maximum allowable current density and its cross-sectional area:

$$I_{max} = J_{max}A \tag{1-2}$$

The maximum current density J_{max} is determined by the maximum conductor temperature that will not damage the conductor or its insulation system. What are the limiting factors on voltage? The fundamental consideration is to provide electrical isolation (or insulation) between adjacent parts that can conduct current—that is, to confine current to the paths through which it was intended to flow. When the voltage exceeds the breakdown strength for a given insulation system, undesirable conduction paths will be created and the system will be either temporarily or permanently disabled. Fluid insulation tends to be "self-healing" (the system will recover from a breakdown if it is deenergized for a short time and then reenergized), whereas solid insulation is permanently damaged by a breakdown. Some common insulating materials are listed in Table 1-6.

Table 1-6 Power insulating materials.

Solids	Liquids
cotton	oil
silk	askarel
paper	
mica	Gasses
glass	air
porcelain	sulphur hexafluoride
quartz	
thermoplastics	Miscellaneous
rubber	silicones
wood	vacuum

The meaning of "ground" is important; we quote from the IEEE Standard Dictionary of Electrical and Electronic Terms [8]:

ground (*earth*) (*electric system*). A conducting connection, whether intentional or accidental, by which an electric circuit or equipment is connected to the earth, or to some conducting body of relatively large extent that serves in place of the earth. Note: It is used for establishing and maintaining the potential of the earth (or of the conducting body) or approximately that potential, on conductors connected to it, and for conducting ground current to and from the earth (or the conducting body).

We understand this to mean that at a given location in the power system, accessible parts of power apparatus and earth constitute an equipotential surface when perfectly grounded. Insulation of conductors from ground is a basic problem.

Let us consider some different schemes for implementing the transmission line indicated in Figure 1-1. For a fair comparison we select constraints that all schemes must satisfy. Although only two conductors are shown in Figure 1-1, we allow any number of conductors to be used, as long as each scheme uses the same amount of conducting material. Given that networks A and B are separated by a fixed physical length, this means that in viewing the lines in cross section we must observe the same cross-sectional conducting area (A) for all schemes. Also, we argue that no conductor shall carry current greater than that constrained by some maximum current density J_0.

We require that at least one conductor be grounded (symbolized as $\perp$) and shall refer to such a conductor as the neutral, designated as "n." If it is not required to conduct any appreciable current, we will not include its cross-sectional area in A. This condition is achieved under certain symmetrical loading conditions, referred to as "balanced" loading, and can be maintained in a practical situation; therefore we allow all schemes to make this assumption. We require that for all schemes, no voltage to ground exceed V_0.

It is assumed that the reader has a background in basic circuit theory. The adjective "dc" essentially means time invariant or constant with time. Recall that the term "ac," which historically stood for "alternating current," in modern usage means "sinusoidal steady state." These terms are used to describe voltages and currents in time invariant (constant) steady-state and sinusoidal steady-state modes. The six schemes we compare will consider dc and ac situations and are tabulated in Figure 1-2, denoted as a, b, c, d, e, and f.

Consider the scheme shown in Figure 1-2(a). One conductor is grounded as required; it is also expected to carry the return current. Therefore we divide our total area A equally between the two conductors. The maximum transmitted power is:

$$P_a = (0.5A)(J_0)V_0 \tag{1-3a}$$

$$= 0.5P_0 \tag{1-3b}$$

where we define

$$P_0 = AJ_0V_0 \tag{1-3c}$$

Scheme (b) is shown in Figure 1-2(b). Because the voltages from the two phases to ground are of opposite polarity, the phase currents will flow in opposite directions. Considering the neutral as a common return path, these

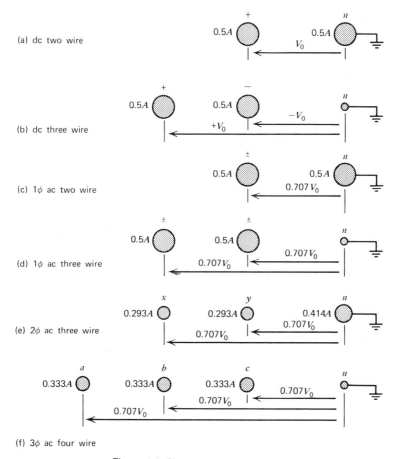

(a) dc two wire

(b) dc three wire

(c) 1ϕ ac two wire

(d) 1ϕ ac three wire

(e) 2ϕ ac three wire

(f) 3ϕ ac four wire

Figure 1-2 Six transmission schemes.

phase currents will subtract from, or cancel, each other. If the phase currents are balanced (equal), the neutral current is zero, allowing the neutral conductor to be small, and using a negligible portion of our conducting cross-section area A. The maximum transmitted power is:

$$P_b = 2(0.5A)J_0V_0 \tag{1-4a}$$

$$= P_0 \tag{1-4b}$$

The improvement in capacity over scheme (a) is dramatic (100%)! This arrangement is similar to that used by Edison in an early 220/110 dc volt system.

15

Introduction

Schemes (a) and (b) were dc. For power transmission in the ac mode the currents and voltages are sinusoidal. Consider

$$v = V_m \cos \omega t \qquad \text{(1-5a)}$$

$$i = I_m \cos \omega t \qquad \text{(1-5b)}$$

The instantaneous power is:

$$p = vi$$

$$= V_m I_m \cos^2 \omega t \qquad \text{(1-6a)}$$

$$= \frac{V_m I_m}{2} [1 + \cos 2\omega t] \qquad \text{(1-6b)}$$

The average power is

$$P = \frac{V_m I_m}{2} \qquad \text{(1-7a)}$$

$$= \frac{V_m}{\sqrt{2}} \left(\frac{I_m}{\sqrt{2}} \right) \qquad \text{(1-7b)}$$

$$= VI \qquad \text{(1-7c)}$$

where V and I are rms (root mean square) values. These ideas are more fully developed in Chapter 2. We are now ready to proceed with the ac schemes.

Equation (1-7) applies directly to scheme (c). The current capacity is dependent on the rms current, whose greatest value is:

$$I = 0.5 A J_0^* \qquad \text{(1-8)}$$

The insulation is limited by the peak or maximum value:

$$V = V_0 / \sqrt{2} \qquad \text{(1-9)}$$

The maximum power transmission is calculated from equation (1-7c), producing:

$$P_c = 0.5 A J_0 [V_0 / \sqrt{2}] \qquad \text{(1-10a)}$$

$$= 0.354 A J_0 V_0 \qquad \text{(1-10b)}$$

$$= 0.354 P_0 \qquad \text{(1-10c)}$$

The instantaneous power p_c is shown in Figure 1-3.

* Assuming uniform current distribution. For purposes of our discussion, this simplifying approximation is reasonable.

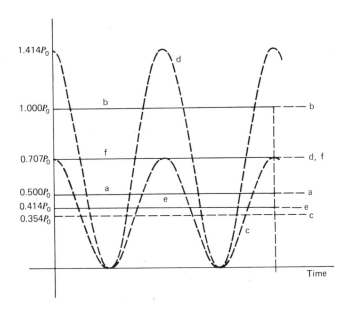

Figure 1-3 Power vs. time for transmission schemes shown in Figure 1-2.

The three wire arrangement for single phase, as shown in Figure 1–2(d), accrues the same advantages as it did for dc. Therefore:

$$P_d = 2P_c \qquad\qquad (1\text{-}11a)$$

$$= 0.707 P_0 \qquad\qquad (1\text{-}11b)$$

The instantaneous power is shown in Figure 1-3(d). Schemes (c) and (d) have a common disadvantage: although the transmitted powers have net positive average values, their instantaneous values are pulsating. Energy is not delivered in a steady flow, but comes in "spurts." This can have undesirable effects on electrical generator prime movers requiring them to likewise deliver energy at a nonuniform rate. Based only on this consideration, the dc form would be preferable.

Scheme (e) is referred to as two phase (2ϕ) three wire. If we require that

$$v_x = V_m \cos \omega t \qquad\qquad (1\text{-}12a)$$

$$i_x = I_m \cos \omega t \qquad\qquad (1\text{-}12b)$$

$$v_y = V_m \sin \omega t \qquad\qquad (1\text{-}12c)$$

$$i_y = I_m \sin \omega t \qquad\qquad (1\text{-}12d)$$

17

then the instantaneous power p_e becomes:

$$p_e = v_x i_x + v_y i_y \tag{1-13a}$$

$$= V_m I_m [\cos^2 \omega t + \sin^2 \omega t] \tag{1-13b}$$

$$= V_m I_m \tag{1-13c}$$

The result given in equation (1-13c) is quite remarkable. Note that the *instantaneous* power p_e is *constant!* Consult Figure 1-3(e). The average value P_e is trivial:

$$P_e = p_e \tag{1-14a}$$

$$= V_m I_m \tag{1-14b}$$

The maximum allowable currents and voltages are:

$$V_m = V_0 \tag{1-15a}$$

$$I_m = J_0(0.293A)\sqrt{2} \tag{1-15b}$$

such that:

$$P_e = 0.414 P_0 \tag{1-16}$$

Note that the neutral conductor for this scheme is larger than the phase conductors since it must carry $\sqrt{2}$ times as much current for balanced operation. This is true since the return phase currents are 90° phase displaced instead of 180°, and therefore do not cancel. However, the pulsating power problem has been overcome.

We continue with the three phase four wire arrangement, scheme (f). If we require that:

$$v_a = V_m \cos \omega t \tag{1-17a}$$

$$v_b = V_m \cos (\omega t - 120°) \tag{1-17b}$$

$$v_c = V_m \cos (\omega t + 120°) \tag{1-17c}$$

$$i_a = I_m \cos \omega t \tag{1-17d}$$

$$i_b = I_m \cos (\omega t - 120°) \tag{1-17e}$$

$$i_c = I_m \cos (\omega t + 120°) \tag{1-17f}$$

the instantaneous power p_f becomes:

$$. p_f = v_a i_a + v_b i_b + v_c i_c \tag{1-18a}$$

$$= V_m I_m [\cos^2 \omega t + \cos^2 (\omega t - 120°) + \cos^2 (\omega t + 120°)] \tag{1-18b}$$

Using the trigonometric identity

$$\cos^2 \theta = \frac{1+\cos 2\theta}{2} \tag{1-19}$$

Equation (1-18b) becomes:

$$p_f = V_m I_m \left[\frac{1+\cos 2\omega t}{2} + \frac{1+\cos(2\omega t - 240°)}{2} + \frac{1+\cos(2\omega t + 240°)}{2} \right] \tag{1-20a}$$

$$= \frac{V_m I_m}{2} [3 + \cos 2\omega t + \cos(2\omega t - 240°) + \cos(2\omega t + 240°)] \tag{1-20b}$$

$$= \frac{3 V_m I_m}{2} \tag{1-20c}$$

Again

$$P_f = p_f \tag{1-21a}$$

$$= \frac{3 V_m I_m}{2} = 3VI \tag{1-21b}$$

The maximum allowable currents and voltages are:

$$V = V_0 / \sqrt{2} \tag{1-22a}$$

$$I = 0.333 A J_0 \tag{1-22b}$$

So that

$$P_f = 3 \frac{V_0}{\sqrt{2}} (0.333 A J_0) \tag{1-23a}$$

$$= 0.707 P_0 \tag{1-23b}$$

Comparative power transmission schemes are summarized in Table 1-7. Suppose we prefer ac schemes because of simpler and cheaper generation and transformation equipment. We reject single phase schemes because of their pulsating power property. Of the two remaing schemes, scheme (f), the three phase four wire arrangement is better because of its greater power transfer capability. Following the same constraints, n phase, $n+1$ wire schemes ($n > 3$) will have the same power capabilities ($0.707 P_0$). Therefore, the three phase scheme is the simplest of the n-phase methods to offer the triple advantages of using the ac mode, constant power flow, and high capacity, and for these reasons it has historically been the standard method used.

Table 1-7 Comparison of transmission schemes.

SCHEME	MODE	MAXIMUM POWER TRANSMITTED
a	dc	$0.500P_0$
b	dc	$1.000P_0$
c	ac	$0.354P_0$
d	ac	$0.707P_0$
e	ac	$0.414P_0$
f	ac	$0.707P_0$

Our comparisons have not been completely fair since they have not accounted for effects such as the additional insulators required for more conductors, differences in tower design, and differences in ac and dc corona loss. However the basic advantages of three phase transmission have been brought out. Note that scheme (b) had the largest transmission capacity, and under some circumstances is the preferred solution. We will discuss dc transmission in Chapter 4 in more detail. Consideration of unbalanced operation will change the relative capacities somewhat (see problem 1-5.

1-4 Electrical Energy Production

Although our concern in this book is with the electrical aspects of electrical power systems, it is important to know what the basic energy sources are that can be used to produce bulk electrical energy. The only practical device currently available for the large-scale conversion of mechanical energy into the electrical form is the electrical generator, a device covered in some detail in Chapter 6. When we claim to "generate" electricity we really mean that we convert energy into the electrical form. Because the generator is a rotational device, mechanical power is supplied to the shaft in the form of an applied torque times the shaft angular velocity. The mechanical torque is produced by a turbine that is one of two basic types: hydraulic (or water driven), or steam driven, of which the latter type is far more common. Steam, of course, is thermally produced and many of the energy sources we shall discuss are simply alternate processes for steam production. We can break down electrical energy sources into two broad categories identified as thermal and nonthermal. A brief discussion of most of the alternatives follows.

THERMAL

Coal. Because we have substantial domestic reserves (some sources claim up to 500 years) and the required technology developed, coal is and will continue as a major energy source. There are important environmental problems involved with coal, related both to its removal from the earth and to its combustion, but such problems are solvable with today's technology. At present coal represents about 45% of the total electrical energy source.

Oil and Natural Gas. Throughout the 1950s and 1960s there was a trend toward greater utilization of these fuels because of their superior combustion properties. However, the cost, scarcity, and competition for petroleum products indicates that while these fuels will remain important for electrical energy production, their contribution expressed as a percentage of the total supply will shrink.

Nuclear Fission of Uranium. As early as 1939 it was observed in some experiments that when certain nuclear particles were subdivided the mass of the component parts totalled less than the original value, with the mass defect appearing as pure energy. There was immediate speculation that such reactions might be sustained and controlled, thereby creating a practical energy source. In 1942 in Chicago a group headed by Enrico Fermi built the first successful nuclear reactor, using uranium as the fissionable material.

Since the 1950s fission reactors have been used commercially for the production of electrical power. Uranium resources in the U.S. are slightly less than the oil and gas reserves. The basic ore contains significant concentrations of uranium oxide (U_3O_8). Only about 0.7% of the uranium is the fissionable light isotope U^{235}; the remainder consists of U^{238}. The conventional fission reactor "burns" U^{235}, requiring refuelling when the supply is exhausted. A second type of reactor whose operation is based on a different nuclear reaction is under development. Called the breeder reactor, it transforms U^{238} into the fissionable product, plutonium (Pu^{239}), which has acceptable nuclear fuel properties. When operational, the breeder reactor will increase the effectiveness of uranium as a nuclear fuel by roughly a factor of 140. The development of a commercially acceptable breeder reactor is not expected before about 1985.

Serious questions regarding environmental impact have accentuated social concern and produced legal and regulatory constraints on the development of this resource. A prudent growth in the importance of this source is expected, concurrent with legitimate and proper concern for environmental impact.

Solar. It is possible to collect solar energy directly and concentrate it on boilers for steam production. The major problems are its diffuse nature, requiring large amounts of land for collectors, and the unreliability caused by atmospheric conditions. At present there are no commercial installations. This source is particularly attractive because no "fuel" is required and because of its nonpolluting characteristics. Roughly 3% of the U.S. land area, at a conversion efficiency of 10%, is sufficient to meet total projected U.S. energy needs in the year 2000.

Nuclear Fusion. It is known that certain nuclear reactions are possible, whereby certain light nuclear particles may be combined, or fused, into heavier particles. Such reactions have end products that exhibit a mass defect, which appears as pure energy. The attractive feature here is that common elements such as isotopes of hydrogen may be used as fuel, making this source essentially inexhaustible. The difficulty is that a sustained fusion reaction requires production of extremely high temperatures and particle concentrations for a sufficient time duration to occur. The associated technical problems are formidable and most experts do not predict a commercial installation until well into the next century.

Geothermal. Heat from the earth's interior and subsurface water combine to produce natural steam, which can be used for electrical energy production. One commercial installation, the Geysers, north of San Francisco, California, is operating in the United States at a capacity of about 400 MW. Total accessible reserves are estimated at up to about half of our total gas and oil reserves. Expectations are that this resource will continue to be developed, but that it will make a minor contribution to the total energy supply.

Biomass. Synthetic gas may be produced from organic material grown expressly for that purpose. The amount of electrical energy produced from this source is negligible at present, and not expected to be significant in the future.

Garbage and Sewage. There are combustible components in garbage that can be used as fuel; these components are separated from noncombustible items and mixed with coal. Sewer gasses are also combustible. In certain situations utilization of these fuels may prove to be economical; however, such installations should be viewed as supplementary, and would contribute only by a small fraction of the total energy supply.

NONTHERMAL

Hydro. Hydroelectric power has historically been an economical and pollution free source of energy. The current contribution to the total energy supply is about 12%. The energy is not "free" since the capital investment in dams, transmission, and generation must be accounted for; nor can we argue that there is no impact on the environment when we consider the destruction of natural white water and the creation of artificial lakes. Still, this is a very attractive energy source; unfortunately, there are few remaining acceptable locations. It is estimated that the hydroelectric contribution to the total energy supply will shrink to about 2% by the year 2000.

The source has the advantage that it is immediately (within seconds at least) available, whereas thermal sources can meet demand at a much slower rate. Hydroelectric sources have the disadvantage that their use is constrained by navigation requirements and actual or predicted rainfall.

Tidal. There are a few sites around the world where it proves economical to convert the change in potential energy caused by tide levels into electrical form. One of the more famous is the French La Rance installation with a rated capacity of 240 MW. The technical problems and aspects of such installations are comparable to those of conventional hydro. The percentage contribution to the total energy supply is quite small, and expected to remain so.

Wind. The wind can be used to drive turbines that in turn drive generators to produce electricity. Because wind is inherently intermittent, such systems must include energy storage devices, such as batteries, or supply loads that are tolerant of unpredictable source interruptions. An example of the latter would be hydrogen production by electrolysis, with the hydrogen used as a fuel, or for making anhydrous ammonia and ammonium nitrate fertilizers. As isolated electric power supplies, such systems are now commercially available in sizes up to about 50 kW. Research on much larger units is currently underway; at present, wind energy is a negligible fraction of the total energy source.

Wave. There have been several experimental machines designed to convert wave kinetic energy into electricity. None as of now appears feasible for large-scale economic electrical energy production.

Direct Solar Conversion. Semiconductors exposed to solar radiation produce free charge carriers for electrical conduction through the so-called photovoltaic effect. It is necessary for the semiconductors to be arranged into large solar radiation collecting panels and placed to intercept a maximum amount of solar energy. Difficulties include the high cost of solar cells and the low efficiencies. Some schemes propose location of such collectors outside of the earth's atmosphere, conversion of the dc energy to microwave frequencies and subsequent beaming to earth. The associated technical problems are substantial, as is the cost of such systems.

There are many schemes for the production of electrical energy, each requiring an associated specialized technology, and meriting a corresponding detailed technical treatment. This is not our objective. Our concern in this book is with a study of the design, modeling, and operation of the electrical power system. It will first be necessary to review some concepts from circuit theory to show how they are conventionally applied to power system analysis.

Bibliography

[1] *A National Plan for Energy Research, Development and Demonstration: Creating Energy Choices for the Future*; Volume 1: The Plan; ERDA-48, U.S. Printing Office, Washington, D.C., 1975.

[2] *A National Plan for Energy Research, Development and Demonstration: Creating Energy Choices for the Future*; Volume 2: Program Implementation; ERDA-48, U.S. Printing Office, Washington, D.C., 1975.

[3] Bright, Jr., Arthur J., *The Electric-Lamp Industry*, The Macmillan Co., New York, 1949.

[4] Dibner, Bern, *Ten Founding Fathers of the Electrical Science*, Burndy Corp., Norwalk, Conn, 1954.

[5] *Edison and Electricity*, Publication APD-53, 7-50 (500 M), General Electric Corp.

[6] Friedlander, G. F., "Energy's Hazy Future," *IEEE Spectrum*, May 1975.

[7] Healy, T. J., *Energy, Electric Power, and Man*, Boyd and Fraser, San Francisco, 1974.

[8] *IEEE Standard Dictionary of Electrical and Electronics Terms*, IEEE Std 100-1972, ANSI C42. 100-1972, Wiley-Interscience, New York, 1972.

[9] Miller, Raymond C., *Kilowatts at Work*, A History of the Detroit Edison Company, Wayne State University Press, Detroit, 1957.

[10] *Reference Manual for SI (Metric) The International System of Units*, Inland Steel Company, April, 1976.

Problems

1-1. Why are power systems ac rather than dc?

1-2. Convert kilowatt-hrs. to joules.

1-3. In the circuit of Figure P1-3 calculate I if:
(a) $R_1 = 10\ \Omega$; $R_2 = \infty$
(b) $R_1 = 10\ \Omega$; $R_2 = 20\ \Omega$
(c) $R_1 = R_2 = 10\ \Omega$

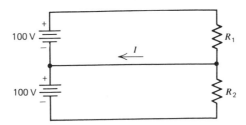

Figure P1-3 Circuit for Problem 1-3.

1-4. Is it preferable to transmit $10\ \text{MW}$ at $V = 100\ \text{V}$ and $I = 100\ \text{kA}$ or $V = 100\ \text{kV}$ and $I = 100\ \text{A}$? Why?

1-5. Following the same logic as that presented in section 1-3, recalculate entries b, d, and f in Table 1-7 if maximum unbalance is anticipated (i.e., that the neutral current may equal the line current).

1-6. Write a short paper (500 words) that provides more information on any one of the energy sources discussed in section 1-4. Use at least two references.

SINUSOIDAL STEADY STATE CIRCUIT CONCEPTS

"As far as possible without surrender be on good terms with all
persons. Speak your truth quietly and clearly; and listen
to others, even to the dull and the ignorant; they
too have their story. Avoid loud and aggressive
persons, they are vexations to the spirit."

Max Ehrmann, DESIDERATA

The normal mode of electrical operation of the power system is ac. Many (in fact, most) of the electrical phenomena of engineering interest can be analyzed using conventional ac circuit methods. It is therefore crucial for a student of power system engineering to be familiar with such methods. Such a background is assumed for readers of this book. However, because of the topic's fundamental importance, the basic concepts of ac circuits are reviewed in this chapter. An additional objective is to explain the author's notation and symbolism.

2-1 Phasor Representation

Consider the general sinusoidal function $f(t)$:

$$f(t) = F_{max} \cos(\omega t + \phi) \tag{2-1}$$

Note that the function has three important parameters:

F_{max} = amplitude, or maximum value

ω = radian frequency

ϕ = phase angle

The parameter F_{max} essentially controls the "strength" of $f(t)$; ω shows the rate at which $f(t)$ is changing; and the parameter ϕ keys the cyclic $f(t)$ to the time origin (or, more to the point, to other sinusoidal functions). Note that any conceivable sinusoidal function can be represented with the proper choice of F_{max}, ω, and ϕ. Recall Euler's identity:

$$e^{j\theta} = \cos\theta + j\sin\theta \tag{2-2}$$

It follows that

$$f(t) = F_{max} \cos(\omega t + \phi) \tag{2-3a}$$

$$= \text{Re}[F_{max} \cos(\omega t + \phi) + jF_{max} \sin(\omega t + \phi)] \tag{2-3b}$$

$$= \text{Re}[F_{max} e^{j(\omega t + \phi)}] \tag{2-3c}$$

$$= \text{Re}[F_{max} e^{j\phi} e^{j\omega t}] \tag{2-3d}$$

$$= \sqrt{2}\text{Re}\left[\frac{F_{max}}{\sqrt{2}} e^{j\phi} e^{j\omega t}\right] \tag{2-3e}$$

We define

$$F = \frac{F_{max}}{\sqrt{2}} e^{j\phi} \tag{2-4}$$

so that

$$f(t) = \sqrt{2}\,\text{Re}[\boldsymbol{F}\,e^{j\omega t}] \tag{2-5}$$

The quantity $\boldsymbol{F}$ is defined as the phasor representation of $f(t)$ and the transformation is defined in equation (2-5). Examine $\boldsymbol{F}$ in equation (2-4) closely. Only two parameters, F_{max} and ϕ, are involved. If we further define

$$F = \frac{F_{\text{max}}}{\sqrt{2}} \tag{2-6}$$

then

$$\boldsymbol{F} = F\,e^{j\phi} \tag{2-7}$$

or

$$\boldsymbol{F} = F\underline{/\phi} \tag{2-8}$$

The notation presented in equation (2-8) is to be understood simply as an alternate way of writing (2-7). Although technically the angle ϕ should be in radians, frequently it is written in degrees because of their familiarity. The value F is the root-mean-square (rms) value of $f(t)$ and is useful because of power considerations (see problem 2-2 for more detail). Note that $\boldsymbol{F}$ contains two-thirds of all the necessary information about the sinusoidal

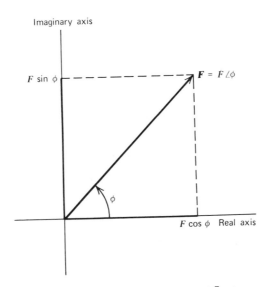

Figure 2-1 Graphical interpretation of $\bar{F}$—the phasor diagram.

29

function $f(t)$ (the frequency ω is not included). If this transformation is applied to a collection of sinusoidal functions of the same frequency, which is independently known, the corresponding phasors contain *all* the essential information. The phasor F is a two dimensional vector from a mathematical viewpoint. The term "phasor" was coined to avoid confusion with spatial vectors: the angular position of the phasor represents position in time, not space. Equation (2-8) is commonly referred to as "polar form"; "rectangular form" is easily produced by applying Euler's identity.

$$F = F\underline{/\phi} \tag{2-9a}$$

$$= F\cos\phi + jF\sin\phi \tag{2-9b}$$

The phasor may be graphically interpreted by plotting the real and imaginary components in conventional cartesian coordinates, with the real and x axes coinciding. Refer to Figure 2-1.

Example 2-1

(a) Transform $f(t) = 100\cos(377t - 30°)$ to the phasor form.

Solution

$$F = \frac{100}{\sqrt{2}}\underline{/-30°}$$

$$= 70.7\underline{/-30°}$$

Observe that ω does not appear.

(b) Transform $F = 100\underline{/+20°}$ to the instantaneous form.

Solution

$$f(t) = 100\sqrt{2}\cos(\omega t + 20°)$$

$$= 141.4\cos(\omega t + 20°)$$

Observe that ω is unknown.

30

(c) Add two sinusoidal functions of the same frequency using phasor methods.

$$a(t) = A\sqrt{2}\cos(\omega t + \alpha)$$

$$b(t) = B\sqrt{2}\cos(\omega t + \beta)$$

$$c(t) = a(t) + b(t)$$

$$= \sqrt{2}[\text{Re}[A\, e^{j(\omega t + \alpha)} + B\, e^{j(\omega t + \beta)}]]$$

$$= \sqrt{2}\,\text{Re}[(A\, e^{j\alpha} + B\, e^{j\beta})\, e^{j\omega t}]$$

$$= \sqrt{2}\,\text{Re}[(A + B)\, e^{j\omega t}]$$

or if:

$$C = A + B$$

$$c(t) = \sqrt{2}\,\text{Re}[C\, e^{j\omega t}]$$

The results of example 2-1(c) are of unexpected importance. Observe that we can add sinusoidal functions of the same frequency by expressing them as phasors and then adding the phasors by the rules of vector algebra. The two basic laws of circuit theory (Kirchhoff's voltage and current laws), when voltages and currents are expressed as phasors, take the following form:

KVL: The sum of all *phasor* voltage drops around any path in a circuit equals zero.
KCL: The sum of all *phasor* currents into any node in a circuit is zero.

The summation of phasors is performed by the rules of ordinary vector addition.

In applying Kirchhoff's laws to circuits in which all voltages and currents are expressed as phasors, it is *absolutely vital* that all voltages and currents *be assigned positive senses*, generally marked on a circuit diagram. It is important to realize that the diagram is necessary to interpret calculated results, and is therefore a part of the solution.

2-2 Impedances of Passive Elements

Table 2-1 summarizes the volt-ampere relations for the three basic passive ideal circuit elements. To understand the basis for these results, consider

Table 2-1 The Three Basic Ideal Passive Circuit Components.

The general case	The ac special case

the inductive case.

$$v = L\frac{di}{dt} \tag{2-10a}$$

If

$$i = \sqrt{2}\text{Re}[\boldsymbol{I}\,e^{j\omega t}]: \tag{2-11}$$

$$v = L\frac{d}{dt}[\sqrt{2}\text{Re}[\boldsymbol{I}\,e^{j\omega t}]] \tag{2-10b}$$

$$= \sqrt{2}\text{Re}\left[L\boldsymbol{I}\frac{d}{dt}\,e^{j\omega t}\right] \tag{2-10c}$$

$$= \sqrt{2}\text{Re}[j\omega L\boldsymbol{I}\,e^{j\omega t}] \tag{2-10d}$$

Examination of equation (2-10d) produces:

$$\boldsymbol{V} = j\omega L\boldsymbol{I} \tag{2-12}$$

If we define impedance as the ratio of phasor voltage to phasor current, for the inductor:

$$Z_L = j\omega L \tag{2-13}$$

follows from (2-12). For the resistor and capacitor we can derive:

$$Z_R = R \tag{2-14}$$

$$Z_C = -j\frac{1}{\omega C} \tag{2-15}$$

Because ω is constant, it is convenient to define reactance for the inductor and capacitor:

$$X_L = \omega L \tag{2-16a}$$

$$X_C = \frac{1}{\omega C} \tag{2-16b}$$

The rules for series and parallel combinations apply to impedances as they do to resistances in the dc case; however, such reductions require complex number arithmetic. It is assumed that the reader has the necessary computational skills; if not, consult any ac circuits textbook, for example, [2] in the Bibliography at the end of the chapter. A more general definition of impedance is illustrated in Figure 2-2. Here:

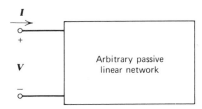

Figure 2-2 General definition of impedance.

$$Z = \frac{V}{I} \tag{2-17a}$$

and

$$R = \text{Re}[Z] \tag{2-17b}$$

$$X = |\text{Im}[Z]| \tag{2-17c}$$

Example 2-2

Given a 100 volt sinusoidal source in series with a 3Ω resistor, an 8Ω inductor, and a 4Ω capacitor,

(a) draw the circuit diagram

The solution is shown in Figure 2-3.

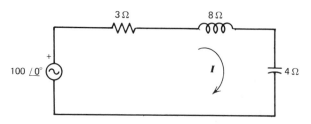

Figure 2-3 Circuit for Example 2-2.

First, the source. "100 volts" is understood to be rms. The angle "0°" is assigned arbitrarily and is a logical choice. We selected the source voltage as phase reference for convenience.

The "8Ω" value assigned to the inductor is its reactance (Its impedance is $+j8$; its inductance is measured in henries, not ohms.)

(b) compute the series impedance:

$$Z = 3 + j8 - j4$$
$$= 3 + j4$$
$$= 5\underline{/+53.1°} \quad \text{(Note that the impedance is a complex number.)}$$

(c) compute the current I:

$$I = \frac{100\underline{/0°}}{5\underline{/53.1°}} = 20\underline{/-53.1°} \text{ amperes}$$

The "20 amperes" is the rms value of the current. The phase angle "$-53.1°$" tells us the current is 53.1° in phase *behind* the source voltage. It is common to say "the current *lags* the voltage by 53.1°."

For some applications an equivalent parameter, admittance, is sometimes used. From equation (2-17a):

$$Y = \text{Complex Admittance} = \frac{I}{V} = \frac{1}{Z} \tag{2-18a}$$

$$G = \text{Conductance} = \text{Re}[Y] \tag{2-18b}$$

$$B = \text{Susceptance} = |\text{Im}[Y]| \tag{2-18c}$$

The units are ohms^{-1} (mhos, Siemens).

2-3 Complex Power in Sinusoidal Steady State Circuits

Consider the situation shown in Figure 2-2. The complex power S flowing into the network is defined as:

$$S = VI^* \tag{2-19a}$$

Consider $V = V\underline{/\alpha}$ and $I = I\underline{/\beta}$. Then

$$S = (V\underline{/\alpha})(I\underline{/\beta})^* \tag{2-19b}$$

$$= VI\underline{/\alpha - \beta} \tag{2-19c}$$

or in rectangular form:

$$S = VI \cos(\alpha - \beta) + jVI \sin(\alpha - \beta) \tag{2-19d}$$

Note that the angle $(\alpha - \beta)$ is the angle by which the current lags the voltage. Concentrate on the real part of equation (2-19d). The reader may recognize this quantity as the average power flowing into the network. That is:

$$v = V\sqrt{2} \cos(\omega t + \alpha) \tag{2-20a}$$

$$i = I\sqrt{2} \cos(\omega t + \beta) \tag{2-20b}$$

and

$$P = \frac{1}{T} \int_0^T vi\, dt \tag{2-21a}$$

$$= VI \cos(\alpha - \beta) \tag{2-21b}$$

where

$$T = \frac{2\pi}{\omega} \tag{2-21c}$$

It is clear from comparing (2-19d) and (2-21b) that:

$$P = \text{Re}[S] \tag{2-22}$$

The other component of S is more difficult to relate to a physical energy flow rate. Let us define it as:

$$Q = \text{Im}[S] = \text{Reactive power} \tag{2-23}$$

A comment about the sign of Q, the reactive power, is in order. Consider $-90° \le (\alpha - \beta) \le 90°$. This will be the case for all passive network terminations. Now notice that the sign of Q depends on the sign of $\sin(\alpha - \beta)$. When $(\alpha - \beta)$ is *positive*, so is Q and vice versa. For what type of load is

$(\alpha - \beta)$ positive? A little thought leads to the conclusion that $(\alpha - \beta)$ is positive when the current *lags* the voltage in phase, which is true for an inductive load. Therefore, inductive loads *absorb* positive reactive power. In power jargon this is referred to as the *lagging* case.

It is clear that the SI units of S, P, and Q must all be the same, namely, joules/s. It is conventional to assign different names to the units of S, P, and Q, and in fact use these unit names synonymously with the quantities themselves. They are:

S: voltamperes (VA), kilovoltamperes (kVA), megavoltamperes (MVA)

P: watts (W), Kilowatts (kW), megawatts (MW)

Q: voltamperes-reactive (var), kilovoltamperes-reactive (kvar), megavoltamperes-reactive (Mvar)

The preceding properties are summarized in the following equations.

$$S = P + jQ = \text{Complex power} \tag{2-24a}$$

where

$$S = VI = \text{apparent power in VA} \tag{2-24b}$$

$$P = VI \cos(\alpha - \beta) = \text{Average (real) power in W} \tag{2-24c}$$

$$Q = VI \sin(\alpha - \beta) = \text{Reactive (imaginary) power in var} \tag{2-24d}$$

It is common practice to refer to the angle $(\alpha - \beta)$ as the "power factor angle (ψ)" and $\cos(\alpha - \beta)$ as the "power factor" (pf).

$$\psi = \text{power factor angle} = \alpha - \beta \tag{2-25a}$$

$$pf = \text{power factor} = \cos(\psi) \tag{2-25b}$$

Notice that ψ is also the angle of the impedance.

$$Z = \frac{V}{I} \tag{2-26a}$$

$$= \frac{V}{I} \underline{/\alpha - \beta} \tag{2-26b}$$

$$= \frac{V}{I} \underline{/\psi} \tag{2-26c}$$

A lagging power factor indicates an inductive impedance and therefore a *positive* value for ψ.

Example 2-3

Calculate the complex power delivered to each of the four elements in the circuit of example 2-2.

Solution

The resistor:

$$S_R = V_R I^* = IRI^* = I^2 R$$
$$= (20^2)3 = 1200 + j0$$

The inductor:

$$S_L = V_L I^* = (jX_L I)I^* = jX_L I^2$$
$$= j(8)(20)^2 = 0 + j3200$$

The capacitor:

$$S_c = V_c I^* = (-jIX_c)I^* = -jX_c I^2$$
$$= -j(4)(20)^2 = 0 - j1600$$

The total load† complex power:

$$S_{\text{Load}} = S_R + S_L + S_c$$
$$= 1200 + j3200 - j1600 = 2000\underline{/+53.1°}$$

The total source complex power:

$$S_v = VI^*$$
$$= (100\underline{/0°})(20\underline{/-53.1°})^*$$
$$= 2000\underline{/+53.1°}$$

Note that $S_v = S_{\text{Load}}$. It is true in general in a given circuit that the net source complex power will equal the net load complex power.

Energy delivered to the network of Figure 2-2 in the time interval $t_2 - t_1$ is

$$W = \int_{t_1}^{t_2} p \, dt \qquad\qquad (2\text{-}27a)$$

† When used in electrical power applications the term "load" does not have a precise technical definition. It can mean current, power, or impedance depending on context. The general idea refers to the component(s) that absorb electrical energy from the system.

For sinusoidal steady state operation:

$$W = P(t_2 - t_1) \tag{2-27b}$$

where P is defined in equation (2-21). If P is in watts and t is in seconds, the energy W is in joules. Sometimes power is given in kilowatts and time in hours, forcing W into kW-hrs. Observe that the kilowatt-hour is not an SI unit, and will require conversion factors in other equations.

2-4 The Ideal Three Phase Source

Consider the circuit shown in Figure 2-4(a). Such a network is by definition an ideal three phase (3ϕ) source. The branches a-n, b-n, and c-n are referred to as the "phases" of the source. This is an ambiguous use of the term "phase," but the terminology is accepted practice. The nodes a, b, and c are referred to as "terminals" or "lines" and the point n as the neutral. The connection is called the wye or star connection. The term "balanced" when applied to 3ϕ voltages or currents means "equal in magnitude, 120 degrees apart in phase."

Because so many variables are involved in 3ϕ circuits it is cumbersome to show them all on the circuit diagram. It is useful to define and use a double subscript notation. For voltages we should understand that the first subscript is defined positive with respect to the second. Thus, changing the order of the subscript produces a 180 degree phase shift in the variable. For currents the subscript order defines the "from–to" direction.

The phasor diagram showing E_{an}, E_{bn}, E_{cn}, and E_{ab} is illustrated in Figure 2-5(a). Let us compute the voltage between terminals a and b (E_{ab}).

$$E_{ab} = E_{an} - E_{bn} \tag{2-28a}$$

$$= E\underline{/0°} - E\underline{/-120°} \tag{2-28b}$$

$$= E\sqrt{3}\underline{/30°} \tag{2-28c}$$

Similarly, it is straightforward to show that:

$$E_{bc} = E\sqrt{3}\underline{/-90°} \tag{2-28d}$$

$$E_{ca} = E\sqrt{3}\underline{/150°} \tag{2-28e}$$

There are two important facts to remember from this result:

$$E_L = \sqrt{3}E\phi \tag{2-29a}$$

where

$$E_L = |E_{ab}| = |E_{bc}| = |E_{ca}| = E\sqrt{3} \tag{2-29b}$$

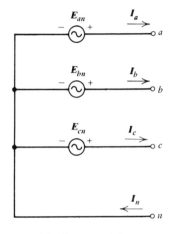

(a) Wye connected

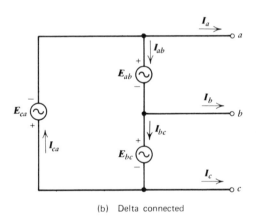

(b) Delta connected

Figure 2-4 The ideal three phase source.

and

$$E_\phi = |E_{an}| = |E_{bn}| = |E_{cn}| = E \qquad (2\text{-}29\text{c})$$

Also, the line voltages (E_L) are not in phase with the phase voltages (E_ϕ). In this particular instance the line voltages are 30° ahead of the phase voltages. The exact phase relations in a given situation depend on the notation used, sign conventions, and choice of voltages. Obviously, the current that flows from a terminal of the generator is the same current that flows in a phase.

39

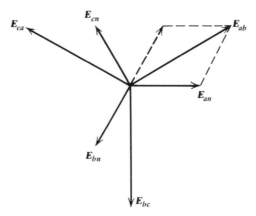

(a)   Phase and line voltages in a wye connection

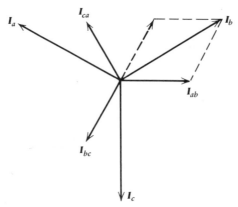

(b)   Phase and line currents in a delta connection

Figure 2-5 Phasor relations in 3ϕ sources.

Study the balanced voltages E_{an}, E_{bn}, and E_{cn}. Relative to E_{an}, there are only two possible locations for E_{bn}: lagging E_{an} by 120°, or leading E_{an} by 120°. These two possibilities are differentiated by a conventional description referred to as "phase sequence." If E_{bn} *lags* E_{an} by 120° the phase sequence is designated as *abc*. The other possible phase sequence is *cba* (E_{bn} *leads* E_{an} by 120°). For the case illustrated, the phase sequence is *abc*.

There is one other possible symmetrical interconnection of the three "phases" that can be made to form a three phase source. See Figure 2-4(b).

40

Such a connection is referred to as a delta, or mesh, connection. Suppose the following:

$$I_{ab} = I\underline{/0°} \qquad (2\text{-}30a)$$

$$I_{bc} = I\underline{/-120°} \qquad (2\text{-}30b)$$

$$I_{ca} = I\underline{/+120°} \qquad (2\text{-}30c)$$

Note that as given here the phase sequence is *abc*. The corresponding phasor diagram is shown in Figure 2-5(b).

By *KCL*

$$I_b = I_{ab} - I_{bc} \qquad (2\text{-}31a)$$

$$= I\underline{/0°} - I\underline{/-120°} \qquad (2\text{-}31b)$$

$$= \sqrt{3}I\underline{/30°} \qquad (2\text{-}31c)$$

Notice that:

$$I_L = \sqrt{3}I_\phi \qquad (2\text{-}32)$$

where I_L is the line current magnitude and I_ϕ is the phase current magnitude in a delta connection. Also notice that the phase and line currents are *not* in phase with each other. Obviously, the phase voltage *is* the line voltage in a delta.

It is possible to find an equivalent delta for every balanced 3ϕ wye connected ideal source, and vice versa. By this we mean that if the two 3ϕ sources were placed in boxes and only three terminals brought out of each (*a*, *b*, and *c*—the neutral is suppressed) there would be no possible electrical measurements that could be used to distinguish the wye from the delta. The conditions for equivalence are that the corresponding line voltages must be equal, in magnitude and phase. We capitalize on this fact by realizing that we may concentrate on the wye connection without loss of generality.

2-5 The Balanced Three Phase Load

Whenever currents that are equal in magnitude and 120° displaced in phase flow as to deliver real power to a three terminal network, the network is viewed as a balanced three phase load. The simplest manifestation of this situation is the case where three passive elements are connected in wye or delta. With regard to impedance, as shown in Figure 2-6(a), "balanced" simply means equal:

$$Z_{an} = Z_{bn} = Z_{cn} = Z_Y \qquad (2\text{-}33)$$

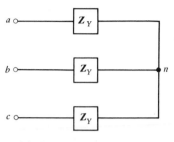

(a)   The wye connected load

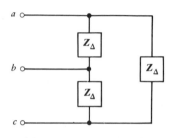

(b)   The delta connected load

Figure 2-6 Balanced three phase loads.

If balanced three phase voltages are applied to the wye connected load of Figure 2-6(a), the currents will be balanced. Observe that

$$I_n = I_a + I_b + I_c = 0 \tag{2-34}$$

If $I_n = 0$, the neutral conductor may be removed with no effect on the system. For the delta connection shown in Figure 2-6(b) "balanced" means:

$$Z_{ab} = Z_{bc} = Z_{ca} = Z_\Delta \tag{2-35}$$

The condition for equivalence between the wye and the delta is:

$$Z_\Delta = 3Z_Y \tag{2-36a}$$

which implies independently that

$$\text{Arg}[Z_\Delta] = \text{Arg}[Z_Y] = \psi \tag{2-36b}$$

and

$$Z_\Delta = 3Z_Y \tag{2-36c}$$

For proof, see problem 2-7. "Equivalence" means that no electrical measurements may be made at the terminals a, b, and c that can be used to distinguish between the two networks. The power factor for the 3ϕ load is

$$pf = \cos \psi \qquad (2\text{-}37)$$

lagging for $0 < \psi \leq \pi/2$

leading for $0 > \psi \geq -\pi/2$

Considering power, "balanced" again means equal:

$$S_{an} = S_{bn} = S_{cn} = S_{1\phi} \qquad (2\text{-}38)$$

For the equivalent delta:

$$S_{ab} = S_{bc} = S_{ca} = S_{1\phi} \qquad (2\text{-}39)$$

Either way:

$$S_{3\phi} = 3S_{1\phi} \qquad (2\text{-}40a)$$

$$= 3V_\phi I_\phi \underline{/\psi} \qquad (2\text{-}40b)$$

where

$V_\phi = $ phase voltage magnitude

$I_\phi = $ phase current magnitude

If we apply equation (2-40b) to the wye load:

$$S_{3\phi} = 3\frac{V_L}{\sqrt{3}}(I_L)\underline{/\psi} \qquad (2\text{-}41a)$$

$$= V_L I_L \sqrt{3}\underline{/\psi} \qquad (2\text{-}41b)$$

where

$V_L = $ line voltage magnitude

$I_L = $ line current magnitude

Identical results occur when applied to the delta load. When we recall:

$$S_{3\phi} = P_{3\phi} + jQ_{3\phi} \qquad (2\text{-}41c)$$

we conclude:

$$S_{3\phi} = V_L I_L \sqrt{3} \qquad (2\text{-}42a)$$

$$P_{3\phi} = V_L I_L \sqrt{3} \cos \psi \qquad (2\text{-}42b)$$

$$Q_{3\phi} = V_L I_L \sqrt{3} \sin \psi \qquad (2\text{-}42c)$$

which are valid independent of whether the load connection is wye or delta.

2-6 Symmetrical Components

The electrical power system normally operates in a balanced three phase sinusoidal steady state mode. However, there are certain situations that can cause unbalanced operation. The most severe of these is the so-called "fault," or short circuit. An example would be a tree in contact with one phase of an overhead transmission line. To protect the system against such contingencies, we need to size protective devices, such as fuses and circuit breakers. For these and other reasons, it is necessary to calculate currents and voltages in the system under such unbalanced operating conditions. Our first impulse is to simply extend our per phase analysis approach to deal with all three phases separately, in effect tripling our work. Unfortunately, it turns out that the work is significantly more than triple that of the balanced case.

In a classical paper, C. L. Fortescue[*] described how arbitrarily unbalanced three phase voltages (or currents) could be transformed into three sets of balanced three phase components, which he called "symmetrical components." The application to power system analysis is of fundamental importance. We can transform an arbitrarily unbalanced condition into symmetrical components, compute the system response by straightforward circuit analysis on simple circuit models, and transform the results back into the original phase variables. This approach proves to be superior to the direct, but much more complicated, method of solving unbalanced problems in the original three phase system.

In a general three phase circuit, instead of a single voltage and current, we must deal with a minimum of three voltages and three currents, which we will loosely refer to as "phase values." Their specific definitions are shown in Figure 2-7. Each is to be considered as sinusoidal steady state and

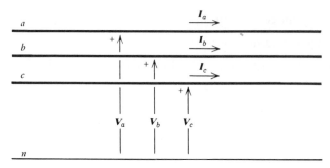

Figure 2-7 Positive definitions of phase voltages and currents in a generalized three phase situation.

[*] C. L. Fortescue, "Method of Symmetrical Coordinates Applied to the Solution of Polyphase Networks," *AIEE Transactions*, Vol. 37, Part 2, 1918.

therefore may be represented with phasor notation. Let us direct our attention to the voltages, realizing that similar statements can and will be made about the currents:

$$V_a = V_a \underline{/\theta_a} \qquad\qquad (2\text{-}43)$$

Note that the phase voltage V_a has two parts: magnitude and phase angle (two degrees of freedom).

To specify three unbalanced voltages we need six numbers:

$$V_a, V_b, V_c, \theta_a, \theta_b, \text{ and } \theta_c$$

Let us think of each phase voltage as having three components such that each voltage is the sum of its components:

$$V_a = V_{a_0} + V_{a_1} + V_{a_2} \qquad\qquad (2\text{-}44\text{a})$$

$$V_b = V_{b_0} + V_{b_1} + V_{b_2} \qquad\qquad (2\text{-}44\text{b})$$

$$V_c = V_{c_0} + V_{c_1} + V_{c_2} \qquad\qquad (2\text{-}44\text{c})$$

Since the phase voltages have only a total of 6 degrees of freedom, these components cannot be completely independent. Suppose we force the components V_{a_1}, V_{b_1}, and V_{c_1} to make up a balanced three phase set with phase sequence *abc*. This will require 2 degrees of freedom (not 6) and accomplish the desirable objective of preserving the balanced case. We refer to this set as the positive sequence components and use the subscript "1" to indicate this.

Let us force the set V_{a_2}, V_{b_2}, and V_{c_2} to be balanced with phase sequence *cba* and refer to these as negative sequence components (use subscript "2"). Again, 2 degrees of freedom are required and we have a second balanced set. The remaining set—V_{a_0}, V_{b_0}, and V_{c_0}—is called the zero sequence set. Note that they cannot be balanced three phase because, if so, they could be combined with either the positive or negative set. Also, the set can only have 2 degrees of freedom. We simply force V_{a_0}, V_{b_0}, and V_{c_0} to be equal in magnitude and phase.

When the components are interrelated as described above, they become "symmetrical components."

2-7 The *a* Operator as used in Symmetrical Component Representation

Recall the operator j. In polar form $j = 1\underline{/90°}$. Note multiplication by j has the effect of rotating a phasor forward 90° without affecting the magnitude.

45

Example 2-4

Compute jA where $A = 10\underline{/60°}$. Refer to Figure 2-8a and b.

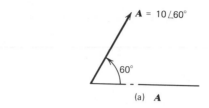

(a) A

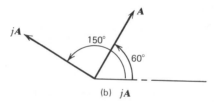

(b) jA

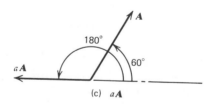

(c) aA

Figure 2-8 j and a effects.

$$jA = 1\underline{/90°}(10\underline{/60°})$$
$$= 10\underline{/150°}$$

Equivalently:

$$jA = j(10\underline{/60°})$$
$$= j(5 + j8.66)$$
$$= j5 - 8.66$$
$$= 10\underline{/150°}$$

46

In a similar manner define $a = 1\underline{/120°}$. Note that multiplication by "*a*" rotates a phasor forward 120° and does not effect the magnitude.

Example 2-5

Compute aA where $A = 10\underline{/60°}$. Refer to Figure 2-8a and c.

$$aA = (1\underline{/120°})(10\underline{/60°})$$

$$= 10\underline{/180°}$$

Note that:

$$a = 1\underline{/120°} \qquad (2\text{-}45\text{a})$$

$$= 1\underline{/-240°}$$

$$a^2 = a \cdot a \qquad (2\text{-}45\text{b})$$

$$= (1\underline{/120°})(1\underline{/120°})$$

$$= 1\underline{/240°}$$

$$= 1\underline{/-120°}$$

$$a^3 = 1\underline{/360°} \qquad (2\text{-}45\text{c})$$

$$= 1\underline{/0°}$$

Now recall the definition of symmetrical components. V_{b_1} *always lags* V_{a_1} by a fixed angle of 120° and always has the same magnitude as V_{a_1}. Similarly, V_{c_1} *leads* V_{a_1} by 120°. It follows then that:

$$V_{b_1} = a^2 V_{a_1} \qquad (2\text{-}46\text{a})$$

$$V_{c_1} = a V_{a_1} \qquad (2\text{-}46\text{b})$$

Similarly, we deduce:

$$V_{b_2} = a V_{a_2} \qquad (2\text{-}46\text{c})$$

$$V_{c_2} = a^2 V_{a_2} \qquad (2\text{-}46\text{d})$$

$$V_{b_0} = V_{a_0} \qquad (2\text{-}46\text{e})$$

$$V_{c_0} = V_{a_0} \qquad (2\text{-}46\text{f})$$

In other words, it is possible to write all nine of our symmetrical components in terms of three, namely those referred to the *a* phase.

Rewriting equations (2-44) and substituting (2-46) we produce:

$$V_a = V_{a_0} + V_{a_1} + V_{a_2} \qquad\qquad (2\text{-}47\text{a})$$

$$V_b = V_{b_0} + V_{b_1} + V_{b_2}$$

$$\quad = V_{a_0} + a^2 V_{a_1} + a V_{a_2} \qquad\qquad (2\text{-}47\text{b})$$

$$V_c = V_{c_0} + V_{c_1} + V_{c_2}$$

$$\quad = V_{a_0} + a V_{a_1} + a^2 V_{a_2} \qquad\qquad (2\text{-}47\text{c})$$

We may simplify the notation as follows. Define:

$$V_0 = V_{a_0} \qquad\qquad (2\text{-}48\text{a})$$

$$V_1 = V_{a_1} \qquad\qquad (2\text{-}48\text{b})$$

$$V_2 = V_{a_2} \qquad\qquad (2\text{-}48\text{c})$$

Remember that the symmetrical components V_0, V_1, and V_2 are referred to the "a" phase. From equations (2-47) we get:

$$V_a = V_0 + V_1 + V_2 \qquad\qquad (2\text{-}49\text{a})$$

$$V_b = V_0 + a^2 V_1 + a V_2 \qquad\qquad (2\text{-}49\text{b})$$

$$V_c = V_0 + a V_1 + a^2 V_2 \qquad\qquad (2\text{-}49\text{c})$$

These equations may be manipulated to solve for V_0, V_1, and V_2 in terms of V_a, V_b, and V_c. Doing this we get:

$$V_0 = 1/3(V_a + V_b + V_c) \qquad\qquad (2\text{-}50\text{a})$$

$$V_1 = 1/3(V_a + a V_b + a^2 V_c) \qquad\qquad (2\text{-}50\text{b})$$

$$V_2 = 1/3(V_a + a^2 V_b + a V_c) \qquad\qquad (2\text{-}50\text{c})$$

Equations (2-50) may be used to convert phase voltages (or currents) to symmetrical component voltages (or currents) and vice versa [equations (2-49)].

Example 2-6

Given $I_a = 1\underline{/60°}$, $I_b = 1\underline{/-60°}$, and $I_c = 0$, find the symmetrical components.

Solution

$$I_1 = 1/3(I_a + a I_b + a^2 I_c)$$

$$= 1/3(1\underline{/60°} + 1\underline{/60°} + 0)$$

$$= 1/3(0.5 + j0.866 + 0.5 + j0.866)$$

$$= 1/3(2\underline{/60°})$$

$$= 0.667\underline{/60°}$$

$$\boldsymbol{I}_2 = 1/3(\boldsymbol{I}_a + a^2\boldsymbol{I}_b + a\boldsymbol{I}_c)$$

$$= 1/3(1\underline{/60°} + 1\underline{/180°})$$

$$= 1/3(0.5 + j0.866 - 1)$$

$$= 0.333\underline{/120°}$$

$$\boldsymbol{I}_0 = 1/3(\boldsymbol{I}_a + \boldsymbol{I}_b + \boldsymbol{I}_c)$$

$$= 1/3(0.5 + j0.866 + 0.5 - j0.866)$$

$$= 0.333\underline{/0°}$$

Note.

$$\boldsymbol{I}_{c_1} = a\boldsymbol{I}_1 = 0.667\underline{/180°}$$

$$\boldsymbol{I}_{c_2} = a^2\boldsymbol{I}_2 = 0.333\underline{/360°}$$

$$\boldsymbol{I}_{c_0} = \boldsymbol{I}_0 = 0.333\underline{/0°}$$

Check.

$$\boldsymbol{I}_c = \boldsymbol{I}_{c_1} + \boldsymbol{I}_{c_2} + \boldsymbol{I}_{c_0}$$

$$= -0.667 + 0.333 + 0.333 \cong 0 \text{ as given.}$$

2-8 Matrix Notation

The transformation equations (2-49) and (2-50) may be written in matrix form:

$$\begin{bmatrix} V_a \\ V_b \\ V_c \end{bmatrix} = \begin{bmatrix} 1 & 1 & 1 \\ 1 & a^2 & a \\ 1 & a & a^2 \end{bmatrix} \begin{bmatrix} V_0 \\ V_1 \\ V_2 \end{bmatrix} \qquad (2\text{-}51)$$

$$\begin{bmatrix} V_0 \\ V_1 \\ V_2 \end{bmatrix} = 1/3 \begin{bmatrix} 1 & 1 & 1 \\ 1 & a & a^2 \\ 1 & a^2 & a \end{bmatrix} \begin{bmatrix} V_a \\ V_b \\ V_c \end{bmatrix} \qquad (2\text{-}52)$$

or more compactly

$$\tilde{V}_{abc} = [T]\tilde{V}_{012} \tag{2-53}$$

$$\tilde{V}_{012} = [T]^{-1}\tilde{V}_{abc} \tag{2-54}$$

where the definitions of $\tilde{V}_{abc}$, $\tilde{V}_{012}$, $[T]$, and $[T]^{-1}$ are obvious comparing equations (2-51) and (2-52) with (2-53) and (2-54).

The same transformation holds for currents:

$$\tilde{I}_{abc} = [T]\tilde{I}_{012} \tag{2-55}$$

$$\tilde{I}_{012} = [T]^{-1}\tilde{I}_{abc} \tag{2-56}$$

2-9 The Effect on Impedance

The effect on impedance must be derived. Suppose we start with:

$$\tilde{V}_{abc} = [Z_{abc}]\tilde{I}_{abc} \tag{2-57}$$

where $[Z_{abc}]$ is a 3×3 matrix giving the self and mutual impedance in and between phases. Substitute (2-53) and (2-55) into (2-57):

$$[T]\tilde{V}_{012} = [Z_{abc}][T]\tilde{I}_{012} \tag{2-58}$$

or

$$\tilde{V}_{012} = [T]^{-1}[Z_{abc}][T]\tilde{I}_{012} \tag{2-59}$$

Define

$$[Z_{012}] = [T]^{-1}[Z_{abc}][T] \tag{2-60}$$

so that

$$\tilde{V}_{012} = [Z_{012}]\tilde{I}_{012} \tag{2-61}$$

The key to understanding the importance of symmetrical components lies in equation (2-60). For typical power system components the matrix $[Z_{abc}]$ is not diagonal, but does possess certain symmetries. These symmetries are such that $[Z_{012}]$ is diagonal, either exactly or approximately. When this is the case, the analysis is greatly simplified. An example illustrates the point.

Example 2-7

Evaulate $[Z_{012}]$ for the line shown in Figure 2-9.

50

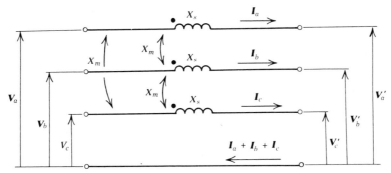

Figure 2-9 Simplified transmission line circuit diagram.

Kirchhoff's Voltage Law produces:

$$V_a - V'_a = jX_s I_a + jX_m I_b + jX_m I_c \tag{2-62a}$$

$$V_b - V'_b = jX_m I_a + jX I_b + jX_m I_c \tag{2-62b}$$

$$V_c - V'_v = jX_m I_a + jX_m I_b + jX_s I_c \tag{2-62c}$$

or in matrix notation:

$$\begin{bmatrix} V_a \\ V_b \\ V_c \end{bmatrix} - \begin{bmatrix} V'_a \\ V'_b \\ V'_c \end{bmatrix} = j \begin{bmatrix} X_s & X_m & X_m \\ X_m & X_s & X_m \\ X_m & X_m & X_s \end{bmatrix} \begin{bmatrix} I_a \\ I_b \\ I_c \end{bmatrix} \tag{2-62d}$$

Even more compactly:

$$\tilde{V}_{abc} - \tilde{V}'_{abc} = [Z_{abc}]\tilde{I}_{abc} \tag{2-62e}$$

Transforming into sequence values

$$\tilde{V}_{012} - \tilde{V}'_{012} = [Z_{012}]\tilde{I}_{012} \tag{2-62f}$$

where

$$[Z_{012}] = [T]^{-1}[Z_{abc}][T] \tag{2-63a}$$

$$= 1/3 \begin{bmatrix} 1 & 1 & 1 \\ 1 & a & a^2 \\ 1 & a^2 & a \end{bmatrix} j \begin{bmatrix} X_s & X_m & X_m \\ X_m & X_s & X_m \\ X_m & X_m & X_s \end{bmatrix} \begin{bmatrix} 1 & 1 & 1 \\ 1 & a^2 & a \\ 1 & a & a^2 \end{bmatrix}$$

$$= j/3 \begin{bmatrix} (X_s + 2X_m) & (X_s + 2X_m) & (X_s + 2X_m) \\ (X_s - X_m) & (aX_s + (1+a^2)X_m) & (a^2 X_s + (1+a)X_m) \\ (X_s - X_m) & (a^2 X_s + (1+a)X_m) & (aX_s + (1+a^2)X_m) \end{bmatrix} \begin{bmatrix} 1 & 1 & 1 \\ 1 & a^2 & a \\ 1 & a & a^2 \end{bmatrix}$$

$$= j \begin{bmatrix} X_s + 2X_m & 0 & 0 \\ 0 & X_s - X_m & 0 \\ 0 & 0 & X_s - X_m \end{bmatrix} \tag{2-63b}$$

We define for this line

$Z_0 =$ Zero sequence impedance $= Z_{00} = j(X_s + 2X_m)$

$Z_1 =$ Positive sequence impedance $= Z_{11} = j(X_s - X_m)$

$Z_2 =$ Negative sequence impedance $= Z_{22} = j(X_s - X_m)$

The sequence networks are shown in Figure 2-10.
Be sure to recognize that the mutual coupling has been eliminated.

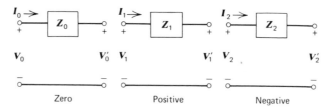

Figure 2-10 Sequence networks.

2-10 Power Considerations

We observe that the total complex power flowing from left to right in the generalized situation shown in Figure 2-7 is:

$$S_{3\phi} = V_a I_a^* + V_b I_b^* + V_c I_c^* \tag{2-64}$$

In matrix notation:

$$S_{3\phi} = \tilde{V}_{abc_t} \tilde{I}_{abc}^* \tag{2-65a}$$

$$= \{[T]\tilde{V}_{012}\}_t \{[T]\tilde{I}_{012}\}^* \tag{2-65b}$$

$$= \tilde{V}_{012_t} [T]_t [T]^* \tilde{I}_{012}^* \tag{2-65c}$$

Now

$$[T]_t [T]^* = \begin{bmatrix} 1 & 1 & 1 \\ 1 & a^2 & a \\ 1 & a & a^2 \end{bmatrix} \begin{bmatrix} 1 & 1 & 1 \\ 1 & a & a^2 \\ 1 & a^2 & a \end{bmatrix}$$

$$= \begin{bmatrix} 3 & 0 & 0 \\ 0 & 3 & 0 \\ 0 & 0 & 3 \end{bmatrix} = 3 \begin{bmatrix} 1 & 0 & 0 \\ 0 & 1 & 0 \\ 0 & 0 & 1 \end{bmatrix} \tag{2-66}$$

Therefore:

$$S_{3\phi} = 3\tilde{V}_{012_t} \tilde{I}_{012}^* \tag{2-67a}$$

or

$$S_{3\phi} = 3[V_0 I_0^* + V_1 I_1^* + V_2 I_2^*] \tag{2-67b}$$

Think about this result. Notice that there are no "cross" terms (such as $V_1 I_0^*$). This is an essential property for the transformation if we are to construct equivalent circuits and analyze with conventional circuit analysis techniques. The factor of three is reasonable when we think in terms of nine components (b and c as well as a phase components). It is possible to remove the three by defining $[T]$ with a $1/\sqrt{3}$ coefficient, which some workers prefer.

Example 2-8

Evaluate $S_{3\phi}$ two ways [equations (2-65a) and (2-67b)], given:

$$\tilde{V}_{abc} = \begin{bmatrix} 100 \\ -100 \\ 0 \end{bmatrix} \quad \text{and} \quad \tilde{I}_{abc} = \begin{bmatrix} j10 \\ -10 \\ -10 \end{bmatrix}$$

Solution

$$\tilde{S}_{3\phi} = \tilde{V}_{abc_t} \tilde{I}_{abc}^*$$

$$= [100 \quad -100 \quad 0] \begin{bmatrix} -j10 \\ -10 \\ -10 \end{bmatrix} = 1000 - j1000$$

Now

$$\tilde{V}_{012} = [T]^{-1} \tilde{V}_{abc}$$

$$= \frac{1}{3} \begin{bmatrix} 1 & 1 & 1 \\ 1 & a & a^2 \\ 1 & a^2 & a \end{bmatrix} \begin{bmatrix} 100 \\ -100 \\ 0 \end{bmatrix} = \frac{1}{\sqrt{3}} \begin{bmatrix} 0 \\ 100\underline{/-30°} \\ 100\underline{/+30°} \end{bmatrix}$$

$$\tilde{I}_{012} = [T]^{-1} \tilde{I}_{abc}$$

$$= \frac{1}{3} \begin{bmatrix} 1 & 1 & 1 \\ 1 & a & a^2 \\ 1 & a^2 & a \end{bmatrix} \begin{bmatrix} j10 \\ -10 \\ -10 \end{bmatrix} = \frac{1}{3} \begin{bmatrix} j10 - 20 \\ j10 + 10 \\ j10 + 10 \end{bmatrix}$$

53

$$S_{3\phi} = 3\tilde{V}_{012_t}\tilde{I}^*_{012}$$

$$= \left[0 \quad \frac{100}{\sqrt{3}}\underline{/-30°} \quad \frac{100}{\sqrt{3}}\underline{/+30°}\right]\begin{bmatrix} -j10-20 \\ 10\sqrt{2}\underline{/-45°} \\ 10\sqrt{2}\underline{/-45°} \end{bmatrix}$$

$$= \frac{1000\sqrt{2}}{\sqrt{3}}[1\underline{/-75°} + 1\underline{/-15°}] = 1000 - j1000$$

2-11 Summary

Power system electrical performance is basically described using sinusoidal steady state (ac) circuit concepts. The phasor method of describing voltages and currents has proven to be of indispensable merit, and is used extensively throughout the rest of this book. Naturally, power and energy calculations are important and were emphasized. It is important to understand the "three kinds" of power (S, P, and Q).

Because the power system is typically three phase, special attention was directed at that topic. Perhaps you were curious as to why the balanced case was emphasized. Many problems in power system analysis are concerned with the system operating in its normal balanced three phase mode and, although it is a special case from a mathematical viewpoint, it is of considerable practical importance. Unbalanced systems are changed through the symmetrical component transformation into three balanced systems. Unexpectedly, balanced concepts are again of prime importance.

We have developed the method of symmetrical components as essentially a mathematical transformation or change of coordinates. We have asserted that circuit models for power system components are quite simple for each set of sequence components. The next step is to discuss such circuit models. We should recognize that when the system is operating in its normal balanced mode, the phase values become the positive sequence values, with the negative and zero sequence quantities evaluating to zero. In that sense, we think of the balanced case as a special example of a more general situation that involves all three sequence networks.

As in all electrical engineering one must be careful not to confuse the physical device with its mathematical model. Thus, when discussing the ideal three phase source, one is tempted to equate it to an actual three phase generator. Actually, these two are dramatically different in many respects; we spend considerable time later developing a more satisfactory model.

Even then the model is different in some important respects from the real thing. There is *always* a trade-off between accuracy and complexity, with the ideal of perfect accuracy unattainable. Where do we stop this process? The answer requires an understanding of what our calculated results are to be used for. This understanding comes only from experience. The models and methods we will investigate represent the combined experience of many engineers and scientists. As our investigation of the power system develops, the student will acquire a growing appreciation of why a thorough understanding of the basics of ac circuits is necessary.

Bibliography

[1] Anderson, Paul M., *Analysis of Faulted Power Systems*, Iowa State Press, Ames, Iowa, 1973.

[2] Brenner, Egon, and Javid, Mansour, *Analysis of Electric Circuits*, McGraw-Hill, Inc., New York, 1967.

[3] Calabrese, G. O., *Symmetrical Components Applied to Electric Power Networks*, Ronald Press, New York, 1959.

[4] Clarke, Edith, *Circuit Analysis of A-C Power Systems*, 2 vols. General Electric Co., Schenectady, N.Y., 1950.

[5] Elgerd, Olle I., *Electric Energy Systems Theory: An Introduction*, McGraw-Hill, New York, 1971.

[6] Fortescue, C. L., *Method of Symmetrical Coordinates Applied to the Solution of Polyphase Networks*, Trans. AIEE 37: pp 1027–1140, 1918.

[7] Hancock, N. N., *Matrix Analysis of Electrical Machinery*, 2nd edition. Pergamon Press, Oxford, 1974.

[8] Hayt, Jr., William H., and Kemmerly, Jack E., *Engineering Circuit Analysis*, 2nd edition, McGraw-Hill, New York, 1971.

[9] Stevenson, Jr., William D., *Elements of Power Systems Analysis*, 3rd edition. McGraw-Hill, New York, 1975.

[10] Van Valkenburg, M. E., *Network Analysis*, 3rd edition. Prentice-Hall, Inc., Englewood Cliffs, N.J., 1974.

[11] Wagner, C. F., and Evans, R. D., *Symmetrical Components*, McGraw-Hill, New York, 1933.

Problems

2-1. (a) For the given $f(t)$ find F.
    (1) $100 \cos(10t - 20°)$
    (2) $-141.4 \sin(100t + 30°)$
  (b) For the given F, find $f(t)$.
    (1) $100\underline{/-80°}$
    (2) $-80 + j60$

2-2. (a) Given that $i(t) = I_{max} \cos(\omega t)$

    prove that the rms value $= I = \dfrac{I_{max}}{\sqrt{2}}$

    by applying the definition:

$$I = \sqrt{\frac{1}{T} \int_0^T i^2(t)\, dt}$$

    where

$$\omega = \frac{2\pi}{T}$$

  (b) Realizing that the instantaneous power p supplied to a resistor is $i^2 R$ and that

$$P = \frac{1}{T} \int_0^T p\, dt$$

    prove that $P = I^2 R$ using the results of (a).

2-3. Consider the given circuit in Figure P2-3.

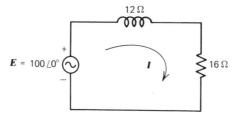

Figure P2-3 Circuit for Problem 2-3.

  (a) Solve for I
  (b) If $\omega = 377$ rad/s, determine L.
  (c) Write $e(t)$ and $i(t)$ as cosine functions.

(d) Prove by direct substitution that the functions $e(t)$ and $i(t)$ satisfy KVL:

$$e = iR + L\frac{di}{dt}$$

2-4 Given that $v(t) = V\sqrt{2}\cos(\omega t + \alpha)$ and $i(t) = I\sqrt{2}\cos(\omega t + \beta)$, show that

$$P = \frac{1}{T}\int_0^T vi\,dt = VI\cos(\alpha - \beta)$$

where

$$\omega = \frac{2\pi}{T}$$

2-5. A wye connected 3ϕ source has $E_{bn} = 120\underline{/0°}$, and a phase sequence of abc. Find E_{ab}, E_{bc}, and E_{ca}.

2-6. A delta connected 3ϕ source has $E_{ab} = 1732\underline{/0°}$. Calculate E_{an}, E_{bn}, and E_{cn} for the wye equivalent source if the phase sequence is to be cba.

2-7. Assume a balanced 3ϕ voltage source (wye or delta) with $E_{ab} = E\sqrt{3}\underline{/0°}$ and phase sequence abc. Compute and locate on a phasor diagram E_{bc}, E_{ca}, E_{an}, E_{bn}, and E_{cn}. Connect the three terminal source (abc) to a balanced delta load such that $Z_\Delta = Z_\Delta\underline{/\psi}$. Compute the load currents I_{ab}, I_{bc}, and I_{ca}. Add these to your diagram. Compute the line currents I_a, I_b, and I_c and locate these on the diagram. Now consider what wye impedance could be used to replace the delta so that the line currents would remain unchanged in both magnitude and phase. The result is:

$$Z_Y = \frac{E_{an}}{I_a}$$

Simplify, relate to Z_Δ, and therefore derive equation (2-36a).

2-8. A three phase load is rated at 2400 volts and 100 kVA, $pf = 0.8$ lagging.
(a) Find the complex phase impedance assuming a wye connection.
(b) Find the complex phase impedance assuming a delta connection.

2-9. Three impedances $Z = 15 - j10$ are connected in wye and connected to a 208 volt 3ϕ 60 Hz supply. Assume $V_{ab} = 208\underline{/0°}$ and a phase sequence of abc. Find complex phasor values for all voltages and currents and show these phasors on a phasor diagram. (i.e., V_{ab}, V_{bc}, V_{ca}, V_{an}, V_{bn}, V_{cn}, I_a, I_b, and I_c).

57

2-10. Repeat problem 2-9 if the same impedances are connected in delta. In addition to the voltages and currents mentioned, compute the phase currents I_{ab}, I_{bc}, and I_{ca}.

2-11. A 3ϕ 500 kVA 2400 volt load has a $pf = 0.8$ lagging. Find the line voltage (V_L), the line current (I_L), 3ϕ real power $(P_{3\phi})$, 3ϕ apparent power $(S_{3\phi})$, and 3ϕ reactive power $(Q_{3\phi})$.

2-12. An ideal wattmeter has a zero impedance current coil and an infinite impedance potential coil, and reads $P = VI \cos(\alpha - \beta)$. Consider the following balanced 3ϕ situation shown in Figure P2-12.

(a) Prove that the top wattmeter reads:

$$P_1 = V_L I_L \cos(\psi + 30°)$$

(b) Prove that the bottom wattmeter reads:

$$P_2 = V_L I_L \cos(\psi - 30°)$$

(c) Now show that

$$P_1 + P_2 = V_L I_L \sqrt{3} \cos \psi$$
$$= P_{3\phi}$$

(d) Likewise, show that

$$P_2 - P_1 = V_L I_L \sin \psi$$

$$= \frac{Q_{3\phi}}{\sqrt{3}}$$

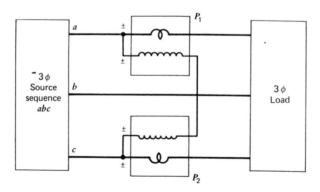

Figure P2-12 Circuit for Problem 2-12.

2-13. In problem 2-12, $P_1 = 50$ kW and $P_2 = 30$ kW.
   (a) Find $P_{3\phi}$
   (b) Find $Q_{3\phi}$
   (c) Find $S_{3\phi}$
   (d) Find pf

2-14. For this problem, four colored pens (or pencils) and a sheet of polar graph paper are recommended. Given that $V_{a_1} = 100\underline{/0°}$, $V_{a_2} = 50\underline{/90°}$, and $V_{a_0} = 20\underline{/30°}$: (a) make a phasor diagram showing all nine symmetrical components, with each sequence in a different color; (b) graphically add the appropriate phasors to obtain the phase voltages, to be shown in a fourth color; (c) evaulate $\tilde{V}_{abc}$ using equation (2-53). Check results against those obtained in part b.

2-15. Given $V_a = V\underline{/\theta}$, $V_b = V\underline{/\theta - 120°}$, $V_c = V\underline{/\theta + 120°}$ evaluate $\tilde{V}_{012}$ using equation (2-54). Are $V_0 = V_2 = 0$? Why?

2-16. Derive equation (2-52) from (2-51).

2-17. Consider the circuit in Figure P2-17. Suppose

$$V_{an} = 100\underline{/0°} \qquad Z_s = 8 + j10$$

$$V_{bn} = 50\underline{/180°} \qquad Z_{ab} = Z_{bc} = Z_{ca} = Z_m = j4$$

$$V_{cn} = 50\underline{/180°}$$

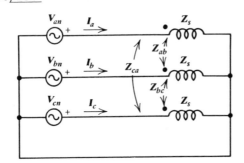

Figure P2-17 Circuit for Problem 2-17.

   (a) Calculate I_a, I_b, and I_c without using symmetrical components.
   (b) Calculate I_a, I_b, and I_c using symmetrical components.

2-18. Given

$$[Z_{abc}] = \begin{bmatrix} Z & 0 & 0 \\ 0 & Z & 0 \\ 0 & 0 & Z \end{bmatrix}$$

Find $[Z_{012}]$.

2-19. Frequently $[T]$ is defined as

$$\frac{1}{\sqrt{3}}\begin{bmatrix} 1 & 1 & 1 \\ 1 & a^2 & a \\ 1 & a & a^2 \end{bmatrix}$$

When this is the case $[T]$ is unitary. (a) Prove it. (A matrix is unitary when its inverse equals its transpose conjugate.) (b) Defining $[T]$ as in part a, prove that $S_{3\phi} = \tilde{V}_{abc_t}\tilde{I}^*_{abc} = \tilde{V}_{012_t}\tilde{I}^*_{012}$ (SI units understood). This result is called the invariance of power condition and makes a strong case for the corresponding definition of $[T]$.

POWER SYSTEM
REPRESENTATION

"If you compare yourself with others, you may become vain
and bitter; for there will always be greater
and lesser persons than yourself."

Max Ehrmann, DESIDERATA

Our attempt to mathematically model the power system is heavily dependent on circuit concepts. The fact that power systems are three phase is a major complication. Another important complicating factor is the large number of components; tens of generators and hundreds of transmission lines and transformers are by no means unusual for typical systems. A third consideration is that transformers partition the system into many different voltage sections. Our methods of representation must particularly deal with these factors, minimizing their complicating effects as much as possible.

Our basic "picture" of the power system is the one line diagram. It communicates the essential interconnection information with maximum simplicity. The per phase equivalent circuit takes advantage of the symmetry inherent in balanced three phase circuits; the per unit system simplifies numerical analysis and eliminates the partitioning effect of transformers. All these representations are very useful in displaying and formulating power system problems and understanding them is necessary for students of power system analysis.

3-1 Per Phase Analysis

When working balanced 3ϕ circuit problems, one realizes that it is only necessary to compute results in one phase and subsequently predict results in the other two phases by exploiting three phase symmetry. The single phase circuit we use is called the per phase equivalent circuit. Such simplifications are demonstrated best with an example.

Example 3-1

A 2400 volt 3ϕ source supplies two parallel loads:
  Load #1: 300 kVA $pf = 0.8$ lagging
  Load #2: 240 kVA $pf = 0.6$ leading
The phase sequence is *abc*. If $V_{an} = \dfrac{2400}{\sqrt{3}} \underline{/0°}$:

(a) Draw a single phase equivalent circuit.
(b) Determine all three source line currents.

Solution

(a) We start by concentrating on the "*a*" phase:

$$S_{a_1} = \tfrac{1}{3} S_{3\phi_1}$$

$$= 100 \underline{/\psi} \, \text{kVA}$$

where

$$\psi = +\cos^{-1}(0.8)$$

$$= +36.9°$$

ψ is positive since the load pf is lagging. Also

$$S_{a_2} = 80\underline{/-53.1°}\,\text{kVA}$$

Now

$$I_{a_1} = \left[\frac{S_{a_1}}{V_{an}}\right]^*$$

$$= \left[\frac{100\underline{/+36.9}}{2.4/\sqrt{3}\underline{/0°}}\right]^*$$

$$= 72.2\underline{/-36.9°}\,\text{A}$$

and

$$I_{a_2} = \left[\frac{S_{a_2}}{V_{an}}\right]^*$$

$$= \left[\frac{80\underline{/-53.1°}}{2.4/\sqrt{3}\underline{/0°}}\right]^*$$

$$= 57.7\underline{/+53.1°}\,\text{A}$$

Now for equivalent wye impedances

$$Z_{an_1} = \frac{V_{an}}{I_{a_1}}$$

$$= \frac{1386\underline{/0°}}{72.2\underline{/-36.9°}}$$

$$= 19.2\underline{/36.9°}$$

$$= j11.52 + 15.36\,\text{ohms}$$

$$Z_{an_2} = \frac{V_{an}}{I_{a_2}}$$

$$= \frac{1386\underline{/0°}}{57.7\underline{/+53.1°}}$$

$$= 24\underline{/-53.1°}$$

$$= 14.4 - j19\cdot2\,\text{ohms}$$

The circuit is shown in Figure 3-1.

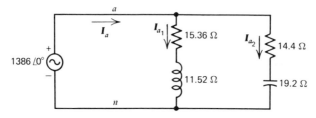

Figure 3-1 Circuit for "a" phase for Example 3-1.

(b) $I_a = I_{a_1} + I_{a_2}$

$\quad = 72.2\underline{/-36.9°} + 57.7\underline{/+53.1°}$

$\quad = 57.7 - j43.3 + 34.6 + j46.2$

$\quad = 92.3 + j2.9$

$\quad = 92.4\underline{/1.8°}$ A

By three phase symmetry:

$I_b = 92.4\underline{/1.8°} - 120° = 92.4\underline{/-118.2°}$ A

$I_c = 92.4\underline{/1.8°} + 120° = 92.4\underline{/121.8°}$ A

It is best to think in terms of equivalent wyes when using per phase analysis; if the actual connection is delta, we may always convert to the equivalent wye. As a consequence, the "bottom line" of our per phase equivalent circuit represents the neutral, the voltages are line to neutral, and the currents are line values.

3-2 The Per Unit System

In many engineering situations it is useful to scale, or normalize, dimensioned quantities. This is commonly done in power system analysis and the standard method used is referred to as the per unit system. Historically this was done to simplify numerical calculations that were done by hand. Although this advantage was eliminated by the use of the computer, other advantages remain, including:

- Device parameters tend to fall in a relatively narrow range, making erroneous values conspicuous.

- The method is defined so as to eliminate ideal transformers as circuit components. Since the typical power system contains hundreds, if not thousands, of transformers this is a nontrivial savings.

- Related to the advantage just mentioned, the voltage throughout the power system is normally close to unity.

Like most things in life, the per unit system is not without its drawbacks, which include:

- The system modifies component equivalent circuits, making them somewhat more abstract. Sometimes phase shifts that are clearly present in the unscaled circuit vanish in the per unit circuit.

- Some equations that hold in the unscaled case are modified when scaled into per unit. Factors such as $\sqrt{3}$ and 3 are removed or added by the method.

It is necessary for power system engineers to become familiar with and facile in the use of the system because of its wide industrial acceptance and use, and also to take advantage of its analytical simplifications. The student is cautioned against a tendency to believe that power circuit analysis is somehow different from "ordinary" circuit theory. The per unit system is simply a scaling method and cannot magically "repeal" any laws of circuit theory.

The basic per unit scaling equation is:

$$\text{Per unit value} = \frac{\text{Actual value}}{\text{Base value}} \tag{3-1}$$

The base value always has the same units as the actual value, forcing the per unit value to be dimensionless. Also, the base value is always a real number, whereas the actual value may be complex. Thinking of a complex value in polar form, the angle of the per unit value is the same as that of the actual value.

Consider complex power:

$$S = VI^* \tag{2-19a}$$

or

$$S \underline{/\psi} = V\underline{/\alpha} I \underline{/-\beta} \tag{2-19b}$$

Suppose we arbitrarily pick a value S_{base}, a real number, with the units of voltamperes. Dividing through by S_{base}:

$$\frac{S\underline{/\psi}}{S_{base}} = \frac{V\underline{/\alpha}I\underline{/-\beta}}{S_{base}} \tag{3-2a}$$

We further define

$$V_{vase}I_{base} = S_{base} \tag{3-3}$$

Either V_{base} or I_{base} may be selected arbitrarily, but not both. Substituting (3-3) into (3-2a) we get:

$$\frac{S\underline{/\psi}}{S_{base}} = \frac{V\underline{/\alpha}I\underline{/-\beta}}{V_{base}I_{base}} \tag{3-2b}$$

$$S_{pu}\underline{/\psi} = \frac{V\underline{/\alpha}}{V_{base}}\frac{I\underline{/-\beta}}{I_{base}} \tag{3-2c}$$

$$S_{pu} = V_{pu}\underline{/\alpha}I_{pu}\underline{/-\beta} \tag{3-2d}$$

$$= V_{pu}I_{pu}{}^* \tag{3-2e}$$

The "pu" subscript implies per unit values. Note that the form of equation (3-2e) is identical to (2-19a). This was not inevitable but resulted from our decision to relate $V_{base}I_{base}$ and S_{base} through (3-3). If we select Z_{base} by

$$Z_{base} = \frac{V_{base}}{I_{base}} \tag{3-4a}$$

$$= \frac{V_{base}{}^2}{S_{base}} \tag{3-4b}$$

We convert Ohm's law

$$Z = \frac{V}{I} \tag{2-26a}$$

into per unit by dividing by Z_{base}.

$$\frac{Z}{Z_{base}} = \frac{V/I}{Z_{base}} \tag{3-5a}$$

$$Z_{pu} = \frac{V/V_{base}}{I/I_{base}} \tag{3-5b}$$

$$= \frac{V_{pu}}{I_{pu}} \tag{3-5c}$$

An example will be helpful.

Example 3-2

Solve the problem of examples 2-2 and 2-3 in per unit on bases of $V_{base} = 100\,V$ and $S_{base} = 500\,VA$.

Solution

$$I_{base} = \frac{S_{base}}{V_{base}}$$

$$= \frac{500}{100} = 5\,A$$

$$Z_{base} = \frac{V_{base}}{I_{base}}$$

$$= \frac{100}{5} = 20\ ohms$$

or

$$Z_{base} = \frac{V_{base}^{2}}{S_{base}}$$

$$= \frac{(100)^{2}}{500} = 20\ ohms$$

(a) We convert the circuit values to per unit:

$$V_{pu} = \frac{V}{V_{base}}$$

$$= \frac{100\underline{/0^\circ}}{100} = 1\underline{/0^\circ}$$

$$R_{pu} = \frac{R}{Z_{base}}$$

$$= \frac{3}{20} = 0.15$$

Power System Representation

$$X_{L_{pu}} = \frac{X_L}{Z_{base}}$$

$$= \frac{8}{20} = 0.40$$

$$X_{C_{pu}} = \frac{X_C}{Z_{base}}$$

$$= \frac{4}{20} = 0.20$$

The circuit is shown in Figure 3-2.

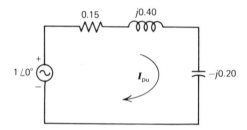

Figure 3-2 Circuit for Example 3-2. All values in per unit.

(b) $Z_{pu} = 0.15 + j(0.4 - 0.2)$

$$= 0.15 + j0.20$$

$$= 0.25\underline{/+53.1°}$$

(c) $I_{pu} = \frac{V_{pu}}{Z_{pu}}$

$$= \frac{1\underline{/0°}}{0.25\underline{/+53.1°}}$$

$$= 4.0\underline{/-53.1°}$$

Per unit solution for example 2-3:

$$S_{R_{pu}} = I^2 R$$

$$= (4)^2(0.15)$$

$$= 2.4 + j0$$

68

$$S_{L_{pu}} = jI^2 X_L$$
$$= j(4)^2(0.4)$$
$$= 0 + j6.4$$
$$S_{C_{pu}} = -jI^2 X_C$$
$$= -j(4)^2(0.2)$$
$$= 0 - j3.2$$
$$S_{pu} = S_{R_{pu}} + S_{L_{pu}} + S_{C_{pu}}$$
$$= 2.4 + j6.4 - j3.2$$
$$= 2.4 + j3.2$$
$$= 4.0\underline{/+53.1°}$$

Refer to Figure 3-2.

Note that the phase angles of the scaled and unscaled complex quantities are the same. Let us convert to actual values.

Z: $0.25(20) = 5$ ohms
I: $4.0(5) = 20$ amperes
S: $4.0(500) = 2000$ VA

To summarize per unit as applied to ac circuits, we realize four quantities are involved: S, V, I, and Z. We may pick base values for any two arbitrarily and calculate base values for the other two from equations (3-3) and (3-4). Per unit values for all quantities are calculated from equation (3-1). Problems are then solved in the scaled circuit by conventional circuit analysis methods. Finally, results are converted back to actual values through equation (3-1).

For power system applications, the base values are arbitrarily selected as S_{base} and V_{base}. The adverb "arbitrarily" is somewhat misleading: in theory this is possible, but in practice we select values that force our results into certain ranges. Thus, for V_{base} a value is chosen such that the system voltage is normally close to unity. Popular power bases used are 1, 10, 100, and 1000 MVA, depending on system size. We examine the effect on transformers later; then we see the important simplifying effect this has on the power system network.

3-3 The Per Unit System Extended to Three Phase Circuits

We recall that in a balanced 3ϕ situation two voltages, the line voltage (V_L), and the line-to-neutral voltage (V_{LN}), are present. It is conventional in 3ϕ

69

power system work to refer to $V_{L_{base}}$ as *the* voltage base. By definition:

$$V_{LN_{base}} = \frac{V_{L_{base}}}{\sqrt{3}}$$

(3-6)

Now consider:

$$V_{L_{pu}} = \frac{V_L}{V_{L_{base}}}$$

(3-7a)

and

$$V_{LN_{pu}} = \frac{V_{LN}}{V_{LN_{base}}}$$

(3-7b)

But

$$V_{LN} = \frac{V_L}{\sqrt{3}}$$

(3-8)

It follows that

$$V_{LN_{pu}} = \frac{V_{LN}}{V_{LN_{base}}}$$

(3-9a)

$$= \frac{V_L/\sqrt{3}}{V_{L_{base}}/\sqrt{3}}$$

(3-9b)

$$= \frac{V_L}{V_{L_{base}}}$$

(3-9c)

$$= V_{L_{pu}}$$

(3-9d)

The result is "simpler" but can cause confusion since the $\sqrt{3}$ factor that clearly was necessary in the unscaled equation has vanished in the per unit equation!

A corresponding result occurs with the power base. The base generally stated is $S_{3\phi_{base}}$. Again by definition

$$S_{1\phi_{base}} = \frac{S_{3\phi_{base}}}{3}$$

(3-10)

70

But $S_{3\phi} = 3S_{1\phi}$ so that:

$$S_{1\phi_{pu}} = \frac{S_{1\phi}}{S_{1\phi_{base}}} \qquad (3\text{-}11a)$$

$$= \frac{S_{3\phi}/3}{S_{3\phi_{base}}/3} \qquad (3\text{-}11b)$$

$$= S_{3\phi_{pu}} \qquad (3\text{-}11c)$$

The wye impedance base is:

$$Z_{Y_{base}} = \frac{V_{LN_{base}}^2}{S_{1\phi_{base}}} \qquad (3\text{-}12a)$$

$$= \frac{[V_{L_{base}}/\sqrt{3}]^2}{S_{3\phi_{base}}/3} \qquad (3\text{-}12b)$$

$$= \frac{V_{L_{base}}^2}{S_{3\phi_{base}}} \qquad (3\text{-}12c)$$

By definition

$$Z_{\Delta_{base}} = 3Z_{Y_{base}} \qquad (3\text{-}13)$$

so that

$$Z_{Y_{pu}} = Z_{\Delta_{pu}} \qquad (3\text{-}14)$$

The line current base $I_{L_{base}}$ is:

$$I_{L_{base}} = \frac{S_{3\phi_{base}}}{V_{L_{base}}\sqrt{3}} \qquad (3\text{-}15)$$

Example 3-3

Rework example 3-1 in per unit on a 300kVA 2400 volt base.

Solution

(a) We understand that

$$S_{3\phi_{base}} = 300 \text{ kVA}$$

$$V_{L_{base}} = 2400 \text{ V}$$

∴ The source voltage = 1.0 per unit.

71

$$S_{1_{pu}} = 1\underline{/+36.9°}$$

$$S_{2_{pu}} = 0.8\underline{/-53.1°}$$

$$\therefore \quad I_{1_{pu}} = 1\underline{/-36.9°}$$

$$I_{2_{pu}} = 0.8\underline{/+53.1°}$$

$$Z_{1_{pu}} = 1\underline{/+36.9°} = 0.8 + j0.6$$

$$Z_{2_{pu}} = 1.25\underline{/-53.1°} = 0.75 - j1.0$$

The circuit is shown in Figure 3-3.

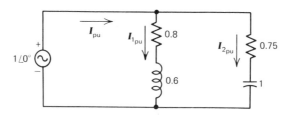

Figure 3-3 The circuit of Example 3-1 in per unit.

(b) $I_{pu} = I_{1_{pu}} + I_{2_{pu}}$

$$= 1\underline{/-36.9°} + 0.8\underline{/+53.1°}$$

$$= 0.8 - j0.6 + 0.48 + j0.64$$

$$= 1.28 + j0.04 = 1.28\underline{/1.8°}$$

$$I_{L_{base}} = \frac{300}{2.4\sqrt{3}}$$

$$= 72.2 \text{ amperes}$$

$$\therefore \quad I_L = (1.28)(72.2)$$

$$= 92.4 \text{ amperes}$$

Note that we added "pu" to variable subscripts to distinguish between scaled and unscaled values. This practice proves to be too cumbersome to continue; we must determine from context whether the per unit system is being used. Although not quite correct, we can think of per unit as being another system of "units" (actually per unit quantities are dimensionless) and state that we are "in per unit", as opposed to using SI units. As we become familiar with it, the system should cause no real difficulties.

3-4 Symmetrical Components Scaled into Per Unit

The equations developed in Chapter 2 assumed that all variables were in SI units. It is useful to interpret the same equations in per unit. Recall equation (2-54):

$$\tilde{V}_{012} = [T]^{-1}\tilde{V}_{abc} \tag{2-54}$$

Divide by $V_{LN_{base}}$:

$$\frac{\tilde{V}_{012}}{V_{LN_{base}}} = [T]^{-1}\frac{\tilde{V}_{abc}}{V_{LN_{base}}} \tag{3-16a}$$

or

$$\tilde{V}_{012_{pu}} = [T]^{-1}\tilde{V}_{abc_{pu}} \tag{3-16b}$$

Similarly

$$\tilde{I}_{012_{pu}} = [T]^{-1}\tilde{I}_{abc_{pu}} \tag{3-17}$$

where the proper current base is $I_{L_{base}}$. Recall equation (2-60):

$$[Z_{012}] = [T]^{-1}[Z_{abc}][T] \tag{2-60}$$

Divide through by $Z_{Y_{base}}$:

$$\frac{1}{Z_{Y_{base}}}[Z_{012}] = \frac{1}{Z_{Y_{base}}}[T]^{-1}[Z_{abc}][T] \tag{3-18a}$$

$$[Z_{012_{pu}}] = [T]^{-1}[Z_{abc_{pu}}][T] \tag{3-18b}$$

Now consider power:

$$S_{3\phi} = 3\tilde{V}_{012_t}\tilde{I}_{012}^* \tag{2-67}$$

Divide through by $S_{3\phi_{base}}$:

$$\frac{S_{3\phi}}{S_{3\phi_{base}}} = \frac{3\tilde{V}_{012_t}\tilde{I}_{012}^*}{S_{3\phi_{base}}} \tag{3-19a}$$

$$S_{pu} = \frac{3\tilde{V}_{012_t}\tilde{I}_{012}^*}{3S_{1\phi base}} \tag{3-19b}$$

$$S_{pu} = \frac{\tilde{V}_{012_t}\tilde{I}_{012}^*}{V_{LN_{base}}I_{L_{base}}} \tag{3-19c}$$

$$S_{pu} = \frac{\tilde{V}_{012_t}}{V_{LN_{base}}}\frac{\tilde{I}_{012}^*}{I_{L_{base}}} \tag{3-19d}$$

$$= \tilde{V}_{012_{pu_t}}\tilde{I}_{012_{pu}}^* \tag{3-19e}$$

73

Note that scaling of voltage, current, and impedance through equations (3-16), (3-17), and (3-18) result in forms identical with those using SI units. The power equation, however, is different in that (3-19e) is missing the factor of three found in (2-67). This is typical of the per unit system because equations that are correct when SI units are used *may require modification when per unit values are substituted.* We could elect to develop and use a more precise, but more cumbersome, notation that distinguishes between SI and per unit values. We will not do so for two reasons. First, for most of the rest of this book we will consistently be in per unit; and second, usage of such a notation consistently will prove discouragingly tedious. We must always be aware of which units are appropriate, taken in context of the specific situation under discussion.

To summarize our results we repeat the proper bases for sequence values:

$$\text{Sequence voltage base} = V_{LN_{\text{base}}} \tag{3-20a}$$

$$\text{Sequence current base} = I_{L_{\text{base}}} \tag{3-20b}$$

$$\text{Sequence impedance base} = Z_{Y_{\text{base}}} \tag{3–20c}$$

$$\text{Sequence power base} = S_{3\phi_{\text{base}}} \tag{3-20d}$$

3-5 The One Line Diagram

One may think of a power system as an extremely complicated electrical network. In fact, the complications are so great that the construction of a conventional circuit diagram is impractical. Yet a diagram that shows interconnections of basic power system components, switch locations, and so on, is the most efficient method of recording that information. Such a diagram is referred to as a one (or single) line diagram. It shows the relative electrical interconnections and locations for generators, transformers, transmission lines, static loads, rotating loads (motors), circuit breakers, reactors, and various types of switches. Sometimes peripheral equipment is also shown, such as instrument transformers, lightning arrestors, and protective relays.

To understand the differences between circuit diagrams and one line diagrams, study Figure 3-4. Power system information is recorded on the one line diagram; most calculations are performed by thinking in terms of the per phase equivalent circuit; and results are extended to the three phase equivalent circuit by using three phase symmetry. Except in the simplest cases, the three phase equivalent circuit is never actually drawn, and the per phase circuit only occasionally constructed. However, the power system engineer *thinks* in terms of these circuit diagrams when formulating the

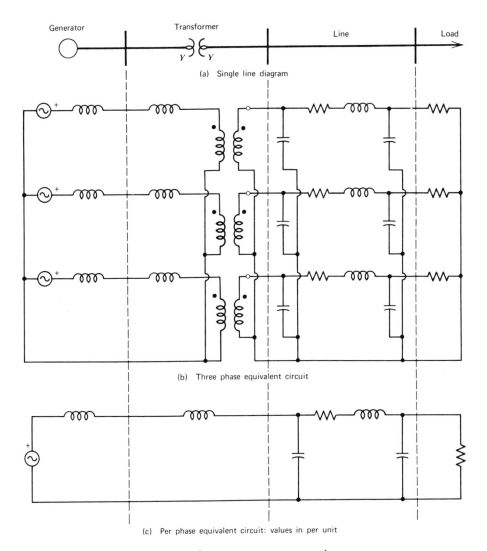

(a) Single line diagram

(b) Three phase equivalent circuit

(c) Per phase equivalent circuit: values in per unit

Figure 3-4 Power system representations.

necessary mathematical relations, and with experience unconsciously converts the single line to the per phase equivalent circuit.

There is no universally accepted set of symbols that are used to represent power system components in one line diagrams; however, variations are usually minor and not difficult to decode.* The symbols used in

* There are standard symbols for electrical devices published by IEEE.

75

one line diagrams in this book are shown in Figure 3-5. The concept of "bus" requires comment. "Bus" in a one line diagram is essentially the same as that of "node" in a circuit diagram, although the three phase nature of power circuits makes the concept somewhat more involved. (Think of bus as a node in the per phase equivalent circuit.) Points along a bus are separated by negligible impedance and therefore are at the same voltage. Busses are to be identified on one line diagrams as short straight lines drawn at right angles to component symbols.

Wye connections are frequently "grounded". The term "ground" used as a verb means to make an electrical connection from the neutral point of

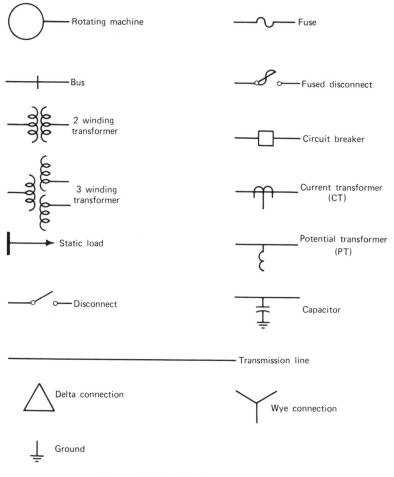

Rotating machine

Fuse

Bus

Fused disconnect

2 winding transformer

Circuit breaker

3 winding transformer

Current transformer (CT)

Static load

Potential transformer (PT)

Disconnect

Capacitor

Transmission line

Delta connection

Wye connection

Ground

Figure 3-5 One line diagram symbols.

the wye to the earth. "Grounded through impedance" means to insert impedance (usually inductance or resistance) between the wye neutral and earth. "Solidly grounded" implies a direct connection from neutral to earth. To insure a good connection to earth at sites of major importance, such as a generator location, a mat of conducting mesh is buried in the earth at the site. Connections to "ground" are then made to this grid. Sometimes a bare conductor referred to as a counterpoise is buried under a transmission line and connected electrically to the tower structure and the overhead neutral conductors. Grounding details are shown on single line diagrams.

3-6 Summary

A typical power system involves interconnections between many components. This information is concisely presented with the one line diagram. This diagram can then be used to construct the per phase equivalent circuit. Under three phase balanced conditions, calculations made on the per phase equivalent circuit can be used to predict three phase performance. The three phase circuit, although technically correct, is too complex to be useful.

Actual calculations are typically made in the per unit system. This system has simplifying features, but can cause confusion occasionally. Appropriate base values must always be clearly understood. For most of our work only four quantities (V, I, P, and Z) are scaled, but in Chapter 12 we extend the method to torque, angular velocity, time, and moments of inertia. Sometimes quantities are expressed in "percent," which simply involves a factor of 100:

$$\text{Percent value} = 100 \times \text{per unit value} \qquad (3\text{-}21)$$

Recall that symmetrical components will transform an unbalanced three phase system into three single phase systems. Calculations are typically done with sequence values scaled into per unit. It is useful to note that what we refer to here as the "per phase equivalent circuit" is identical to the positive sequence equivalent circuit.

Problems

3-1. A three phase load is rated at 13.8 kV, 100 MVA, $pf = 0.8$ lagging. Find
  (a) Z_Y in ohms
  (b) Z_Δ in ohms

(c) Z_{load} suitable for use in a per phase equivalent circuit in per unit, assuming the rated values are used for base values.

3-2. The following equation is sometimes used to convert an impedance in pu on one base to a second base:

$$Z_{\text{pu}_2} = Z_{\text{pu}_1} \left[\frac{V_{L_{\text{base 1}}}}{V_{L_{\text{base 2}}}} \right]^2 \left[\frac{S_{3\phi_{\text{base 2}}}}{S_{3\phi_{\text{base 1}}}} \right]$$

Prove its validity.

3-3. For the system shown in Figure P3-3, draw a per phase equivalent circuit, using the circuit models shown in Figure 3-4(c).

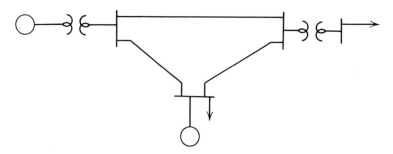

Figure P3-3 Diagram for Problem 3-3.

3-4. Draw a three phase equivalent circuit for the system of problem 3-3.

3-5. Examine the situation in Figure P3-5. (a) Compute appropriate Z_Y and Z_Δ values for each of the loads in ohmns. (b) Convert results in (a) to per unit on bases of $S_{3\phi_{\text{base}}} = 1$ MVA and $V_{L_{\text{base}}} = 4.16$ kV.

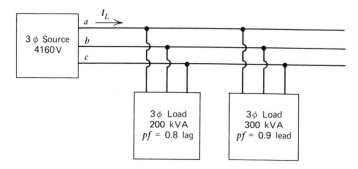

Figure P3-5 Circuit for Problems 3-5 and 3-6.

3-6. Continue problem 3-5 as follows
  (a) Draw an appropriate per phase equivalent circuit, with all values in per unit.
  (b) Compute the source line current (I_L) in per unit.
  (c) Convert I_L to amperes.

3-7. Work problem 2-11 in per unit on bases of $V_{L\text{base}} = 2.4$ kV and $S_{3\phi_{\text{base}}} = 500$ kVA.

3-8. Examine the single line system diagram of Figure P3-8. The loads are:

$S_1 = 3.0 + j0.8$ $S_4 = 2.0 - j0.6$

$S_2 = 2.0 + j0.4$ $S_5 = 2.5 + j0.6$

$S_3 = 1.5 + j0.3$

Bus	Voltage (kV)
1, 2	12
3–10	46
11–15	4.16

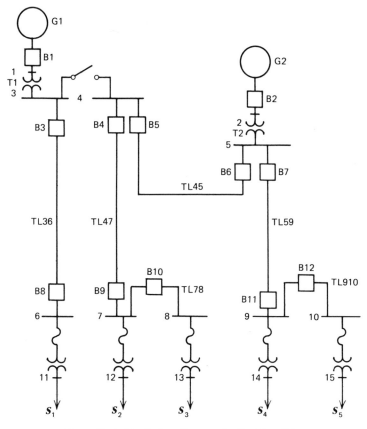

Figure P3-8 Single line diagram for Problem 3-8.

Values are in per unit referred to a 10 MVA base. Assume all components are lossless and that each generator must be sized large enough to carry the total load. The disconnect between busses 3 and 4 may be opened or closed.

(a) What is the minimum MVA rating of each generator?

(b) What is the maximum current in amperes to be expected in line TL78?

(c) What is the maximum current to be expected in line TL45?

4
TRANSMISSION
LINES

"Enjoy your achievements as well as your plans. Keep interested in your own career, however humble; it is a real possession in the changing fortunes of time."

Max Ehrmann, DESIDERATA

The overhead three phase power transmission line is the main energy corridor in a power system. One might assume that the circuit model would be trivial (ideal conductors), but three different phenomena produce effects that cannot reasonably be ignored. In order of importance, they are the series voltages induced by the magnetic fields surrounding the conductors, the shunt displacement currents resulting from the electric fields between conductors, and the ohmic resistance of the conductor material. A fourth, and minor, effect is the leakage conduction current that flows through contamination films on the insulators, which we will neglect.

Figure 4-1 500 kV transmission line. (Photograph courtesy of Alabama Power Company.)

A typical overhead power transmission line is shown in Figure 4-1. The overhead neutrals are electrically in contact with the tower and therefore grounded. They primarily exist to provide lightning shielding for the phase conductors and also to carry zero sequence and harmonic currents that help to maintain balanced sinusoidal voltages. They are usually steel or aluminum and are small (diameter about 1 cm.). The phase conductors are much larger (diameter about 5 cm.), and are typically stranded aluminum surrounding a stranded steel cable (for increased tensile strength). Sometimes more than one (a "bundle") comprise a phase. All are bare (no insulating covering) for heat dissipation reasons; the phase conductors are insulated from each other and the tower by suspension from insulator strings.

We might touch on some of the reasons bearing on the decision to build a line. A common situation is that load growth in an area has brought existing lines near their thermal or stability limits. Studies might have shown that system reliability at a specific location has fallen below acceptable levels. Additional lines might enhance the transient stability characteristics of generators. New generation sites at remote locations require additional lines. Additional lines allow for more flexibility in system operation.

Having decided to build a line, basic considerations are the line capacity (power rating) and voltage to be used. As we shall see, the line capacity is related to the length and voltage rating. For a fixed length the capacity varies as the square of the voltage, whereas the line cost varies roughly linearly with the voltage. Only standard voltage levels are to be considered because of the availability of equipment. For a certain power level and line length, a specific voltage class will prove to be economically optimum. The greater the required capacity or line length, the higher the optimum voltage class. For very great lengths dc transmission is preferred (see section 4-10).

Suppose the voltage and power ratings have been chosen for a proposed line of known length. What further decisions remain? The number, size, and spacing of conductors per phase bundle must be selected. The decisive criteria here are corona and line impedance effects. The phase to phase spacing must be chosen. The number, location, and conductor type for overhead neutrals must be chosen, lightning shielding being the major consideration. The level of insulation must be selected, deciding how many suspension discs are to be used in a string.

When the weight of the line is essentially fixed, we direct our attention to the tower design. Local weather conditions must be considered; specifically, the worst icing and wind conditions within reason must be estimated because these factors also relate to the tower load. Architectural aesthetics should be considered, particularly in populated areas. Tower width relates to available right-of-way, and the decision to use guyed or free standing towers must be made. Overhead minimum clearances for railroads, highways, structures, vegetation, and the earth must be established; these dimensions fix tower height, along with tower spacing and permissible conductor sag. Wind excited conductor motions must be anticipated and minimized.

83

All the above factors must be considered and satisfactorily accounted for with an economically acceptable design. We will not conduct a comprehensive investigation in all these areas, but will look closer at some of the electrical characteristics important in the design and operation of transmission lines.

Our first objective is to derive a circuit model for this device and we start with the most important parameter, the line inductance. A notational problem arises that could potentially cause confusion. The nature of the model is that line impedance and admittance expressions per unit length are first derived, with the effect of the line length dealt with later. It is awkward to develop symbolism that distinguishes per unit length in all cases. Thus we find ourselves using "Z" for a parameter measured in ohms/metre. No confusion will result if we read carefully; to help the reader make the distinction, the units are given to the equation's right when necessary. Also, the necessary equations we derive are rigorously applicable only to a line of infinite length. However, for power applications, end effects are negligible.

4-1 Line Inductance

We recall from electromagnetic field theory that an infinitely long conductor will have a radial E field and concentric H field. Examine Figure 4-2, which

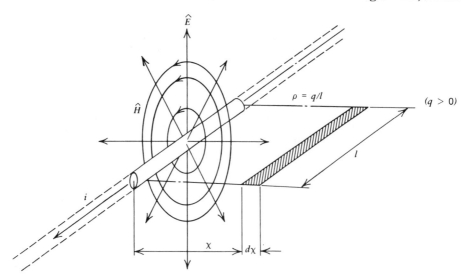

Figure 4-2 Magnetic and electric fields surrounding a conductor segment l of an infinitely long conductor.

shows a conductor segment of length l of an infinitely long conductor. The conductor is of uniform circular cross section and radius r.

Our first concern is with the magnetic effects. Ampere's law relates H to i:

$$\int_{\text{line}} \hat{H} \cdot dl = i_{\text{enclosed}} \tag{4-1}$$

At a distance χ external to the conductor this relation simplifies to:

$$H = \frac{i}{2\pi\chi} \tag{4-2}$$

When H exists in a constant permeability medium:

$$B = \mu H \tag{4-3}$$

In air $\mu \cong \mu_0 = 4\pi \times 10^{-7}$ H/m.

$$\therefore \quad B = \frac{\mu i}{2\pi\chi} \tag{4-4}$$

The external flux linking the conductor segment l out to a distance $\chi = D$ is obtained by integrating from the conductor surface ($\chi = r$, $r =$ conductor radius) out to location $\chi = D$.

$$\lambda' = \int_r^D Bl \, d\chi \tag{4-5a}$$

$$= \frac{\mu i l}{2\pi} \ln \frac{D}{r} \tag{4-5b}$$

A thin walled conductor will have no internal flux linkage since the internal magnetic field is zero. It is possible to compute the radius of an equivalent thin walled conductor (r') for any conductor geometry. It is "equivalent" in the sense that it has the same *total* flux linkages as the original conductor. This value r' is referred to as the geometric mean radius (GMR). Consult the Appendix for more detail as to how r' is determined. Therefore, total flux linkage out to a distance D is:

$$\lambda' = \frac{\mu i l}{2\pi} \ln \frac{D}{r'} \text{ webers} \tag{4-6}$$

Define the flux linkage per unit length to be λ.

$$\lambda = \frac{\lambda'}{l} \text{ webers/m} \tag{4-7a}$$

$$= \frac{\mu i}{2\pi} \ln \frac{D}{r'} \tag{4-7b}$$

Now consider the array of n conductors shown in Figure 4-3, each carrying a current (i_1, i_2, etc.) out of the page, given that:

$$i_1 + i_2 + \ldots + i_n = 0 \tag{4-8}$$

For the ith conductor the flux linkages per unit length would be, by superposition,

$$\lambda_i = \lambda_{i1} + \lambda_{i2} + \ldots + \lambda_{in} \tag{4-9}$$

where the subscript notation λ_{ij} means the flux linkage per unit length around the ith conductor due to current flowing in the jth conductor.

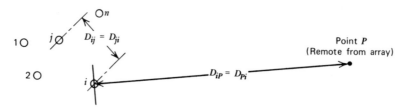

Figure 4-3 Array of n current carrying conductors. Positive currents are assumed out of the paper.

Computing flux linkage out to some point P remote from the array we get from (4-7b):

$$\lambda_{i1} = \frac{\mu i_1}{2\pi} \ln \frac{D_{1P}}{D_{i1}} \tag{4-10a}$$

$$\lambda_{i2} = \frac{\mu i_2}{2\pi} \ln \frac{D_{2P}}{D_{i2}} \tag{4-10b}$$

$$\vdots$$

$$\lambda_{ii} = \frac{\mu i_i}{2\pi} \ln \frac{D_{iP}}{r_i'} \tag{4-10i}$$

$$\vdots$$

$$\lambda_{in} = \frac{\mu i_n}{2\pi} \ln \frac{D_{nP}}{D_{in}} \tag{4-10n}$$

where $D_{ij} = D_{ji} =$ center to center distance between conductors i and j. ($D_{ii} = r_i'$)

Substituting back into (4-9):

$$\lambda_i = \frac{\mu}{2\pi}\left[i_1 \ln\frac{D_{1P}}{D_{i1}} + i_2 \ln\frac{D_{2P}}{D_{i2}} + \ldots + i_n \ln\frac{D_{nP}}{D_{in}}\right] \tag{4-11a}$$

$$= \frac{\mu}{2\pi}\left[i_1 \ln\frac{1}{D_{i1}} + i_2 \ln\frac{1}{D_{i2}} + \ldots + i_n \ln\frac{1}{D_{in}} + i_1 \ln D_{1P} + i_2 \ln D_{2P}\right.$$

$$\left. + \ldots + i_n \ln D_{nP}\right] \tag{4-11b}$$

To compute the total flux linkage per unit length the point P must approach ∞. As it does, $D_{1P} \cong D_{2P} \cong \ldots \cong D_{nP} \cong D$.
Then:

$$\lim_{D \to \infty}(i_1 + i_2 + \ldots + i_n)\ln D = 0 \tag{4-12}$$

and equation (4-11b) simplifies to:

$$\lambda_i = \frac{\mu}{2\pi}\left[i_1 \ln\frac{1}{D_{i1}} + i_2 \ln\frac{1}{D_{i2}} + \ldots + i_n \ln\frac{1}{D_{in}}\right] \tag{4-13}$$

Similar equations may be written for all conductors. The results may be summarized in a matrix formulation:

$$\tilde{\lambda} = [L]\tilde{i} \tag{4-14a}$$

where

$\tilde{\lambda}$ is an $n \times 1$ vector containing $\lambda_1\lambda_2 \ldots \lambda_n$

$\tilde{i}$ is an $n \times 1$ vector containing $i_1 i_2 \ldots i_n$

$[L]$ is an $n \times n$ matrix

whose general entry is:

$$l_{ij} = \frac{\mu}{2\pi}\ln\frac{1}{D_{ij}} \quad \text{henry/m} \tag{4-14b}$$

$$i, j = 1, 2, \ldots n \tag{4-14c}$$

$$D_{ii} = r_i' \tag{4-14d}$$

To understand how this relates to a three phase line we apply equation (4-14) to an array of 4 conductors (3 phases and a neutral). Think of $\tilde{\lambda}$ as (4×1), $\tilde{i}$ as (4×1), $[L]$ as (4×4), and the index parameters i, j as running over the values a, b, c, and n. At this point, to develop a circuit model for a short line section of length l, we prefer to think in terms of voltage rather than flux linkage. From Faraday's law:

$$v = \frac{d\lambda}{dt} \tag{4-15a}$$

$$= L\frac{di}{dt} \tag{4-15b}$$

and for sinusoidal steady state

$$V = j\omega L I \tag{4-15c}$$

The line section may be modeled by the equivalent circuit shown in Figure 4-4.

Applying Kirchhoff's voltage law we produce

$$\begin{bmatrix} V_{aa'} \\ V_{bb'} \\ V_{cc'} \\ V_{nn'} \end{bmatrix} = l\{[R] + j\omega[L]\} \begin{bmatrix} I_a \\ I_b \\ I_c \\ I_n \end{bmatrix} \tag{4-16}$$

where the double subscript notation V_{ij} means that i is understood positive with respect to j. Also:

$$[R] = \begin{bmatrix} R_a & 0 & 0 & 0 \\ 0 & R_b & 0 & 0 \\ 0 & 0 & R_c & 0 \\ 0 & 0 & 0 & R_n \end{bmatrix} \quad \text{ohm/m} \tag{4-17}$$

More compactly we write an equivalent expression for (4-16):

$$\tilde{V}_{ii'} = l[Z]\tilde{I} \tag{4-18}$$

where $[Z]$ is 4×4 with entries

$$Z'_{ii} = R_i + j\omega l_{ii} \quad \text{ohm/m} \tag{4-19a}$$

$$Z'_{ij} = j\omega l_{ij}(i \neq j) \quad \text{ohm/m} \tag{4-19b}$$

$$l_{ij} = \frac{\mu}{2\pi}\ln\frac{1}{D_{ij}} \quad \text{henry/m} \tag{4-19c}$$

$$i, j = a, b, c, n$$

A common voltage reference point is desirable. Consider the fourth equation of the set (4-18):

$$V_{nn'} = [j\omega l_{na}I_a + j\omega l_{nb}I_b + j\omega l_{nc}I_c + (R_n + j\omega l_{nn})I_n]l \tag{4-20}$$

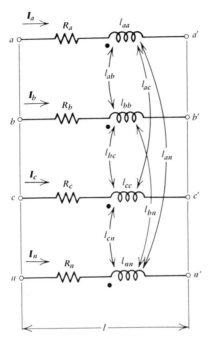

Figure 4-4 General equivalent circuit for a four conductor line section of length l considering serial inductive and resistive effects.

Consider the consequences of subtracting (4-20) from the top three equations of the set (4-18). The serial voltage $V_{nn'}$ would be "absorbed" (i.e., accounted for) in the remaining three equations. By KVL we produce:

$$V_{aa'} - V_{nn'} = V_{an} - V_{a'n'} \tag{4-21a}$$

$$V_{bb'} - V_{nn'} = V_{bn} - V_{b'n'} \tag{4-21b}$$

$$V_{cc'} - V_{nn'} = V_{cn} - V_{c'n'} \tag{4-21c}$$

To accomplish the operations discussed we perform the subtraction on the right side of (4-18) and perform the necessary simplifying algebra. Also we divide through by the length l. The result is:

$$\frac{1}{l}(V_{an} - V_{a'n'}) = \left[R_a + j\frac{\omega\mu}{2\pi} \ln \frac{D_{na}}{r'_a} \right] I_a + \left[j\frac{\omega\mu}{2\pi} \ln \frac{D_{nb}}{D_{ab}} \right] I_b$$

$$+ \left[j\frac{\omega\mu}{2\pi} \ln \frac{D_{nc}}{D_{ac}} \right] I_c + \left[-R_n + j\frac{\omega\mu}{2\pi} \ln \frac{r'_n}{D_{an}} \right] I_n \tag{4-22a}$$

89

$$\frac{1}{l}(V_{bn} - V_{b'n'}) = j\left[\frac{\omega\mu}{2\pi}\ln\frac{D_{na}}{D_{ba}}\right]I_a + \left[R_b + j\frac{\omega\mu}{2\pi}\ln\frac{D_{nb}}{r_b'}\right]I_b$$

$$+ \left[j\frac{\omega\mu}{2\pi}\ln\frac{D_{nc}}{D_{bc}}\right]I_c + \left[-R_n + j\frac{\omega\mu}{2\pi}\ln\frac{r_n'}{D_{bn}}\right]I_n \qquad (4\text{-}22\text{b})$$

$$\frac{1}{l}(V_{cn} - V_{c'n'}) = \left[j\frac{\omega\mu}{2\pi}\ln\frac{D_{na}}{D_{ca}}\right]I_a + \left[j\frac{\omega\mu}{2\pi}\ln\frac{D_{nb}}{D_{cb}}\right]I_b$$

$$+ \left[R_c + j\frac{\omega\mu}{2\pi}\ln\frac{D_{nc}}{r_c'}\right]I_c + \left[-R_n + j\frac{\omega\mu}{2\pi}\ln\frac{r_n'}{D_{cn}}\right]I_n \qquad (4\text{-}22\text{c})$$

To eliminate I_n we use equation (4-8). Simplifying, we get:

$$\frac{1}{l}(V_{an} - V_{a'n'}) = \left[R_a + R_n + j\frac{\omega\mu}{2\pi}\ln\frac{D_{na}^2}{r_a'r_n'}\right]I_a$$

$$+ \left[R_n + j\frac{\omega\mu}{2\pi}\ln\frac{D_{an}D_{nb}}{D_{ab}r_n'}\right]I_b + \left[R_n + j\frac{\omega\mu}{2\pi}\ln\frac{D_{an}D_{nc}}{D_{ac}r_n'}\right]I_c$$

$$(4\text{-}23\text{a})$$

$$\frac{1}{l}(V_{bn} - V_{b'n'}) = \left[R_n + j\frac{\omega\mu}{2\pi}\ln\frac{D_{bn}D_{na}}{D_{ba}r_n'}\right]I_a$$

$$+ \left[R_b + R_n + j\frac{\omega\mu}{2\pi}\ln\frac{D_{nb}^2}{r_b'r_n'}\right]I_b + \left[R_n + j\frac{\omega\mu}{2\pi}\ln\frac{D_{bn}D_{nc}}{D_{bc}r_n'}\right]I_c$$

$$(4\text{-}23\text{b})$$

$$\frac{1}{l}(V_{bn} - V_{b'n'}) = \left[R_n + j\frac{\omega\mu}{2\pi}\ln\frac{D_{cn}D_{na}}{D_{ca}r_n'}\right]I_a$$

$$+ \left[R_n + j\frac{\omega\mu}{2\pi}\ln\frac{D_{cn}D_{nb}}{D_{cb}r_n'}\right]I_b + \left[R_c + R_n + j\frac{\omega\mu}{2\pi}\ln\frac{D_{cn}^2}{r_c'r_n'}\right]I_c$$

$$(4\text{-}23\text{c})$$

or writing (4-23) in matrix form:

$$\frac{1}{l}(\tilde{V}_{abc} - \tilde{V}_{abc}') = [Z_{abc}]\tilde{I}_{abc} \qquad (4\text{-}24\text{a})$$

where (4-23) serves to define the entries in $[Z_{abc}]$. Transforming equation (4-24a) to sequence quantities we get

$$\frac{1}{l}(\tilde{V}_{012} - \tilde{V}_{012}') = [Z_{012}]\tilde{I}_{012} \qquad (4\text{-}24\text{b})$$

where

$$[Z_{012}] = [T]^{-1}[Z_{abc}][T] \qquad \text{ohm/m} \qquad (2\text{-}60)$$

A general computation of $[Z_{012}]$ is formidable and should be executed by computer. For educational purposes, let us examine a symmetrical arrangement of conductors referred to as the equilateral line, a case simple enough to develop by hand.

Example 4-1

Given a transmission line, the phase conductors of which are located at the vertices of an equilateral triangle, and a single neutral at its centroid, evaluate $[Z_{012}]$ if $r'_a = r'_b = r'_c = r'_n = r'$, $R_a = R_b = R_c = R_n = R$, and $D_{ab} = D_{bc} = D_{ca} = D$.

Solution

Equation (4-24a) becomes:

$$\frac{1}{l}\begin{bmatrix} V_{an} \\ V_{bn} \\ V_{cn} \end{bmatrix} - \frac{1}{l}\begin{bmatrix} V_{a'n'} \\ V_{b'n'} \\ V_{c'n'} \end{bmatrix} = \begin{bmatrix} Z_s & Z_m & Z_m \\ Z_m & Z_s & Z_m \\ Z_m & Z_m & Z_s \end{bmatrix}\begin{bmatrix} I_a \\ I_b \\ I_c \end{bmatrix} \qquad (4\text{-}25a)$$

where

$$Z_s = R_s + jX_s \quad \text{ohm/m} \qquad (4\text{-}25b)$$

$$Z_m = R_m + jX_m \quad \text{ohm/m} \qquad (4\text{-}25c)$$

$$R_s = 2R \quad \text{ohm/m} \qquad (2\text{-}25d)$$

$$R_m = R \quad \text{ohm/m} \qquad (4\text{-}25e)$$

$$X_s = \frac{\mu\omega}{2\pi}\ln\frac{(D/\sqrt{3})^2}{r'^2} \quad \text{ohm/m} \qquad (4\text{-}25f)$$

$$X_m = \frac{\mu\omega}{2\pi}\ln\frac{(D/\sqrt{3})^2}{Dr'} \quad \text{ohm/m} \qquad (4\text{-}25g)$$

We have encountered this particular symmetry before in example 2-7. Recall that application of (2-60)

$$[Z_{012}] = [T]^{-1}[Z_{abc}][T] \qquad (2\text{-}60)$$

produces the result:

$$[Z_{012}] = \begin{bmatrix} Z_0 & 0 & 0 \\ 0 & Z_1 & 0 \\ 0 & 0 & Z_2 \end{bmatrix} \qquad (4\text{-}26a)$$

where

$$Z_0 = R_0 + jX_0 \tag{4-26b}$$

$$Z_1 = R_1 + jX_1 \tag{4-26c}$$

$$Z_2 = R_2 + jX_2 \tag{4-26d}$$

$$R_0 = R_s + 2R_m = 4R \tag{4-26e}$$

$$R_1 = R_2 = R_s - R_m = R \tag{4-26f}$$

$$X_0 = X_s + 2X_m = \frac{\mu\omega}{2\pi} \ln \frac{D^4}{27r'^4} \tag{4-26g}$$

$$X_1 = X_2 = X_s - X_m = \frac{\mu\omega}{2\pi} \ln \frac{D}{r'} \tag{4-26h}$$

Again be sure to note the diagonalization of $[Z_{012}]$. These results were derived for the equilateral line. If the line is completely transposed, Equation (4-26h) can still be used to calculate X_1 (see Problem 4-24 for proof). Transposition means that each phase position allocated for phase conductors is occupied for equal lengths by phases a, b, and c. Specifically, for the transposed line:

$$X_1 = \frac{\mu\omega}{2\pi} \ln \frac{D}{r'} \tag{4-26h}$$

where

$$D = \sqrt[3]{D_{ab}D_{bc}D_{ca}}$$

A corresponding expression is derived for X_0 for the transposed line in problem (4-24).

Example 4-2

Continuing example 4-1, find numerical values for Z_0, Z_1, and Z_2 for $\mu = \mu_0$, $f = 60$ Hz, $D = 10$ metres, and $r' = 4$ cm. Also take R to be 0.06 milliohms/m.

Solution

$$R_0 = 4R \tag{4-26e}$$

$$= 0.24 \text{ milliohms/m}$$

$$R_1 = R \tag{4-26f}$$

$$= 0.06 \text{ milliohms/m}$$

and

$$R_2 = 0.06 \text{ milliohms/m}$$

$$X_0 = \frac{\mu\omega}{2\pi} \ln \frac{D^4}{27r'^4} \tag{4-26g}$$

$$= \frac{(4\pi \times 10^{-7})(377)}{2\pi} \ln \frac{10^4}{27(0.04)^4}$$

$$= 1.42 \text{ milliohms/m}$$

$$X_1 = \frac{\mu\omega}{2\pi} \ln \frac{D}{r'} \tag{4-26h}$$

$$= \frac{(4\pi \times 10^{-7})(377)}{2\pi} \ln \frac{10}{0.04}$$

$$= 0.42 \text{ milliohms/m}$$

$$\therefore \quad X_2 = 0.42 \text{ milliohms/m}$$

Therefore

$$Z_0 = 0.24 + j1.42 \text{ milliohms/m}$$

$$Z_1 = 0.06 + j0.42 \text{ milliohms/m}$$

$$Z_2 = 0.06 + j0.42 \text{ milliohms/m}$$

This approach can be extended to any number of neutrals by similar steps. Also, the effect of ground is important, and can be included in the calculations. More detail on these, and related topics, is given in the Appendix. Fortunately, more complicated lines produce a $[Z_{012}]$ that is approximately diagonal; that is, the off-diagonal entries are relatively small. For most practical applications, the assumption of a diagonal $[Z_{012}]$ is reasonable.

4-2 Line Capacitance

Refer back to Figure 4-2 and recall the structure of the electric field surrounding a charged, infinitely long conductor with line charge density ρ.

Consider Gauss' Law:

$$\int_{\text{Surface}} \hat{D} \cdot ds = q_{\text{enclosed}} \qquad (4\text{-}27)$$

In this simple geometry, (4-27) simplifies to:

$$E = \frac{\rho}{2\pi\varepsilon\chi} \qquad (4\text{-}28a)$$

where

$$\varepsilon = \varepsilon_0 = \frac{1}{36\pi} \times 10^{-9} \text{ farad/m} \qquad (4\text{-}28b)$$

The equipotential surfaces are concentric cylinders surrounding the conductor. The potential difference between cylinders, from the position χ_a to χ_b, is:

$$v_{ab} = \int_{\chi_a}^{\chi_b} E \, d\chi \qquad (4\text{-}29a)$$

$$v_{ab} = \int_{\chi_a}^{\chi_b} \frac{\rho}{2\pi\varepsilon\chi} \, d\chi \qquad (4\text{-}29b)$$

$$= \frac{\rho}{2\pi\varepsilon} \ln \frac{\chi_b}{\chi_a} \qquad (4\text{-}29c)$$

A word of explanation on the positive sense of variables is in order. The notation v_{ab} implies the voltage drop from "a" relative to "b,"—that is, "a" is understood positive relative to "b." The charge density ρ carries its own sign (positive charge produces positive ρ and vice versa). Whenever $\chi_b > \chi_a$, $\ln(\chi_b/\chi_a) > 0$, and v_{ab} is positive for positive ρ, which is physically correct. Equation (4-29c) is valid for χ external to the conductor and holds for instantaneous quantities. We apply equation (4-29c) to calculate the voltage between two conductors (i, j) due to charge on a third (k).

$$v_{ij}\rfloor_{\rho_k} = v_{ik}\rfloor_{\rho_k} + v_{kj}\rfloor_{\rho_k} \qquad (4\text{-}30a)$$

$$= \frac{\rho_k}{2\pi\varepsilon} \left[\ln \frac{r_k}{D_{ik}} + \ln \frac{D_{jk}}{r_k} \right] \qquad (4\text{-}30b)$$

$$= \frac{\rho_k}{2\pi\varepsilon} \ln \frac{D_{jk}}{D_{ik}} \qquad (4\text{-}30c)$$

where r_k is the radius of conductor k and D_{ij} is the center to center distance between the ith and jth conductors, the same notation that was used in section 4-1.

We now need to develop an equation for the voltage between two conductors in the presence of several charged conductors. Consider the n conductor array as shown in Figure 4-3. By superposition:

$$v_{ij} = v_{ij}\rfloor_{\rho_1} + v_{ij}\rfloor_{\rho_2} + \ldots + v_{ij}\rfloor_{\rho_i} + v_{ij}\rfloor_{\rho_j} + \ldots + v_{ij}\rfloor_{\rho_n} \tag{4-31a}$$

$$= \frac{1}{2\pi\varepsilon}\left[\rho_1 \ln\frac{D_{1j}}{D_{1i}} + \rho_2 \ln\frac{D_{2j}}{D_{2i}} + \ldots + \rho_i \ln\frac{D_{ij}}{r_i} + \rho_j \ln\frac{r_i}{D_{ji}} + \ldots + \rho_n \ln\frac{D_{nj}}{D_{ni}}\right] \tag{4-31b}$$

or in a more compact notation:

$$v_{ij} = \frac{1}{2\pi\varepsilon}\sum_{k=1}^{n}\rho_k \ln\frac{D_{kj}}{D_{ki}}; \qquad D_{kk} = r_k \tag{4-31c}$$

We apply this work only to situations where charge is conserved:

$$\rho_1 + \rho_2 + \ldots + \rho_n = 0 \tag{4-32}$$

Equations (4-31) and (4-32) constitute the bases for handling any conductor configuration for an arbitrary three phase line. We are now forced to choose between generality with its necessary complexities and a specific simple case with its benefits in understanding. The problem is solved in some generality in the Appendix. We choose to investigate a four conductor array (three phases and a neutral). In our notation we use a, b, c, and n to designate phase and neutral conductors, respectively. Equation (4-31c), with "j" selected as "n," the neutral, becomes:

$$v_{in} = \frac{1}{2\pi\varepsilon}\sum_{k=a}^{n}\rho_k \ln\frac{D_{kn}}{D_{ki}} \tag{4-33a}$$

where

$$i = a, b, c, n \tag{4-33b}$$

$$k = a, b, c, n \tag{4-33c}$$

$$D_{ii} = r_i = \text{radius of } i\text{th conductor} \tag{4-33d}$$

Note what happens to the fourth equation in the set (4-33a). For $i = n$, $\ln(D_{kn}/D_{kn}) = \ln(1) = 0$ is a factor of every term on the right side and of course $v_{nn} = 0$ to produce the trivial result $0 = 0$.

The remaining three equations of the set (4-33a) are nontrivial. However, we still have four variables on the right side (ρ_a, ρ_b, ρ_c, and ρ_n). We eliminate ρ_n, using (4-32):

$$\rho_n = -(\rho_a + \rho_b + \rho_c) \tag{4-34}$$

Using the "a" phase equation from (4-33a):

$$v_{an} = v_a = \frac{1}{2\pi\varepsilon}\left[\rho_a \ln \frac{D_{an}}{r_a} + \rho_b \ln \frac{D_{bn}}{D_{ba}} + \rho_c \ln \frac{D_{cn}}{D_{ca}} - (\rho_a + \rho_b + \rho_c)\ln \frac{r_n}{D_{na}}\right]$$

(4-35)

Simplifying:

$$v_a = \frac{1}{2\pi\varepsilon}\left[\rho_a \ln \frac{D_{an}^2}{r_a r_n} + \rho_b \ln \frac{D_{bn}D_{na}}{D_{ba}r_n} + \rho_c \ln \frac{D_{cn}D_{na}}{D_{ca}r_n}\right]$$

(4-36a)

Similarly:

$$v_b = \frac{1}{2\pi\varepsilon}\left[\rho_a \ln \frac{D_{an}D_{nb}}{D_{ab}r_n} + \rho_b \ln \frac{D_{bn}^2}{r_b r_n} + \rho_c \ln \frac{D_{cn}D_{nb}}{D_{cb}r_n}\right]$$

(4-36b)

$$v_c = \frac{1}{2\pi\varepsilon}\left[\rho_a \ln \frac{D_{an}D_{nc}}{D_{ac}r_n} + \rho_b \ln \frac{D_{bn}D_{nc}}{D_{bc}r_n} + \rho_c \ln \frac{D_{cn}^2}{r_c r_n}\right]$$

(4-36c)

Converting to matrix notation:

$$\tilde{v}_{abc} = [F_{abc}]\tilde{\rho}_{abc}$$

(4-37a)

where $[F_{abc}]$ is 3×3 and has the general entry:

$$f_{ij} = \frac{1}{2\pi\varepsilon}\ln \frac{D_{in}D_{nj}}{D_{ij}r_n}$$

(4-37b)

$$i, j = a, b, c$$

(4-37c)

For sinusoidal steady state analysis both voltage and charge density may be represented by phasors. Therefore:

$$\tilde{V}_{abc} = [F_{abc}]\tilde{P}_{abc}$$

(4-38)

It follows that

$$\tilde{P}_{abc} = [F_{abc}]^{-1}\tilde{V}_{abc}$$

(4-39a)

$$= [C_{abc}]\tilde{V}_{abc}$$

(4-39b)

where

$$[C_{abc}] = [F_{abc}]^{-1}$$

(4-39c)

where the units of $[C_{abc}]$ are farads/m.

Converting to shunt admittance:

$$[Y_{abc}] = j\omega[C_{abc}]$$

(4-40)

Although it is straightforward to write general entries for $[F_{abc}]$, the matrix inversion step prevents obtaining simple expressions for $[C_{abc}]$ or $[Y_{abc}]$. Transforming $[Y_{abc}]$ to sequence values:

$$[Y_{012}] = [T]^{-1}[Y_{abc}][T] \tag{4-41}$$

Comparing (4-40) with (4-41) we could as readily write

$$[C_{012}] = [T]^{-1}[C_{abc}][T] \text{ farads/m} \tag{4-42a}$$

if we define:

$$[C_{012}] = 1/j\omega[Y_{012}] \tag{4-42b}$$

Similarly

$$[F_{012}] = [T]^{-1}[F_{abc}][T] \tag{4-43}$$

We have our choice from equations (4-41), (4-42), or (4-43) for converting to sequence values. We may prefer to compute $[F_{012}]$ first (if it is diagonal) since it would be simple to invert $[F_{012}]$ to obtain $[C_{012}]$. As in the inductance calculation in the nonsymmetric situation, it is unreasonable to calculate by hand. To clarify the procedure, an example will help. Again we resort to the equilateral line.

Example 4-3

The equilateral line of example 4-1 is given, with $r_a = r_b = r_c = r_n = r$. Evaluate $[C_{012}]$.

Solution

We realize $D_{ab} = D_{bc} = D_{ca} = D$

and

$$D_{an} = D_{bn} = D_{cn} = D/\sqrt{3}$$

$$\therefore \; [F_{abc}] = \frac{1}{2\pi\varepsilon} \begin{bmatrix} \ln\dfrac{(D/\sqrt{3})^2}{r^2} & \ln\dfrac{(D/\sqrt{3})^2}{Dr} & \ln\dfrac{(D/\sqrt{3})^2}{Dr} \\[3mm] \ln\dfrac{(D/\sqrt{3})^2}{Dr} & \ln\dfrac{(D/\sqrt{3})^2}{r^2} & \ln\dfrac{(D/\sqrt{3})^2}{Dr} \\[3mm] \ln\dfrac{(D/\sqrt{3})^2}{Dr} & \ln\dfrac{(D/\sqrt{3})^2}{Dr} & \ln\dfrac{(D/\sqrt{3})^2}{r^2} \end{bmatrix} \tag{4-44a}$$

or

$$[F_{abc}] = \begin{bmatrix} f_s & f_m & f_m \\ f_m & f_s & f_m \\ f_m & f_m & f_s \end{bmatrix} \tag{4-44b}$$

where

$$f_s = \frac{1}{2\pi\varepsilon} \ln \frac{D^2}{3r^2} \tag{4-44c}$$

$$f_m = \frac{1}{2\pi\varepsilon} \ln \frac{D}{3r} \tag{4-44d}$$

Again we recognize this special symmetry so that:

$$[F_{012}] = [T]^{-1}[F_{abc}][T] \tag{4-45a}$$

$$= \begin{bmatrix} f_0 & 0 & 0 \\ 0 & f_1 & 0 \\ 0 & 0 & f_2 \end{bmatrix} \tag{4-45b}$$

where

$$f_0 = f_s + 2f_m = \frac{1}{2\pi\varepsilon} \ln \frac{D^4}{27r^4} \tag{4-45c}$$

$$f_1 = f_s - f_m = \frac{1}{2\pi\varepsilon} \ln \frac{D}{r} \tag{4-45d}$$

$$f_2 = f_s - f_m = \frac{1}{2\pi\varepsilon} \ln \frac{D}{r} \tag{4-45e}$$

Inverting $[F_{012}]$ is simple:

$$[C_{012}] = [F_{012}]^{-1} \tag{4-46a}$$

$$= \begin{bmatrix} 1/f_0 & 0 & 0 \\ 0 & 1/f_1 & 0 \\ 0 & 0 & 1/f_2 \end{bmatrix} \tag{4-46b}$$

$$= \begin{bmatrix} C_0 & 0 & 0 \\ 0 & C_1 & 0 \\ 0 & 0 & C_2 \end{bmatrix} \tag{4-46c}$$

where

$$C_0 = \frac{2\pi\varepsilon}{\ln(D^4/27r^4)} \tag{4-46d}$$

$$C_1 = \frac{2\pi\varepsilon}{\ln(D/r)} \qquad\qquad (4\text{-}46e)$$

$$C_2 = \frac{2\pi\varepsilon}{\ln(D/r)} \qquad\qquad (4\text{-}46f)$$

The units are farads/metre. The sequence admittance matrix $[Y_{012}]$ is

$$[Y_{012}] = j \begin{bmatrix} B_0 & 0 & 0 \\ 0 & B_1 & 0 \\ 0 & 0 & B_2 \end{bmatrix} \qquad\qquad (4\text{-}47a)$$

where

$$B_0 = \omega C_0 = \frac{2\pi\varepsilon\omega}{\ln(D^4/27r^4)} \qquad\qquad (4\text{-}47b)$$

$$B_1 = \omega C_1 = \frac{2\pi\varepsilon\omega}{\ln(D/r)} \qquad\qquad (4\text{-}47c)$$

$$B_2 = \omega C_2 = \frac{2\pi\varepsilon\omega}{\ln(D/r)} \qquad\qquad (4\text{-}47d)$$

The units are siemens/metre.

Example 4-4

Continuing Example 4-3, find numerical values for B_0, B_1, and B_2 for $\varepsilon = \varepsilon_0$, $f = 60$ Hz, $D = 10$ metres, and $r = 5$ cm.

$$
\begin{aligned}
B_0 &= \frac{2\pi\varepsilon\omega}{\ln(D^4/27r^4)} \\[2mm]
&= \frac{2\pi(1/36\pi)\times 10^{-9}(377)}{\ln((10)^4/27(0.05)^4)} \qquad\qquad (4\text{-}47b) \\[2mm]
&= 1.17\times 10^{-9} \text{ siemen/m}
\end{aligned}
$$

$$
\begin{aligned}
B_1 = B_2 &= \frac{2\pi\varepsilon\omega}{\ln(D/r)} \\[2mm]
&= \frac{2\pi(1/36\pi)\times 10^{-9}(377)}{\ln(10/0.05)} \qquad\qquad (4\text{-}47c) \\[2mm]
&= 3.95\times 10^{-9} \text{ siemen/m}
\end{aligned}
$$

As mentioned before, a more general treatment is given in the Appendix. Again $D = \sqrt[3]{D_{ab}D_{bc}D_{ca}}$ may be used in equation (4-47c) for the completely transposed line as an approximation.

4-3 Line Resistance

Virtually all overhead power transmission lines utilize bare aluminum conductors because of their economy, good electrical conduction properties, and light weight. Some design types include a steel or metal alloy cable in the center for high tensile strength surrounded by stranded aluminum cable for high electrical conductivity. Conductor types are designated as follows:

AAC—All Aluminum Conductor

AAAC—All Aluminum Alloy Conductor

ACSR—Aluminum Conductor Steel Reinforced

ACAR—Aluminum Conductor Alloy Reinforced

Such conductors are sized by their cross-sectional area, frequently given in "circular mils." One circular mil is defined as the area of a circle of diameter 10^{-3} inch. One thousand circular mils is designated by the abbreviation "MCM." Thus, a 1000 MCM stranded aluminum conductor has a diameter of one inch (an approximation since "diameter" implies a perfect circular cross section, which stranding prevents).

Computation or measurement of conductor resistance would seem to be simple. However, there are several effects that complicate the problem, the most important of which are temperature, skin effect, and spiralling. The dc resistance of a conductor of uniform material and cross-sectional area is:

$$R_{dc} = \frac{\rho l}{A} \qquad (4\text{-}48)$$

where

R_{dc} = dc conductor resistance in ohms

A = conductor cross-sectional area in m^2

l = conductor length in m

ρ = conductor resistivity in metre-ohms

$= 2.83 \times 10^{-8}$ metre-ohm for aluminum at 20°C.

Observe that the resistivity ρ is given at a specific temperature. Variation of ρ with temperature is approximately linear and may be calculated from:

$$\rho_2 = \rho_1 \frac{T_2 + T_0}{T_1 + T_0}$$

where $T_0 = 228$ for aluminum and ρ_1 and ρ_2 are resistivities at temperatures T_1 and T_2, measured in degrees Celsius.

Skin effect is the tendency for an ac current to concentrate at the conductor's surface, thereby increasing the effective resistance. The effect increases with frequency and is observable at 60 Hz. Recognizing this, we find that "ac resistances" are greater than dc values. Because large power conductors are stranded, and the strands wound in a spiral fashion around the conductor center, each strand is somewhat longer than the finished conductor. This too slightly increases the resistance.

Accounting for these effects typically increases the value calculated from (4-48) by a few percent. For a thorough treatment of the topic, the reader is referred to the Bibliography at the end of the chapter [1]. We shall find it sufficient to simply look up suitable values in tables supplied for such purposes for our work. Refer to Table 4-1 for typical values; consult [1] for more complete data.

Table 4-1 Data for selected commercial ACSR bare electrical conductors.

CONDUCTOR SIZE (MCM)	STRANDING AL/ST	AC RESISTANCE 75° C (OHMS/KM)	RADIUS (CM)	GMR (CM)	APPROXIMATE CURRENT RATING (AMPERES)
266.8	26/7	0.2568	0.815	0.661	460
556.5	24/7	0.1240	1.161	0.933	730
795	45/7	0.0880	1.350	1.073	900
954	54/7	0.0732	1.519	1.225	1010
1590	54/19	0.0448	1.962	1.594	1380

Closely related to the problem of conductor resistance is deciding what the conductor current rating should be. This is obviously important because this directly determines the line's energy transporting capacity, which, after all, is the line's only reason for existence. Since the line temperature is the critical factor, the current and its time of duration both are significant. Elevated temperature will cause the line to fail by producing excessive conductor sagging. The problem is that critical clearances fail to be met and arcs can flash over from the phase conductors to ground points. The point to be avoided occurs when the elastic limit of the conductors is reached; that is, the limit is not exceeded when the elongated conductors can shrink to their original length when cooled. Among the significant factors that affect current ratings are ambient temperature, ambient wind velocity, weathered condition of conductors, and line clearances. Table 4-1 gives some approximate current ratings for selected ACSR power conductors.

4-4 Consideration of Line Length

We have observed that the results $[Z_{012}]$ and $[Y_{012}]$ are impedances and admittances per unit length. These are distributed parameters; that is, the effects discussed are distributed uniformly along the length of the line. To achieve our goal of finding an equivalent circuit for the line we must properly account for the line length. Consider the circuit shown in Figure 4-5, which depicts a general distributed parameter two conductor line.

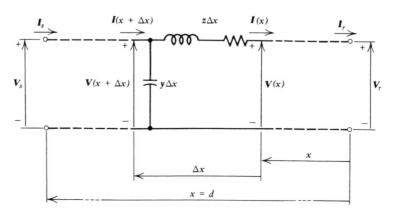

Figure 4-5 Distributed parameter transmission line.

We define the following:

$x =$ position along the line, measured positive from the right (receiving end) toward the left (sending end), in metres

$V(x) =$ phasor voltage at location x on the line

$I(x) =$ phasor current at location x on the line

$z = R + j\omega L =$ Series impedance per unit length in ohms/m

$y = j\omega C =$ Shunt admittance per unit length in siemens/m

102

$V_s = V(d) =$ Sending end voltage

$V_r = V(0) =$ Receiving end voltage

$I_s = I(d) =$ Sending end current

$I_r = I(0) =$ Receiving end current

$d =$ line length

From Kirchhoff's voltage law:

$$V(x + \Delta x) = V(x) + z\,\Delta x I(x) \tag{4-49a}$$

Rearranging

$$\frac{V(x + \Delta x) - V(x)}{\Delta x} = zI(x) \tag{4-49b}$$

Taking the limit as Δx goes to zero:

$$\lim_{\Delta x \to 0} \frac{V(x + \Delta x) - V(x)}{\Delta x} = zI(x) \tag{4-49c}$$

$$\frac{dV(x)}{dx} = zI(x) \tag{4-49d}$$

Now by Kirchhoff's current law:

$$I(x + \Delta x) = I(x) + y\,\Delta x V(x + \Delta x) \tag{4-50a}$$

Rearranging:

$$\frac{I(x + \Delta x) - I(x)}{\Delta x} = yV(x + \Delta x) \tag{4-50b}$$

Taking the limit as Δx goes to zero:

$$\lim_{\Delta x \to 0} \frac{I(x + \Delta x) - I(x)}{\Delta x} = \lim_{\Delta x \to 0} yV(x + \Delta x) \tag{4-50c}$$

$$\frac{dI(x)}{dx} = yV(x) \tag{4-50d}$$

Differentiate (4-49d) and (4-50d) with respect to x, producing

$$\frac{d^2 V(x)}{dx^2} = z\frac{dI}{dx}(x) \tag{4-49e}$$

$$\frac{d^2 I(x)}{dx^2} = y\frac{dV}{dx}(x) \tag{4-50e}$$

Eliminate $dI(x)/dx$ and $dV(x)/dx$ from equations (4-49e) and (4-50e) using (4-50d) and (4-49d), producing:

$$\frac{d^2 V(x)}{dx^2} = zy V(x) \tag{4-51a}$$

$$\frac{d^2 I(x)}{dx} = zy I(x) \tag{4-51b}$$

We find it expedient at this point to define

$$\gamma^2 = zy \tag{4-52a}$$

or

$$\gamma = \sqrt{zy} \tag{4-52b}$$

What are the units of γ? Recall that z and y are ohms/m and siemens/m. The propagation constant, γ, must be in per meter. Also note that γ is complex. We further define

$$\alpha = \text{Re}[\gamma] = \text{attenuation constant} \tag{4-53a}$$

$$\beta = \text{Im}[\gamma] = \text{phase constant} \tag{4-53b}$$

so that

$$\gamma = \alpha + j\beta \tag{4-53c}$$

We wish to solve equation (4-51a) with zy replaced by its equivalent, γ^2.

$$\frac{d^2 V(x)}{dx^2} = \gamma^2 V(x) \tag{4-54a}$$

Using p for $d/dx(\ )$, the characteristic equation is:

$$p^2 - \gamma^2 = 0 \tag{4-54b}$$

$$\therefore \quad p = \pm\gamma \tag{4-54c}$$

and

$$V(x) = V^+ e^{\gamma x} + V^- e^{-\gamma x} \tag{4-54d}$$

The complex constants V^+ and V^- arise from solving the second order equation (4-54a). Substitute this result into (4-49d)

$$zI(x) = \frac{d}{dx}[V^+ e^{\gamma x} + V^- e^{-\gamma x}] \tag{4-55a}$$

$$= \gamma V^+ e^{\gamma x} - \gamma V^- e^{-\gamma x} \tag{4-55b}$$

or

$$\frac{z}{\gamma} I(x) = V^+ e^{\gamma x} - V^- e^{-\gamma x} \tag{4-55c}$$

We find it expedient at this point to define the characteristic impedance Z_c:

$$Z_c = \sqrt{\frac{z}{y}} \tag{4-56}$$

What are the units of Z_c? Ohms! (Note the per unit lengths cancel). Observe that in general Z_c is complex, since z and y are also. In the special, but important, case where the series resistance is zero (called the lossless case) an extraordinary result occurs. Since $z = j\omega L$ and $y = j\omega C$, in (4-56) the "$j\omega$'s" cancel and Z_c reduces to $\sqrt{L/C}$, which is real! We note this only in passing; for our work it is not necessary to neglect the resistance because we are concerned only with sinusoidal steady state performance. Returning our attention to (4-55c):

$$\frac{z}{\gamma} = \frac{z}{\sqrt{zy}} \tag{4-57a}$$

$$= \sqrt{\frac{z}{y}} \tag{4-57b}$$

$$= Z_c \tag{4-57c}$$

Equation (4-55c) becomes:

$$Z_c I(x) = V^+ e^{+\gamma x} - V^- e^{-\gamma x} \tag{4-58}$$

We wish to replace the constants V^+ and V^- with V_r and I_r, since these are more conveniently evaluated. We aptly equations (4-54d) and (4-58) at the receiving end ($x = 0$).

$$V(0) = V_r = V^+ + V^- \tag{4-59}$$

$$Z_c I(0) = Z_c I_r = V^+ - V^- \tag{4-60}$$

Add (4-59) to (4-60) and divide by two:

$$V^+ = \frac{V_r + Z_c I_r}{2} \tag{4-61a}$$

Similarly, subtract (4-60) from (4-59) and divide by two:

$$V^- = \frac{V_r - Z_c I_r}{2} \tag{4-61b}$$

We can now eliminate V^+ and V^- from equations (4-54d) and (4-58). We get:

$$V(x) = \frac{V_r + I_r Z_c}{2} e^{\gamma x} + \frac{V_r - I_r Z_c}{2} e^{-\gamma x} \tag{4-62a}$$

$$Z_c I(x) = \frac{V_r + I_r Z_c}{2} e^{\gamma x} - \frac{V_r - I_r Z_c}{2} e^{-\gamma x} \tag{4-62b}$$

Rearranging:

$$V(x) = \left[\frac{e^{\gamma x} + e^{-\gamma x}}{2}\right] V_r + \left[\frac{e^{\gamma x} - e^{-\gamma x}}{2}\right] Z_c I_r \tag{4-63a}$$

$$Z_c I(x) = \left[\frac{e^{\gamma x} - e^{-\gamma x}}{2}\right] V_r + \left[\frac{e^{\gamma x} + e^{-\gamma x}}{2}\right] Z_c I_r \tag{4-63b}$$

Recognizing the hyperbolic functions sinh and cosh we can equivalently write:

$$V(x) = \cosh \gamma x V_r + Z_c \sinh \gamma x I_r \tag{4-64a}$$

$$I(x) = \frac{1}{Z_c} \sinh \gamma x V_r + \cosh \gamma x I_r \tag{4-64b}$$

Equation (4-65) is completely adequate to calculate line performance as knowing Z_c and γ for the line, and the receiving end quantities. To obtain sending end values we simply set x equal to d.

$$V_s = V(x)]_{x=d} = \cosh \gamma d V_r + Z_c \sinh \gamma d I_r \tag{4-65a}$$

$$I_s = I(x)]_{x=d} = \frac{1}{Z_c} \sinh \gamma d V_r + \cosh \gamma d I_r \tag{4-65b}$$

Equation (4-65) is completely adequate to calculate line performance as sensed at its sending and receiving end terminals. However, an equivalent circuit representation is more suitable for our purposes, since we must interconnect the line with other components. Consider the pi network shown in Figure 4-6. Kirchhoff's voltage law produces:

$$V_s = Z\left(I_r + \frac{Y}{2} V_r\right) + V_r \tag{4-66a}$$

Rearranging:

$$V_s = \left(1 + \frac{ZY}{2}\right) V_r + Z I_r \tag{4-66b}$$

Kirchhoff's current law produces

$$I_s = \frac{Y}{2} V_s + I_r + \frac{Y}{2} V_r \qquad (4\text{-}67\text{a})$$

Using (4-66b) to eliminate V_s:

$$I_s = \frac{Y}{2}\left[\left(1+\frac{ZY}{2}\right)V_r + ZI_r\right] + I_r + \frac{Y}{2} V_r \qquad (4\text{-}67\text{b})$$

Rearranging:

$$I_s = \left(Y + \frac{ZY^2}{4}\right)V_r + \left(1+\frac{ZY}{2}\right)I_r \qquad (4\text{-}67\text{c})$$

Compare (4-65a) and (4-65b) with (4-66b) and (4-67c). They will become identical if we force:

$$Z = Z_c \sinh \gamma d \qquad (4\text{-}68\text{a})$$

$$1 + \frac{ZY}{2} = \cosh \gamma d \qquad (4\text{-}68\text{b})$$

Eliminate Z with (4-68a) in (4-68b):

$$1 + \sinh \gamma d \left[\frac{Z_c Y}{2}\right] = \cosh \gamma d \qquad (4\text{-}69\text{a})$$

$$\frac{Z_c Y}{2} = \frac{\cosh \gamma d - 1}{\sinh \gamma d} \qquad (4\text{-}69\text{b})$$

$$= \frac{e^{\gamma d} + e^{-\gamma d} - 2}{e^{\gamma d} - e^{-\gamma d}} \qquad (4\text{-}69\text{c})$$

$$\frac{Z_c Y}{2} = \frac{(e^{\gamma d/2} - e^{-\gamma d/2})^2}{(e^{\gamma d/2} + e^{-\gamma d/2})(e^{\gamma d/2} - e^{-\gamma d/2})} \qquad (4\text{-}69\text{d})$$

$$= \frac{e^{\gamma d/2} - e^{-\gamma d/2}}{e^{\gamma d/2} + e^{-\gamma d/2}} \qquad (4\text{-}69\text{e})$$

$$= \tanh \frac{\gamma d}{2} \qquad (4\text{-}69\text{f})$$

or

$$\frac{Y}{2} = \frac{1}{Z_c} \tanh \frac{\gamma d}{2} \qquad (4\text{-}69\text{g})$$

We are nearing our objective. The Z and Y components used in Figure 4-6 are to be calculated from equations (4-68a) and (4-69g). To apply this to

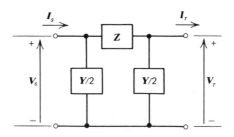

Figure 4-6 Pi network.

the three phase line we think in terms of symmetrical components, and realize we need *three* networks, one for each sequence. Therefore we need three γ's and three Z_c's. Review the definitions for z and y. We further define:

$$z_0 = Z_{00} \tag{4-70a}$$

$$z_1 = Z_{11} \tag{4-70b}$$

$$z_2 = Z_{22} \tag{4-70c}$$

$$y_0 = Y_{00} \tag{4-70d}$$

$$y_1 = Y_{11} \tag{4-70e}$$

$$y_2 = Y_{22} \tag{4-70f}$$

where the Z_{ii} and Y_{ii} values are the main diagonal entries in $[Z_{012}]$ and $[Y_{012}]$. See equations (2-60) and (4-41). It follows that:

$$\gamma_0 = \sqrt{z_0 y_0} \tag{4-71a}$$

$$\gamma_1 = \sqrt{z_1 y_1} \tag{4-71b}$$

$$\gamma_2 = \sqrt{z_2 y_2} \tag{4-71c}$$

and

$$Z_{c_0} = \sqrt{z_0/y_0} \tag{4-72a}$$

$$Z_{c_1} = \sqrt{z_1/y_1} \tag{4-72b}$$

$$Z_{c_2} = \sqrt{z_2/y_2} \tag{4-72c}$$

Finally we apply (4-68a) and (4-69g):

$$Z_0 = Z_{c_0} \sinh \gamma_0 d \tag{4-73a}$$

$$Z_1 = Z_{c_1} \sinh \gamma_1 d \tag{4-73b}$$

$$Z_2 = Z_{c_2} \sinh \gamma_2 d \qquad\qquad (4\text{-}73\text{c})$$

and

$$\frac{Y_0}{2} = \frac{1}{Z_{c_0}} \tanh \frac{\gamma_0 d}{2} \qquad\qquad (4\text{-}74\text{a})$$

$$\frac{Y_1}{2} = \frac{1}{Z_{c_1}} \tanh \frac{\gamma_1 d}{2} \qquad\qquad (4\text{-}74\text{b})$$

$$\frac{Y_2}{2} = \frac{1}{Z_{c_2}} \tanh \frac{\gamma_2 d}{2} \qquad\qquad (4\text{-}74\text{c})$$

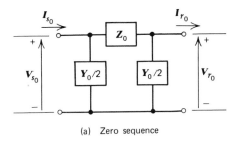

(a)   Zero sequence

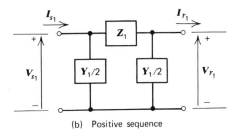

(b)   Positive sequence

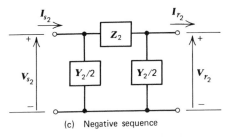

(c)   Negative sequence

Figure 4-7 Transmission line sequence networks.

109

The corresponding circuits are shown in Figure 4-7. One last detail before we apply this to an example. If $\gamma d \ll 1$:

$$Z = Z_c \sinh \gamma d \tag{4-75a}$$

$$Z \cong Z_c(\gamma d) \tag{4-75b}$$

$$Z \cong \sqrt{\frac{z}{y}}\sqrt{zy}\,d \tag{4-75c}$$

$$\cong zd \tag{4-75d}$$

and

$$\frac{Y}{2} = \frac{1}{Z_c}\tanh\frac{\gamma d}{2} \tag{4-76a}$$

$$\cong \frac{1}{Z_c}(\gamma d/2) \tag{4-76b}$$

$$\cong \sqrt{\frac{y}{z}}\sqrt{zy}\,d/2 \tag{4-76c}$$

$$\cong \frac{yd}{2} \tag{4-76d}$$

When equations (4-75d) and (4-76d) are used, the resultant circuit model is referred to as the "nominal" pi equivalent circuit. Generally, the approximation is good for transmission lines of length up to 200 km. When the calculations are done by computer, there is no reason to use equations (4-75d) and (4-76d) instead of (4-71) through (4-74).

Example 4-5

The line of examples 4-1, 2, 3, and 4 is 200 km long. Compute values for the sequence circuit models shown in Figure 4-7.

Solution

From examples 4-2 and 4-3:

$$z_0 = 0.24 + j1.42 \text{ milliohms/m}$$

$$z_1 = 0.06 + j0.42 \text{ milliohms/m}$$

$$y_0 = j1.17 \ \text{nanosiemens/m}$$

$$y_1 = y_2 = j3.95 \ \text{nanosiemens/m}$$

We calculate

$$Z_{c_0} = \sqrt{\frac{z_0}{y_0}} \tag{4-72a}$$

$$= \sqrt{\frac{(0.24 + j1.42) \times 10^{-3}}{j1.17 \times 10^{-9}}}$$

$$= 1109\underline{/-4.8°} \ \text{ohms}$$

$$Z_{c_1} = Z_{c_2} = \sqrt{\frac{z_1}{y_1}} \tag{4-72b}$$

$$= \sqrt{\frac{(0.06 + j0.42) \times 10^{-3}}{j3.95 \times 10^{-9}}}$$

$$= 328\underline{/-4.07°} \ \text{ohms}$$

$$\gamma_0 = \sqrt{z_0 y_0} \tag{4-71a}$$

$$= \sqrt{(0.24 + j1.42)(j1.17) \times 10^{-12}}$$

$$= (0.109 + j1.29) \times 10^{-6} \ \text{per metre}$$

$$\gamma_1 = \gamma_2 = \sqrt{z_1 y_1} \tag{4-71b}$$

$$= \sqrt{(0.06 + j0.42)(j3.95) \times 10^{-12}}$$

$$= (0.092 + j1.29) \times 10^{-6} \ \text{per metre}$$

As an approximation
Let

$$\gamma_0 = \gamma_1 = j1.29 \times 10^{-6}$$

so that

$$\gamma_0 d = \gamma_1 d = (j1.29 \times 10^{-6})(2 \times 10^5)$$

$$= j0.258$$

also let

$$Z_{c_0} = 1109 \ \Omega$$

and

$$Z_{c_1} = 328 \ \Omega$$

Then

$$Z_0 = Z_{c_0} \sinh \gamma_0 d \tag{4.73a}$$

$$= 1109 \sinh j0.258$$

$$= j283 \text{ ohms}$$

$$Y_0/2 = \frac{1}{Z_{c_0}} \tanh \frac{\gamma_0 d}{2} \tag{4-74a}$$

$$= \frac{1}{1109} \tanh \frac{j0.258}{2}$$

$$= j1.17 \times 10^{-4} \text{ siemen}$$

$$Z_1 = Z_{c_1} \sinh \gamma_1 d \tag{4-73b}$$

$$= (328) \sinh j0.258$$

$$= j83.7 \text{ ohms}$$

$$Y_1/2 = \frac{1}{Z_{c_1}} \tanh \frac{\gamma_1 d}{2} \tag{4-74b}$$

$$= \frac{1}{328} \tanh \frac{j0.258}{2} = j3.95 \times 10^{-4} \text{ siemen}$$

4-5 Maximum Power Loading for Power Transmission Lines

It is of fundamental importance to consider the question of how much power a transmission line is capable of transmitting. There are two basic limits: first, the line thermal limit, imposed by the current carrying capacity of the phase conductors; and second, the steady state stability limit, imposed by the line's impedance values. The line is assumed operating in its normal balanced three phase sinusoidal steady state mode at its rated voltage. Only the positive sequence equivalent circuit is required. The thermal limit is:

$$S_{3\phi_{\text{rated}}} = V_{L_{\text{rated}}} I_{L_{\text{rated}}} \sqrt{3} \tag{4-77}$$

where the units are SI (not per unit).

There is some difficulty in deciding just what the rated line current should be. Since the problem is conductor overheating, both the ambient temperature and wind velocity are important. The problem is not insignificant when one realizes that each ampere at $500\,\text{kV}$ represents

866 kVA of transmitted power. Obviously the winter rating of conductors should exceed the summer rating.

Only for relatively short lines will equation (4-77) represent the actual power limit. Recall the transmission line positive sequence network as represented in Figure 4-7(b). The corresponding performance equations are:

$$V_s = \cosh \gamma d V_r + Z_c \sinh \gamma d I_r \tag{4-65a}$$

$$I_s = \frac{1}{Z_c} \sinh \gamma d V_r + \cosh \gamma d I_r \tag{4-65b}$$

Interpret the voltages as line to neutral, the currents as line values, and the impedances as wye connected. The units are SI, not per unit. Equations (4-65) are sometimes written as

$$V_s = A V_r + B I_r \tag{4-78a}$$

$$I_s = C V_r + D I_r \tag{4-78b}$$

where

$$A = A\underline{/\alpha} = \cosh \gamma d \tag{4-78c}$$

$$B = B\underline{/\beta} = Z_c \sinh \gamma d \tag{4-78d}$$

$$C = \frac{1}{Z_c^2} B \tag{4-78e}$$

$$D = A \tag{4-78f}$$

$$V_s = V_s\underline{/\delta} \tag{4-78g}$$

$$V_r = V_r\underline{/0} \tag{4-78h}$$

From (4-78a):

$$I_r = \frac{V_s}{B} - \frac{A V_r}{B} \tag{4-79a}$$

$$I_r = \frac{V_s}{B}\underline{/\delta - \beta} - \frac{A V_r}{B}\underline{/\alpha - \beta} \tag{4-79b}$$

and

$$I_r^* = \frac{V_s}{B}\underline{/\beta - \delta} - \frac{A V_r}{B}\underline{/\beta - \alpha} \tag{4-80}$$

113

The receiving end complex power S is:

$$S_{3\phi_r} = 3V_r I_r^* \tag{4-81a}$$

$$= \frac{3V_s V_r}{B} \underline{/\beta - \delta} - \frac{3AV_r^2}{B} \underline{/\beta - \alpha} \tag{4-81b}$$

For constant V_s and V_r, the only variable in equation (4-81b) is δ, the power angle. Let us display equation (4-81) graphically as shown in Figure 4-8. The locus of S_r in the P_r, Q_r plane as δ varies is a circle. When the receiving end power is zero, δ is small (point $\cdot a$). As we load the line δ

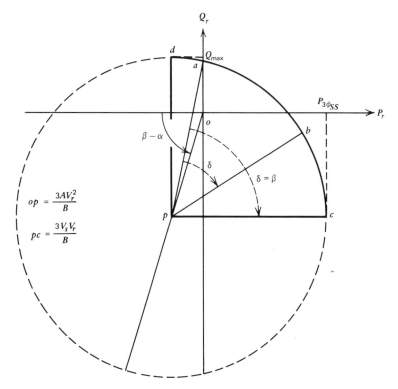

Figure 4-8 Receiving end circle diagram.

increases (point b). We can continue to load the line to the steady state stability limit $P_{3\phi ss}$. This point (c) is reached when $\delta = \beta$. The maximum reactive power that could be transmitted occurs at point d corresponding to $\delta = \beta - 90°$. For real power flow from the sending to receiving end, only operation over the range abc is important. It is not advisable to operate too close to $P_{3\phi ss}$, with a minimum margin of about 20% recommended—that is, $P_{3\phi_r} \leq 0.8 P_{3\phi ss}$. From equation (4-81b):

$$P_{3\phi ss} = R_e \left[\frac{3V_s V_r}{B} \underline{/\beta - \delta} - \frac{3AV_r^2}{B} \underline{/\beta - \alpha} \right]_{\substack{V_s = V_r = V_{L_\text{rated}}/\sqrt{3} \\ \delta = \beta}} \tag{4-82a}$$

$$P_{3\phi ss} = \frac{V_{L_\text{rated}}^2}{B} [1 - A \cos(\beta - \alpha)] \tag{4-82b}$$

As a given line length increases this limit becomes the decisive factor. The corresponding reactive power value is:

$$Q_{3\phi ss} = -\frac{AV_{L_\text{rated}}^2}{B} \sin(\beta - \alpha) \tag{4-82c}$$

and the corresponding apparent power is:

$$S_{3\phi ss} = \sqrt{P_{3\phi ss}^2 + Q_{3\phi ss}^2} \tag{4-82d}$$

$$= \frac{V_{L_\text{rated}}^2}{B} \sqrt{1 + A^2 - 2A \cos(\beta - \alpha)} \tag{4-82e}$$

This limit becomes decisive when $S_{3\phi ss} < S_{3\phi \text{rated}}$. We apply these ideas to an example line in the Appendix.

4-6 Line Voltage Regulation and Compensation

The line impedance elements can have an effect on the voltage at all points along the line, the effect varying with line loading. Serious over-voltages can occur at light loads and unacceptable low voltages can occur at heavy loads. The problem becomes more serious as the line length increases.

This effect can severely limit the acceptable operating range of practical long lines and justifies the cost of installing additional equipment to correct the situation. We observe that the problem in the loaded case is essentially caused by the series inductance. The insertion of series capacitance would have a cancellation effect and relieve the difficulty to some degree. Such a modification is referred to as "series compensation." The overvoltage condition is caused by the line shunt capacitance and this may at least partially be eliminated by the installation of shunt reactors (inductance). This is referred to as "shunt compensation." To investigate the effect of

115

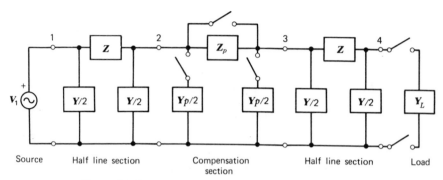

Figure 4-9 Transmission line with mid point compensation.

compensation consider the situation illustrated in Figure 4-9. We define:

$$Z_p = -jX_p \qquad (4\text{-}83a)$$

= Series compensating capacitor impedance; inserted during heavy load conditions, bypassed under light load conditions

$$Y_p = -jB_p \qquad (4\text{-}83b)$$

= Shunt compensating inductor admittance; inserted under light load conditions, removed under heavy load conditions

These elements may be switched in and out of the line as required.

The compensating elements should be located at points along the line spaced so as to keep the line voltage within required tolerances at all points. The economics of such installations must be considered as a part of the line cost. The compensating elements may be inserted or bypassed depending on the switch positions. To investigate the effect let us place a source at one end of the line and a load (passive admittance) at the other. The results achieved for a particular set of line values are presented in Example 4-6.

Example 4-6

For the situation illustrated in Figure 4-9 the following values are given:

$$V_1 = 1\underline{/0} \qquad Y_L = 1.0 + j0$$

$$Z = +j0.5$$

$$Y = j0.10$$

Example 4-6

(a) Calculate the voltage magnitude at nodes one through four as a function of $Z_p(Y_p = 0)$ for loaded conditions.

(b) Repeat (a) for $Y_p(Z_p = 0)$ for the no load condition.

Solution

The circuit solution was programmed and results are presented in Table 4-2. In the loaded condition observe the initial rise in line voltage for increasing series compensation. The no load voltage, which is initially high, is decreased by adding shunt compensation.

Table 4-2 Results for Example 4-6

XP	BP	V1	V2	V3	V4	GLOAD	BLOAD
0.000	0.000	1.000	0.825	0.825	0.753	1.000	0.000
−0.100	0.000	1.000	0.840	0.865	0.790	1.000	0.000
−0.200	0.000	1.000	0.860	0.906	0.827	1.000	0.000
−0.300	0.000	1.000	0.886	0.946	0.864	1.000	0.000
−0.400	0.000	1.000	0.918	0.985	0.899	1.000	0.000
−0.500	0.000	1.000	0.955	1.021	0.932	1.000	0.000
−0.600	0.000	1.000	0.997	1.053	0.961	1.000	0.000
−0.700	0.000	1.000	1.042	1.078	0.983	1.000	0.000
−0.800	0.000	1.000	1.089	1.094	0.998	1.000	0.000
−0.900	0.000	1.000	1.134	1.101	1.005	1.000	0.000
−1.000	0.000	1.000	1.177	1.098	1.002	1.000	0.000

XP	BP	V1	V2	V3	V4	GLOAD	BLOAD
0.000	0.000	1.000	1.082	1.082	1.110	0.000	0.000
0.000	−0.020	1.000	1.070	1.070	1.098	0.000	0.000
0.000	−0.040	1.000	1.059	1.059	1.086	0.000	0.000
0.000	−0.060	1.000	1.048	1.048	1.075	0.000	0.000
0.000	−0.080	1.000	1.037	1.037	1.064	0.000	0.000
0.000	−0.100	1.000	1.026	1.026	1.053	0.000	0.000
0.000	−0.120	1.000	1.016	1.016	1.042	0.000	0.000
0.000	−0.140	1.000	1.006	1.006	1.031	0.000	0.000
0.000	−0.160	1.000	0.996	0.996	1.021	0.000	0.000
0.000	−0.180	1.000	0.986	0.986	1.011	0.000	0.000
0.000	−0.200	1.000	0.976	0.976	1.001	0.000	0.000

4.7 Line Insulation and Lightning

There are basically three factors to consider when selecting the insulation level for a power transmission line:

- The 60 Hz power voltage

- Surge voltages, caused by lightning

- Surge voltages, caused by switching

Surge voltages provide the most stringent test and supply the rationale for the standard impulse voltage waveform; that is, if the line is properly insulated to withstand surges, it can accommodate the highest expected 60 Hz voltages.

The problem of selecting the appropriate line insulation to withstand voltage surges is not as straightforward as one might assume. This is chiefly due to the nature of insulation breakdown. In general, one cannot claim that if the voltage across a particular insulator is some specific value the insulator will (or won't) with certainty breakdown. A critical factor is the *duration* of the impressed voltage. Insulators are more tolerant of short duration overvoltage than sustained values. Even when this parameter is fixed, if back-to-back tests are made with identical voltage waveforms, the insulator may fail one test and withstand the test voltage in the other.

For purposes of impulse testing a standard waveform is defined, as shown in Figure 4-10. The waveform is referred to as "$T_1 \times T_2$" where both values are conventionally given in microseconds. The standard wave is

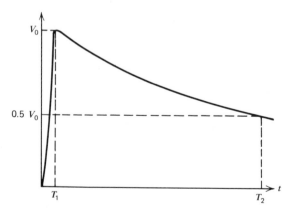

Figure 4-10 Standard voltage impulse waveform.

usually 1.2×50. The peak value of the waveform is V_0, which is defined as the B.I.L. (basic impulse insulation level). We define the "withstand voltage" as that B.I.L. that can be repeatedly applied to an insulator without flashover, disruptive discharge, puncture, or other electric failure, under specified test conditions. Frequently this withstand voltage is expressed in per unit; when this is the case the base value is understood to be the maximum line to neutral value. For example, for 500 kV, a V_0 of 2.7 pu. equates to $2.7 (500/\sqrt{3})\sqrt{2} = 1102$ kV. Manufacturers have met IEEE standards so that equipment designed for a nominal system voltage will have a B.I.L. rating that is adequate for most reasonable estimated switching and lightning surges. The precise B.I.L. value selected is a matter of engineering opinion; the trend is to select values somewhat lower (in pu.) for the higher voltage lines. Values selected for 500 kV generally fall in the range 2.5 to 3.0 pu.

Lightning is a natural electrical phenomenon that consists of a high current short time discharge that neutralizes an accumulation of charge in the atmosphere. The discharge path can be between two locations in a cloud, two clouds, a cloud and the earth, or a cloud and structures attached to the earth. The mechanisms by which such charge accumulations form is not completely understood, but is related to the motions of large air masses that meet certain conditions of humidity, temperature, and pressure. The charge build up in the cloud can be of either polarity, but is typically negative.

When voltage gradients within the cloud build to the order of 5 to 10 kV per cm an ionized path, or "leader," begins to form, moving from the cloud to the earth. The path taken is irregular, unpredictable, and proceeds relatively slowly, with the front moving at a typical velocity of 10^5 m/s. An opposite charge accumulation (typically positive) occurs under the cloud, and when the leader is close enough, a similar streamer from the earth rises to meet it. If and when the two streamers intersect, a lightning discharge occurs, neutralizing the charge centers.

The current involved in a lightning stroke is formidable, and its waveform parameters can range over extreme limits, with a peak value from one to 240 kA, rise times from 1 to 10 microseconds, and duration times from 100 to 1000 microseconds. A typical pulse is one of a series, consisting typically of three to five, but occasionally including as many as forty. These discharges are generally separated by about 40 milliseconds.

We are interested in lightning because of its potentially harmful effects on the power system. A direct stroke on a phase conductor will elevate the voltage to ground to abnormally high values. This will cause flashover across suspension insulators on the line and insulation breakdown in transformers and generators. These breakdowns in turn cause mechanical damage and fire. System protective devices will operate, temporarily removing equipment from service. Even a near miss can induce unusually high voltages and currents into the system.

The system is protected from lightning by several methods. Lines have neutral conductors that are electrically connected to the towers and therefore to ground. These conductors are strung at the highest point and will effectively shield objects from direct strokes within a certain sector beneath them. Field experience shows that in a 60° sector (30° from the vertical) beneath a ground conductor the chances of an object being hit by a direct stroke is reduced by about a factor of 1000. Because the neutrals are connected to the towers, connection to the ground is made at the tower footings. It is therefore important that there be low resistance between the tower base and a remote earth point. Care is taken to insure that this is true; sometimes bare conductors are buried under the line and connected to sequential towers.

Although the overhead neutrals effectively shield the power conductors from direct lightning strokes, such strokes can be expected to hit the neutrals. When this happens the enormous currents that flow down the tower to ground will produce IZ voltages that may be well in excess of the breakdown voltage of the line insulation. A phenomenon called "backflash" then occurs: arcing from the neutrals at temporary high voltage to the phase conductors. Once the breakdown occurs the fault will be sustained by the 60 Hz power voltage and can only be removed by de-energizing the line. Also, large currents and voltages can be induced into power circuits by strokes that are near misses.

To prevent damage to power equipment from these effects, air (rod and horn) gaps, surge diverters, and lightning arresters are used. Generally, such devices are connected from phase conductors to ground. The simplest device, the rod gap, is a preset air gap designed to flashover first in the event of unacceptable overvoltage. It suffers from the disadvantages that once it arcs over, it cannot clear itself and will present a fault to the 60 Hz voltages, and the electrodes are damaged in the arcing process. Surge suppressors are essentially nonlinear resistors, which draw excessive current at high voltage. They operate on the principle that elevated voltages will require substantial current flow through the suppressor, and that the accompanying IZ drop in the system will limit the voltage to manageable levels. The material used in such resistors is usually silicon carbide.

The surge suppressor continuously draws current, and therefore constitutes a steady power loss, and if the permissible ceiling voltage is too close to the rated value this loss is prohibitive. The next step in sophistication is the lightning arrestor that incorporates the features of both gap protection and nonlinear resistance. The arrestor must function to some extent like a circuit breaker, in that after breaking down and shunting surge current, it must later interrupt the 60 Hz follow current.

Overhead power transmission line phase conductors are bare; insulation would greatly reduce their current-carrying capacity. Also the use of insulated conductors would substantially increase the line cost. The phase conductors rely on the insulating properties of the surrounding air and are insulated from the towers by the hardware used at the point of attachment. The most common method of attachment to the towers employs strings of interlocking discs. These discs are designed to have the necessary mechanical and electrical strength; they are constructed of porcelain or glass and steel. The discs retain their insulating properties even when exposed to rain, snow, and dust contamination. Increasing the insulation strength of a line is simply a matter of adding more discs to a suspension string.

4.8 Transient Analysis of Transmission Lines

We have so far confined our study to sinusoidal steady state balanced three phase performance. Lightning and switching surge require some attention to modeling the line under transient conditions. This is a complex situation, and we should note some of the complications. Even for a two wire line, if losses are considered, the general transient response cannot be expressed in closed form. Modeling the effect of the earth for transients is formidable; the concept of "three phase" is now meaningless, but the interactions between the line conductors must still be properly accounted for.

To serve as an introduction to the problem let us make the simplifying assumption of ignoring line losses. This development parallels that presented in section 4-4, defining:

$$z(s) = sL \tag{4-84a}$$

    = Series transform impedance per unit length in ohms/m.

      Resistance is neglected.

$$y(s) = sC \tag{4-84b}$$

    = Shunt transform admittance per unit length in siemens/m.

    s = Laplace Transform operator $\tag{4-84c}$

The equation corresponding to (4-52) defines the lossless transform propagation constant:

$$\gamma = \sqrt{zy} \tag{4-85a}$$

$$= s\sqrt{LC} \tag{4-85b}$$

From equation (4-56) the characteristic impedance is:

$$Z_c = \sqrt{\frac{z}{y}} \qquad\qquad (4\text{-}86\text{a})$$

$$= \sqrt{\frac{L}{C}} \qquad\qquad (4\text{-}86\text{b})$$

In section 4-4 it was irrelevant which way position along the line was measured since the variable x was to be eliminated; also it was convenient to use the receiving end as reference ($x = 0$). Now, however, we are interested in which way waveforms are traveling along the line. For this reason we shall now define x as shown in Figure 4-11. Note that $x = 0$ is the sending end and the direction of increasing x is from sending to receiving end.

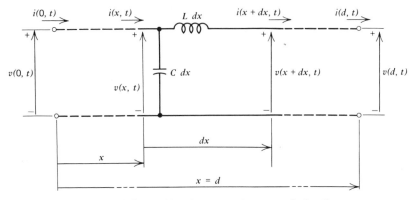

Figure 4-11 General lossless two wire transmission line.

Therefore:

$$\frac{\partial v(x, t)}{\partial x} = -L \frac{\partial i(x, t)}{\partial t} \qquad\qquad (4\text{-}87\text{a})$$

$$\frac{\partial i(x, t)}{\partial x} = -C \frac{\partial v(x, t)}{\partial t} \qquad\qquad (4\text{-}87\text{b})$$

Taking the Laplace Transform with respect to t produces:

$$\frac{dV(x, s)}{dx} = -L[sI(x, s) - i(x, 0)] \qquad\qquad (4\text{-}88\text{a})$$

$$\frac{dI(x, s)}{dx} = -C[sV(x, s) - v(x, 0)] \qquad\qquad (4\text{-}88\text{b})$$

If the line has zero initial stored energy, then $v(x, 0) = i(x, 0) = 0$, producing:

$$\frac{dV(x, s)}{dx} = -sLI(x, s) \qquad (4\text{-}89a)$$

$$\frac{dI(x, s)}{dx} = -sCV(x, s) \qquad (4\text{-}89b)$$

If we differentiate with x:

$$\frac{d^2V(x, s)}{dx^2} = -sL\frac{dI(x, s)}{dx} \qquad (4\text{-}90a)$$

$$\frac{d^2I(x, s)}{dx^2} = -sC\frac{dV(x, s)}{dx} \qquad (4\text{-}90b)$$

Substituting (4-89) into (4-90):

$$\frac{d^2V(x, s)}{dx^2} = s^2LCV(x, s) \qquad (4\text{-}91a)$$

Similarly:

$$\frac{d^2I(x, s)}{dx^2} = s^2LCI(x, s) \qquad (4\text{-}91b)$$

Again defining the p operator as d/dx ( ), the characteristic equation is:

$$p^2 - s^2LC = 0 \qquad (4\text{-}92a)$$

whose roots are:

$$p = \pm s\sqrt{LC} \qquad (4\text{-}92b)$$

The solutions to the differential equations are therefore of the form

$$V(x, s) = V^+(s)\, e^{-s\sqrt{LC}x} + V^-(s)\, e^{+s\sqrt{LC}x} \qquad (4\text{-}93a)$$

$$I(x, s) = I^+(s)\, e^{-s\sqrt{LC}x} + I^-(s)\, e^{+s\sqrt{LC}x} \qquad (4\text{-}93b)$$

The "constants" that arise must be recognized as possible functions of s since the second independent variable, time, is involved. The association of $V^+(s)$ with $e^{-s\sqrt{LC}x}$ is consistent with the notation of section 4-4, and will prove to be logical. Recall that the positive sense of x has been reversed. The time shifting (or delay) theorem from Laplace Transforms is:

$$\mathcal{L}[f(t - \tau)u(t - \tau)] = \dot{F}(s)\, e^{-\tau s} \qquad (4\text{-}94a)$$

123

where

$$F(s) = \mathcal{L}[f(t)] \tag{4-94b}$$

$u(t - \tau) = $ Shifted unit step function

$$= 1; \; t \geq \tau$$
$$= 0; \; t < \tau \tag{4-94c}$$

$\tau = $ Time shift

Applying the theorem to the problem of computing the inverse transforms of equations (4-93) we get:

$$v(x, t) = v^+(t - \sqrt{LC}x)u(t - \sqrt{LC}x) + v^-(t + \sqrt{LC}x)u(t + \sqrt{LC}x) \tag{4-95a}$$

$$i(x, t) = i^+(t - \sqrt{LC}x)u(t - \sqrt{LC}x) + i^-(t + \sqrt{LC}x)u(t + \sqrt{LC}x) \tag{4-95b}$$

Examine a function of the form

$$f^+(t - kx)u(t - kx) \tag{4-96}$$

as shown in Figure 4-12. As we move to an observation point further down the line (i.e., x more positive) the function f^+ arrives at a *later* time (specifically, at $t = kx$ seconds). For this reason f^+ is referred to as a "traveling wave" (specifically, a *forward* traveling wave because its direction of motion is from sending to receiving; f^- indicates a *backward* traveling wave, since its direction of travel is in the negative x direction). Also observe that the units of k must be s/m or $1/k$ must be m/s, that of velocity. The voltage and currents travel on our lossless line with velocity $1/\sqrt{LC}$. The four factors $V^+(s)$, $V^-(s)$, $I^+(s)$, $I^-(s)$, represent the transform forward and reverse traveling waves. We choose to eliminate the currents. Using equation (4-89a):

$$I(x, s) = \frac{1}{-sL}[-s\sqrt{LC}V^+(s) \, e^{-s\sqrt{LC}x} + s\sqrt{LC}V^-(s) \, e^{+s\sqrt{LC}x}] \tag{4-97a}$$

$$= \frac{V^+(s)}{Z_c} e^{-s\sqrt{LC}x} - \frac{V^-(s)}{Z_c} e^{+s\sqrt{LC}x} \tag{4-97b}$$

revealing that

$$I^+(s) = \frac{V^+(s)}{Z_c} \tag{4-97c}$$

$$I^-(s) = \frac{-V^-(s)}{Z_c} \tag{4-97d}$$

We are primarily interested in the voltage, realizing we could develop

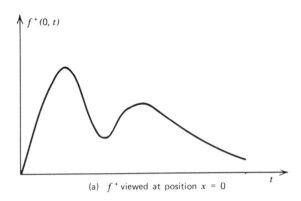

(a) f^+ viewed at position $x = 0$

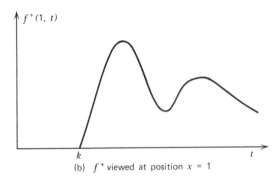

(b) f^+ viewed at position $x = 1$

Figure 4-12 Traveling wave on lossless line.

similar equations to compute current. Consider the line terminated in Z_r. At the receiving end ($x = d$):

$$V_r = V(d, s) = V^+(s) \, e^{-s\sqrt{LC}d} + V^-(s) \, e^{+s\sqrt{LC}d} \tag{4-98a}$$

$$I_r = I(d, s) = \frac{V^+(s)}{Z_c} \, e^{-s\sqrt{LC}d} - \frac{V^-(s)}{Z_c} \, e^{+s\sqrt{LC}d} \tag{4-98b}$$

Dividing (4-98a) by (4-98b):

$$\frac{V_r}{I_r} = Z_r = Z_c \left[\frac{V^+(s) \, e^{-s\sqrt{LC}d} + V^-(s) \, e^{+s\sqrt{LC}d}}{V^+(s) \, e^{-s\sqrt{LC}d} - V^-(s) \, e^{+s\sqrt{LC}d}} \right] \tag{4-99}$$

Manipulating 4-99) we get:

$$V^-(s) = \frac{Z_r - Z_c}{Z_r + Z_c} \, V^+(s) \, e^{-2s\sqrt{LC}d} \tag{4-100}$$

125

Returning to equation (4-93a):

$$V(x, s) = V^+(s)\, e^{-s\sqrt{LC}x} + \frac{Z_r - Z_c}{Z_r + Z_c} V^+(s)\, e^{+s\sqrt{LC}(x-2d)} \tag{4-101}$$

A study of equation (4-101) provides an interesting physical insight into the problem. The equation says that reverse traveling waves are scaled versions of forward traveling waves (the scaling factor is $(Z_r - Z_c)/(Z_r + Z_c)$ and delayed $2\sqrt{LC}(d-x)$ seconds, the time required for a waveform traveling at velocity $1/\sqrt{LC}$ to move from position x to the end of the line and return. We then can visualize the reverse traveling wave as a "reflection" of the forward traveling wave and in this sense define:

$\Gamma_r^{?}$ = Receiving end reflection coefficient

$$= \frac{Z_r - Z_c}{Z_r + Z_c} \tag{4-102a}$$

In a similar manner, define:

Γ_s = Sending end reflection coefficient

$$= \frac{Z_s - Z_c}{Z_s + Z_c} \tag{4-102b}$$

where

Z_s is the sending end Thevenin equivalent impedance.

We are now prepared to deal with equation (4-93a):

$$V^+(s)\, e^{-s\sqrt{LC}x} = \text{Sum of all forward traveling waves}$$

$$= V_1\, e^{-s\sqrt{LC}x} + V_3\, e^{-s\sqrt{LC}x} + \ldots \tag{4-103a}$$

$$V^-(s)\, e^{+s\sqrt{LC}x} = \text{Sum of all reverse traveling waves}$$

$$= V_2\, e^{+s\sqrt{LC}x} + V_4\, e^{+s\sqrt{LC}x} + \ldots \tag{4-103b}$$

where

$$V_1\, e^{-s\sqrt{LC}x} = V_0\, e^{-s\sqrt{LC}x} = \text{Forward traveling initial incident wave} \tag{4-103c}$$

$$V_2\, e^{+s\sqrt{LC}x} = \Gamma_r V_1\, e^{-s\sqrt{LC}x} = \text{Reverse traveling reflection of } V_1\, e^{-s\sqrt{LC}x} \tag{4-103d}$$

$$V_3\, e^{-s\sqrt{LC}x} = \Gamma_s V_2\, e^{+s\sqrt{LC}x} = \text{Forward traveling reflection of } V_2\, e^{+s\sqrt{LC}x} \tag{4-103e}$$

$$V_4\,e^{+s\sqrt{LC}x} = \Gamma_r V_3\,e^{-s\sqrt{LC}x} = \text{Reverse traveling reflection of } V_3\,e^{-s\sqrt{LC}x}$$

$$(4\text{-}103\text{f})$$

and so on.

Substitution of the above into (4-103a) and (4-103b) and substitution of the results into (4-93a) produces:

$$V(x, s) = V_0[e^{-s\sqrt{LC}x} + \Gamma_r e^{-s\sqrt{LC}(2d-x)} + \Gamma_s\Gamma_r e^{-s\sqrt{LC}(2d+x)}$$

$$+ \Gamma_r\Gamma_s\Gamma_r e^{-s\sqrt{LC}(4d-x)} + \Gamma_s\Gamma_r\Gamma_s\Gamma_r e^{-s\sqrt{LC}(4d+x)} + \dots] \qquad (4\text{-}104)$$

The problem now centers on the evaluation of V_0. Assume a generalized sending end circuit modeled by its transform Thevenin equivalent. Further suppose that $Z_r = Z_c$ so that $\Gamma_r = 0$. We observe that $V^-(s)\,e^{+s\sqrt{LC}x}$ is zero and that $V^+(s)\,e^{-s\sqrt{LC}x}$ consists only of $V_0\,e^{-s\sqrt{LC}x}$, the incident voltage. Similar conditions hold for the current. Therefore:

$$V(x, s) = V_0\,e^{-s\sqrt{LC}x} \qquad (4\text{-}105\text{a})$$

$$I(x, s) = \frac{V_0}{Z_c}\,e^{-s\sqrt{LC}x} \qquad (4\text{-}105\text{b})$$

Specifically at the sending end ($x = 0$):

$$V(0, s) = V_0 \qquad (4\text{-}106\text{a})$$

$$I(0, s) = \frac{V_0}{Z_c} \qquad (4\text{-}106\text{b})$$

Dividing $V(0, s)$ by $I(0, s)$:

$$\frac{V(0, s)}{V(0, s)} = \frac{V_s}{I_s} = Z_c \qquad (4\text{-}107)$$

The meaning of (4-107) is deeper than we might first appreciate. It says that "looking in" to line at the sending end we "see" the impedance Z_c. This is true *in general initially* whether $Z_r = Z_c$ or not, since it will take a finite time for the incident wave to travel down the line and produce the first reflection. Therefore at first ($t = 0^+$; it is assumed the source is energized at $t = 0$) the circuit of Figure 4-13 is a proper model. By the voltage divider:

$$V_0 = \frac{Z_c}{Z_s + Z_c} E_s \qquad (4\text{-}108)$$

We are now prepared to solve equation (4-104). For simple problems the Bewley lattice diagram, shown in Figure 4-14, is a useful organizer and visual aid. The diagonal lines represent the traveling waves. The reflections

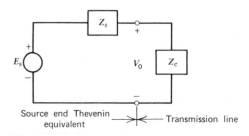

Figure 4-13 Source line circuit model for calculating the initial incident voltage V_0.

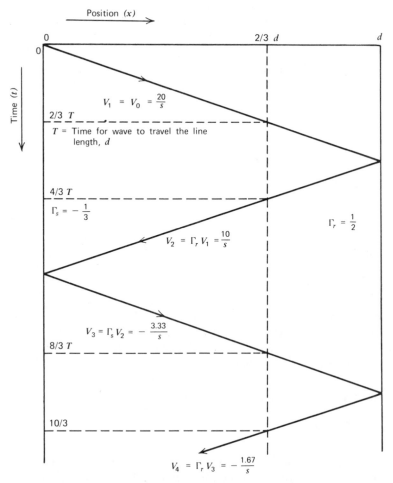

Figure 4-14 Bewley lattice diagram—numerical values are for Example 4-7.

are calculated by multiplying the incident waves by the appropriate reflection coefficient. The voltage at any point in space or time is determined by summing all terms directly above the space, time point. An example will demonstrate.

Example 4-7

A two wire lossless transmission line has the following parameters:

$d = 1$ meter

$L = 1$ H/m

$C = 1$ F/m

Sending end termination:

$$E_s = \frac{30}{s}$$

$$Z_s = 1/2$$

Receiving end termination: $Z_r = 3$

Find and sketch $v(x, t)$ at location $x = 2/3m$.

Solution

First calculate:

$$Z_c = \sqrt{\frac{L}{C}} = 1$$

$$\therefore V_0 = \frac{1}{1+1/2}\left[\frac{30}{s}\right] = \frac{20}{s}$$

$$\Gamma_r = \frac{3-1}{3+1} = \frac{1}{2}$$

$$\Gamma_s = \frac{1/2-1}{1/2+1} = -\frac{1}{3}$$

Results are presented on the lattice diagram shown in Figure 4–14. The voltage $v(2/3, t)$ is plotted in Figure 4–15. The final value may be calculated by the voltage divider:

129

$$v\left(\frac{2}{3}, \infty\right) = \frac{3}{3+1/2}(30)$$

$$= 25.7 \text{ volts}$$

We could also use equation (4-104) directly:

$$V\left(\frac{2}{3}, s\right) = \frac{20}{s}\left[e^{-\frac{2}{3}s} + \frac{1}{2}e^{-\frac{4}{3}s} + \left(-\frac{1}{3}\right)\left(\frac{1}{2}\right)e^{-\frac{8}{3}s}\right.$$

$$\left. + \left(\frac{1}{2}\right)\left(-\frac{1}{3}\right)\left(\frac{1}{2}\right)e^{-\frac{10}{3}s} + \left(-\frac{1}{3}\right)\left(\frac{1}{2}\right)\left(-\frac{1}{3}\right)\left(\frac{1}{2}\right)e^{-\frac{14}{3}s} + \dots\right]$$

$$\therefore \quad v\left(\frac{2}{3}, t\right) = 20\left[u\left(t-\frac{2}{3}\right) + 0.5u\left(t-\frac{4}{3}\right) - 0.167u\left(t-\frac{8}{3}\right)\right.$$

$$\left. - 0.0833u\left(t-\frac{10}{3}\right) + 0.0278u\left(t-\frac{14}{3}\right) + \dots\right]$$

which plots as shown in Figure 4-15.

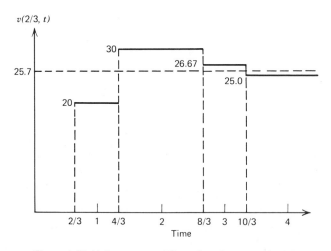

Figure 4-15 Voltage at $x = 2/3$ vs time for example 4-7.

4-9 Application to the Three Phase Line

Our main concern in this section is to extend the techniques of the previous section to the three phase line. As we already know, there are many complicated mutual couplings involved in 3ϕ lines, a fact that previously led us into the symmetrical components method. Unfortunately, symmetrical components are basically defined for sinusoidal steady state conditions.* What we need is a similar transformation that will diagonalize the phase impedance matrix and also be suitable for transient analysis.

A suitable transformation, referred to as the modal transformation, is given in equation (4-109):

$$\begin{bmatrix} v_a \\ v_b \\ v_c \end{bmatrix} = \begin{bmatrix} 1/\sqrt{3} & 1/\sqrt{6} & 1/\sqrt{2} \\ 1/\sqrt{3} & -2/\sqrt{6} & 0 \\ 1/\sqrt{3} & 1/\sqrt{6} & -1/\sqrt{2} \end{bmatrix} \begin{bmatrix} v_0 \\ v_\alpha \\ v_\beta \end{bmatrix} \qquad \text{(4-109a)}$$

or

$$\tilde{v}_{abc} = [L]\tilde{v}_{0\alpha\beta} \qquad \text{(4-109b)}$$

where the $0\alpha\beta$ subscripts identify modal quantities.
The matrix $[L]$ is unitary. Therefore:

$$[L]^{-1} = [L]_t \qquad \text{(4-110)}$$

See problem 4-19 for proof. We define the same transformation to hold for currents:

$$\tilde{i}_{abc} = [L]\tilde{i}_{0\alpha\beta} \qquad \text{(4-111)}$$

The transformation is power invariant (see problem 4-20 for proof), that is:

$$\tilde{v}_{abc_t}\tilde{i}_{abc} = \tilde{v}_{0\alpha\beta_t}\tilde{i}_{0\alpha\beta} \qquad \text{(4-112)}$$

Taking the Laplace Transforms of equations (4-109) and (4-111) produces:

$$\tilde{V}_{abc} = [L]\tilde{V}_{0\alpha\beta} \qquad \text{(4-113a)}$$

$$\tilde{I}_{abc} = [L]\tilde{I}_{0\alpha\beta} \qquad \text{(4-113b)}$$

where

$$\tilde{V}_{abc} = \begin{bmatrix} V_a(s) \\ V_b(s) \\ V_c(s) \end{bmatrix} \quad \text{(4-113c)}; \qquad \tilde{V}_{0\alpha\beta} = \begin{bmatrix} V_0(s) \\ V_\alpha(s) \\ V_\beta(s) \end{bmatrix} \qquad \text{(4-113d)}$$

$$\tilde{I}_{abc} = \begin{bmatrix} I_a(s) \\ I_b(s) \\ I_c(s) \end{bmatrix} \quad \text{(4-113e)}; \qquad \tilde{I}_{0\alpha\beta} = \begin{bmatrix} I_0(s) \\ I_\alpha(s) \\ I_\beta(s) \end{bmatrix} \qquad \text{(4-113f)}$$

---

* Some authors also use symmetrical components for transient analysis.

The transformation on impedance is easily developed. We start from:

$$\tilde{V}_{abc} = [Z_{abc}]\tilde{I}_{abc} \tag{4-114}$$

where the entries in $[Z_{abc}]$ are transform values. Transforming the voltages and currents

$$[L]\tilde{V}_{0\alpha\beta} = [Z_{abc}][L]\tilde{I}_{0\alpha\beta} \tag{4-115a}$$

$$\tilde{V}_{0\alpha\beta} = [L]^{-1}[Z_{abc}][L]\tilde{I}_{0\alpha\beta} \tag{4-115b}$$

$$\tilde{V}_{0\alpha\beta} = [Z_{0\alpha\beta}]\tilde{I}_{0\alpha\beta} \tag{4-115c}$$

where

$$[Z_{0\alpha\beta}] = [L]^{-1}[Z_{abc}][L] \tag{4-115d}$$

$$= [L]_t[Z_{abc}][L] \tag{4-115e}$$

We apply this result to the situation described in Example 4-1.

Example 4-8

Evaluate $[Z_{0\alpha\beta}]$ for the line described in example 4-1.

Solution

$$[Z_{abc}] = \begin{bmatrix} Z_s & Z_m & Z_m \\ Z_m & Z_s & Z_m \\ Z_m & Z_m & Z_s \end{bmatrix} \tag{4-116a}$$

where

$$Z_s = R_s + sL_s \tag{4-116b}$$

$$Z_m = R_m + sL_m \tag{4-116c}$$

$$R_s = 2R \tag{4-116d}$$

$$R_m = R \tag{4-116e}$$

$$L_s = \frac{\mu}{2\pi}\ln\frac{(D/\sqrt{3})^2}{r'^2} \tag{4-116f}$$

$$L_m = \frac{\mu}{2\pi}\ln\frac{(D/\sqrt{3})^2}{Dr'} \tag{4-116g}$$

Now we are ready to evaluate $[Z_{0\alpha\beta}]$

$$[Z_{0\alpha\beta}] = [L]_t[Z_{abc}][L] \tag{4-115e}$$

$$= \begin{bmatrix} 1/\sqrt{3} & 1/\sqrt{3} & 1/\sqrt{3} \\ 1/\sqrt{6} & -2/\sqrt{6} & 1/\sqrt{6} \\ 1/\sqrt{2} & 0 & -1/\sqrt{2} \end{bmatrix} \begin{bmatrix} Z_s & Z_m & Z_m \\ Z_m & Z_s & Z_m \\ Z_m & Z_m & Z_s \end{bmatrix}$$

$$\begin{bmatrix} 1/\sqrt{3} & 1/\sqrt{6} & 1/\sqrt{2} \\ 1/\sqrt{3} & -2/\sqrt{6} & 0 \\ 1/\sqrt{3} & 1/\sqrt{6} & -1/\sqrt{2} \end{bmatrix} \tag{4-117a}$$

$$= \begin{bmatrix} Z_0 & 0 & 0 \\ 0 & Z_\alpha & 0 \\ 0 & 0 & Z_\beta \end{bmatrix} \tag{4-117b}$$

where

$$Z_0 = Z_s + 2Z_m \tag{4-117c}$$

$$Z_\alpha = Z_\beta = Z_s - Z_m \tag{4-117d}$$

Observe that the modal impedances equate to corresponding sequence values (except that the modal values are transform impedances and the sequence values are complex $[s = j\omega]$).

In the lossless case (all line resistance neglected) for the equilateral line (a diagonal sequence impedance matrix) it is possible to directly determine the modal series impedances (and shunt admittances) from the sequence values:

$$Z_0 = \frac{s}{j\omega} Z_0 \tag{4-118a}$$

$$Z_\alpha = \frac{s}{j\omega} Z_1 \tag{4-118b}$$

$$Y_0 = \frac{s}{j\omega} Y_0 \tag{4-118c}$$

$$Y_\alpha = \frac{s}{j\omega} Y_1 \tag{4-118d}$$

When the positive and negative sequence values are equal (which is always the case in our work here) the beta values are equal to the alpha values. The "zero" voltage and current components of the $0\alpha\beta$ transformation should not be confused with the "zero" voltage and current components of the 012 transformation; they are similar but differ by a factor of $\sqrt{3}$.

We are now prepared to discuss a general approach to line transient analysis. We first transform terminal conditions into $0\alpha\beta$ components. Using the $0\alpha\beta$ decoupled line circuit models, we compute the line transient response independently in each of the three modes by the methods of section 4-8. Finally we transform our $0\alpha\beta$ results back into abc phase quantities.

Example 4-9

A lossless version of the equilateral line of example 4-2 is terminated as shown in Figure 4-16. Determine the general transient response, $v_a(x, t)$, $v_b(x, t)$, and $v_c(x, t)$ for $t \geq 0$; $0 \leq x \leq d$. The source voltages are:

$$e_a = V_m \cos(\omega t - 60°)u(t)$$

$$e_b = V_m \cos(\omega t - 180°)u(t)$$

$$e_c = V_m \cos(\omega t + 60°)u(t)$$

where

$$\omega = 377 \text{ rad/s}$$

Also

$$d = \text{Line length}$$

and

$$R_g = 325 \text{ Ohms}$$

Solution

We start by calculating the $0\alpha\beta$ source voltages:

$$\tilde{e}_{0\alpha\beta} = \begin{bmatrix} 1/\sqrt{3} & 1/\sqrt{3} & 1/\sqrt{3} \\ 1/\sqrt{6} & -2/\sqrt{6} & 1/\sqrt{6} \\ 1/\sqrt{2} & 0 & -1/\sqrt{2} \end{bmatrix} \begin{bmatrix} V_m \cos(\omega t - 60°)u(t) \\ V_m \cos(\omega t - 180°)u(t) \\ V_m \cos(\omega t + 60°)u(t) \end{bmatrix}$$

$$= \begin{bmatrix} 0 \\ \sqrt{\tfrac{3}{2}}V_m \cos(\omega t)u(t) \\ \sqrt{\tfrac{3}{2}}V_m \sin(\omega t)u(t) \end{bmatrix}$$

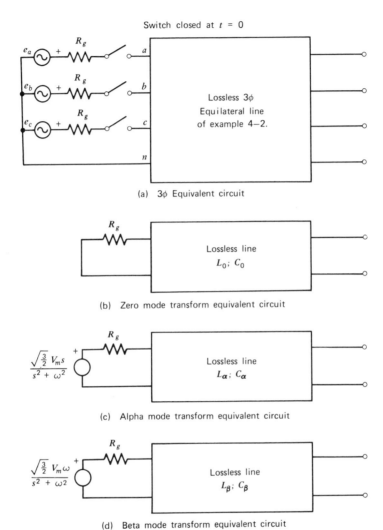

Switch closed at $t = 0$

(a) 3ϕ Equivalent circuit

(b) Zero mode transform equivalent circuit

(c) Alpha mode transform equivalent circuit

(d) Beta mode transform equivalent circuit

Figure 4-16 Lossless line of example 4-9.

The transform values are

$$\tilde{E}_{0\alpha\beta} = \sqrt{\tfrac{3}{2}}V_m \begin{bmatrix} 0 \\ \dfrac{s}{s^2 + \omega^2} \\ \dfrac{\omega}{s^2 + \omega^2} \end{bmatrix}$$

135

The source impedances are

$$[Z_{0\alpha\beta}] = [L]^{-1}[Z_{abc}][L]$$

$$= [L]^{-1} \begin{bmatrix} 325 & 0 & 0 \\ 0 & 325 & 0 \\ 0 & 0 & 325 \end{bmatrix} [L]$$

$$= \begin{bmatrix} 325 & 0 & 0 \\ 0 & 325 & 0 \\ 0 & 0 & 325 \end{bmatrix}$$

From examples 4-2 and 4-3:

$$L_0 = \frac{1.42}{377} = 3.77 \ \mu\text{H/m}$$

$$L_1 = \frac{0.42}{377} = 1.11 \ \mu\text{H/m}$$

$$C_0 = \frac{1.17 \times 10^{-9}}{377} = 3.10 \ \text{pF/m}$$

$$C_1 = \frac{3.95 \times 10^{-9}}{377} = 10.5 \ \text{pF/m}$$

It follows that

$$L_\alpha = L_\beta = L_1 = 1.11 \ \mu\text{H/m}$$

$$C_\alpha = C_\beta = C_1 = 10.5 \ \text{pF/m}$$

We need not calculate zero components since there is no zero mode source.

$$Z_c = Z_{c_\alpha} = \sqrt{\frac{L_\alpha}{C_\alpha}}$$

$$= \sqrt{\frac{1.11 \times 10^{-6}}{10.5 \times 10^{-12}}}$$

$$= 325 \ \text{ohms}$$

$$\sqrt{LC} = \sqrt{L_\alpha C_\alpha} = 3.41 \times 10^{-9} \ \text{s/m}$$

We calculate the reflection coefficients:

$$\Gamma_r = \Gamma_{r_\alpha} = \lim_{Z_{r_\alpha} \to -\infty} \left[\frac{Z_{r_\alpha} - Z_{c_\alpha}}{Z_{r_\alpha} + Z_{c_\alpha}} \right] = 1$$

$$\Gamma_s = \Gamma_{s_\alpha} = \frac{325 - 325}{325 + 325} = 0$$

The β values are identical.
We need to evaluate V_{0_α}:

$$V_{0_\alpha} = \frac{1}{2}E_\alpha = \frac{V_m}{2}\sqrt{\frac{3}{2}}\left(\frac{s}{s^2 + \omega^2}\right)$$

$$V_{0_\beta} = \frac{1}{2}E_\beta = \frac{V_m}{2}\sqrt{\frac{3}{2}}\left(\frac{\omega}{s^2 + \omega^2}\right)$$

Equation (4-104) applies directly:

$$V_\alpha(x, s) = \frac{V_m}{2}\sqrt{\frac{3}{2}}\left(\frac{s}{s^2 + \omega^2}\right)[e^{-s\sqrt{LC}x} + (1)\, e^{-s\sqrt{LC}(2d-x)}]$$

$$V_\beta(x, s) = \frac{V_m}{2}\sqrt{\frac{3}{2}}\left(\frac{\omega}{s^2 + \omega^2}\right)[e^{-s\sqrt{LC}x} + (1)\, e^{-s\sqrt{LC}(2d-x)}]$$

To substitute numerical values it is expedient to measure time in milliseconds and position in kilometers. Therefore:

$$V_\alpha(x, s) = \frac{0.612V_m s}{s^2 + 0.142}[e^{-0.00341sx} + e^{-0.00341s(2d-x)}]$$

$$V_\beta(x, s) = \frac{0.231V_m}{s^2 + 0.142}[e^{-0.00341sx} + e^{-0.00341s(2d-x)}]$$

To transform to the time domain:

$$v_\alpha(x, t) = 0.612V_m \cos\,[0.377(t - 0.00341x)]u(t - 0.00341x)$$
$$+0.612V_m \cos[0.377(t - 0.00341(2d - x))]$$
$$\cdot u(t - 0.00341(2d - x))$$

$$v_\beta(x, t) = 0.612V_m \sin[0.377(t - 0.00341x)]u(t - 0.00341x)$$
$$+0.612V_m \sin[0.377(t - 0.00341(2d - x))]$$
$$\cdot u(t - 0.00341(2d - x))$$

$$v_0(x, t) = 0$$

The final step requires transformation back to the phase (abc) values. Using equation (4-109) we calculate:

$$v_a(x, t) = 0.5V_m \cos[0.377(t - 0.00341x) - 60°] \cdot u(t - 0.00341x)$$
$$+0.5V_m \cos[0.377(t - 0.00341(2d - x)) - 60°]$$
$$\cdot u(t - 0.00341(2d - x))$$

$$v_b(x, t) = 0.5V_m \cos[0.377(t-0.00341x) - 180°] \cdot u(t-0.00341x)$$

$$+0.5V_m \cos[0.377(t-0.00341(2d-x)) - 180°]$$

$$\cdot u(t-0.00341(2d-x))$$

$$v_c(x, t) = 0.5V_m \cos[0.377(t-0.00341x) + 60°] \cdot u(t-0.00341x)$$

$$+0.5V_m \cos[0.377(t-0.00341(2d-x)) + 60°]$$

$$\cdot u(t-0.00341(2d-x))$$

The analytical complications are discouraging. Even in this simple case (i.e., the lossless, equilateral line) the equations were quite involved. Also, we were terminated in the characteristic impedance at the sending end, forcing Γ_s to zero, and reducing equation (4-104) to only two terms. Furthermore, we did not have to deal with the zero mode response. Obviously, in a more general situation, the complexities of the equations prohibit a closed form analytical approach. However, a general approach to the problem is now apparent, and can be extended by numerical methods to more general situations.

4-10 Dc Transmission

At very high power levels and very long distances dc transmission lines are preferable to ac lines. This is mainly because although the ac/dc conversion equipment is quite expensive, the cost per unit length for lines of equal capacities is less for dc. Therefore there is a critical distance at which dc transmission is more economical than ac; at present this value is about 600 km. The voltage regulation problem is much less serious for dc since only the "IR" drop is involved. ("IX" is zero since $\omega = 0$.) For the same reason steady state stability is no longer a major consideration. The dc line is an asynchronous link and can even interconnect systems of different frequencies.

For a single dc line between two converter stations, circuit breakers are unnecessary since control of the converters can be used to block current flow in case of a short circuit. The typical arrangement is to have two phases (also called poles) at $\pm V_{dc}$ with respect to ground; this scheme has a reliability comparable to a double circuit ac line. Also, dc cables are considerably cheaper than their ac equivalents.

The dc line is not without its disadvantages. As mentioned, the ac/dc conversion equipment is complicated and expensive. Dc switchgear is more expensive. Component reliability has not reached the level attained by ac devices, although it has and will continue to improve.

A major point of interest is the ac/dc converter and its basic component the electronic switching device. This switching device is one of two physical types: the ionic valve or the thyristor. Although these devices operate on different physical principles, their terminal characteristics are similar. The switching device is essentially a diode with the important modification that conduction cannot occur until the control circuitry provides a strong positive pulse at the gate (or grid). It is possible, through this control action, to allow dc power to flow in either direction: when the flow direction is from ac to dc the converter is described as a "rectifier," and from dc to ac, an "inverter." The controlled diodes are arranged in six or twelve phase rectifier circuits, with the source produced by appropriate delta wye transformer connections. Required voltage and current ratings are achieved by placing the necessary number of units in series and in parallel.

4-11 Corona

The high voltages at which transmission lines operate produce electric field strengths of sufficient intensity to ionize the air near the phase conductors. This effect, called corona, is detectable audibly as a buzzing, hissing sound and visually as a faint bluish aura surrounding the conductors. The critical field intensity E_c at which this ionization begins for dry air is:

$$E_c = 30\delta m \left[1 + \frac{0.3}{\sqrt{\delta r}} \right] \text{ kV/cm} \tag{4-119}$$

where

δ = relative air density = $\dfrac{3.92b}{T}$

b = atmospheric pressure in cm Hg

T = absolute temperature in degrees Kelvin

m = stranding factor $(0 < m < 1)$

    $m = 1$, smooth cylinder

    $m = 0.9$, weathered ACSR

r = conductor radius, cm

Using bundled conductors per phase tends to produce a larger effective phase radius and therefore reduces electric field intensity levels in the conductor vicinity.

Corona has two undesirable features: power loss and radio interference. An expression for fair weather single phase corona loss was given by Peterson [30] as:

$$P = \frac{3.37 \times 10^{-5} f V^2 F}{[\log_{10}(2s/d)]^2} \text{ kw/phase/mile} \tag{4-120}$$

where

V = rms line to neutral voltage in kV

f = frequency in Hertz

F = corona factor determined by test

s = phase spacing

d = conductor diameter

The power loss is small, computing to about 1 to 2 kw per km for 500 kV, 3 conductors per phase bundle. However, the corona losses increase dramatically when the line encounters precipitation in any form, with frost being the worst situation. Losses can run as high as 30 kw/km, with an average of about 2.4 kw/km expected for line design similar to our 500 kV example located in the Southeastern United States.

Radio interference is also a problem, occurring essentially over a frequency range from 0.2 to 4 MHz, centered around $f_0 = 0.8$ MHz. Precipitation increases RF interference, as does high humidity. As conductors age, RF interference levels tend to decrease.

Formulation of general equations that account for all relevant variables and give accurate results is a difficult problem. Results are obtained using empirical relations and statistical methods applied to impressive amounts of recorded data. Corona power loss and corona RF interference are two more factors that must be considered when evaluating a line design.

4-12 Summary

We have observed how the magnetic and electric fields surrounding the transmission line produce serial voltage drops and shunt current paths, creating the need for inductive and capacitive elements in the line circuit models. We learned that even in a completely symmetrical situation (the equilateral case), the phase impedance and admittance matrices were not diagonal, requiring mutual elements. Transformation to sequence values diagonalized the impedance and admittance matrices, eliminating mutual elements, and resulting in our final equivalent circuits, shown in Figure 4-7.

Note that the positive and negative networks are identical, a fact that is intuitively predictable (for balanced three phase excitation, the line response is indifferent to phase sequence).

We are somewhat frustrated by the fact that we are unable to derive simple exact circuit models for more practical line geometries, including arbitrary numbers of conductors and the effect of ground. For the benefit of students who wish to pursue the subject in more depth, the Appendix extends our work to consider some of these effects. The student is fore-warned that for these cases it is practical to obtain accurate numerical results only by using the computer. Furthermore, the matrices $[Z_{012}]$ and $[Y_{012}]$ are not truly diagonal, although they are approximately so $(Z_{ii} \gg Z_{ij})$. Sometimes lines are designed so that each phase conductor occupies each position in the conductor geometrical arrangement for one-third the line length. This technique, called transposition, has the effect of cancelling asymmetries and making $[Z_{012}]$ and $[Y_{012}]$ more nearly diagonal.

It is surprising to find that for a completely transposed line equations (4-26h) and (4-47c) may be used for X_1 and B_1, if D is calculated as $D = \sqrt[3]{D_{ab}D_{bc}D_{ca}}$. Even when the line is not transposed, X_1 and B_1 are sometimes calculated this way as an approximation. Unfortunately, the corresponding expressions for X_0 and B_0 can be in serious error since they depend heavily on neutral and ground effects.

There are many interesting and important problems associated with the operation and design of power transmission lines. Of basic importance is the line transmission capacity. There are two limits to observe: the thermal rating and the steady state stability limit. Also, the impedance effects of the line can cause the line voltage to vary outside of acceptable limits, with high voltage encountered at light load and low voltage encountered at rated load. The situation may be relieved by the insertion of series and shunt compensating elements.

Line insulation is basically determined by consideration of 60 Hz voltage levels, lightning induced transients, and switching induced transients. Such levels are referred to as "basic impulse insulation levels" (B.I.L.) and relate to the crest value of a standard voltage pulse waveform. Lightning is the most common cause of line faults (short circuits) and, as such, is a proper object for study. Two important techniques for reducing the harmful effects of lightning are the placement of overhead neutrals that shield the phase conductors, and maintenance of a low resistance to ground at the tower footings.

Line transient response is a complicated analytical problem that until recently was dealt with almost exclusively on an analog device referred to as the "transient network analyzer" (TNA). The TNA is a scaled laboratory circuit model that can simulate simple systems (a few lines and trans-formers), and includes components whose nonlinearities were comparable to those of the actual system devices. Examples of the worst case switching conditions can be quickly isolated by skillful operators, and therefore provide useful information for line operation and design. It is possible to handle certain simplified situations analytically and, using the same approach, extend these methods to more practical and complicated cases.

Dc transmission is practical and reasonable when long distances are to be spanned. Such lines are far more common outside the United States, since it is not unusual to find generation sites that are widely separated from load centers. Corona effects are undesirable as they constitute a power loss and a source of radio interference. The use of larger and bundled conductors will reduce these undesirable effects to some extent.

Bibliography

[1] *Aluminum Electrical Conductor Handbook*, The Aluminum Association, New York, 1971.

[2] Anderson, Paul M., *Analysis of Faulted Power Systems*, Iowa State Press, Ames, Iowa, 1973.

[3] Anderson, P. M., Bowen, D. W., and Shah, A. P., "An Indefinite Admittance Network Description for Fault Computation," *Trans. IEEE*, PAS-89 (July/August): pp. 1215–19, 1970.

[4] Beck, Edward, *Lightning Protection for Electric Systems*, McGraw-Hill, New York, 1954.

[5] Carson, John R., Wave Propagation in Overhead Wires with Ground Return, *Bell System Tech. J.* 5: 539–54, 1926.

[6] Clarke, Edith, *Circuit Analysis of A-C Power Systems, Vol. I*, Wiley, New York: 1943.

[7] Clarke, Edith, *Circuit Analysis of A-C Power Systems, Vol. II*, General Electric Co., Schenectady, New York, 1950.

[8] *EHV Transmission Line Reference Book*, Edison Electric Institute, New York, 1968.

[9] *Electrical Transmission and Distribution Reference Book*, Westinghouse Electric Corporation, East Pittsburg, PA, 1964.

[10] Elgerd, Olle I., *Electric Energy Systems Theory: An Introduction*, McGraw-Hill Inc., New York, 1971.

[11] General Electric Company. Project EHV. *EHV Transmission Line Reference Book*, Edison Electric Institute, New York, 1968.

[12] Greenwood, Allan, *Electrical Transients in Power Systems*, Wiley-Interscience, New York, 1971.

[13] Gross, C. A., "Modal Impedance of Non-Transposed Power Transmission Lines," *Conference Record of the Midwest Power Symposium*, Rolla, Mo., 1974.

[14] Gross, E. T. B., and Hesse, M. H., "Electromagnetic Unbalance of Untransposed Lines," *Trans. AIEE* Vol. 72: pp. 1323–36, 1953.

[15] Gross, E. T. B., and Nelson, S. W., "Electromagnetic Unbalance of Untransposed Transmission Lines. II. Single Lines with Horizontal Conductor Arrangement," *Trans. AIEE* Vol. 74: pp. 887–93, 1954.

[16] Gross, E. T. B., Drinnan, J. H., and Jochum, E., "Electromagnetic Unbalance of Untransposed Transmission Lines. III. Double Circuit Lines," *Trans. AIEE* Vol. 79: pp. 1362–71, 1959.

[17] Hedman, D. E., "Propagation on Overhead Transmission Lines: I– Theory of Modal Analysis," *IEEE Transactions on PAS*, March 1965, pp. 200–205.

[18] Hedman, D. E., "Propagation on Overhead Transmission Lines: II– Earth-conduction Effects and Practical Results," *IEEE Transactions on PAS*, March 1965, pp. 205–211.

[19] Hesse, M. H., "Electromagnetic and Electrostatic Transmission Line Parameters by Digital Computer," *Trans. IEEE* Vol. PAS-82: pp. 282–91, 1963.

[20] Holley, H., Colemen, D., and Shipley, R. B., "Untransposed EHV Line Computations," *Trans. IEEE* Vol. PAS-83: pp. 291–96, 1964.

[21] *IEEE Standard Dictionary of Electrical and Electronics Terms*, Wiley-Interscience, New York, 1972.

[22] Javid, Mansour, and Brenner, Egon, *Analysis, Transmission and Filtering of Signals*, McGraw-Hill, New York, 1963.

[23] Nekrasov, A. M., and Rokotyan, S. S., *500 kV Long Distance Electric Transmission*, Published for the U.S. Dept. of Interior and NSF by the Israel Program for Scientific Translations, 1966.

[24] Neuenswander, John R., *Modern Power Systems*, International Textbook Co., Scranton, 1971.

[25] Nieman, L. R., Glinternik, S. R., Emel'yanov, A. V., and Nouitski, V. C., *Dc Transmission in Power Systems*, Published for the U.S. Dept. of Interior and NSF by the Israel Program for Scientific Translations, 1967.

[26] Saums, H. L., and Pendleton, W. W., *Materials for Electrical Insulating and Dielectric Functions*, Hayden, Rochelle Park, N.J., 1973.

[27] Stevenson, Jr., William D., *Elements of Power Systems Analysis*, 3rd edition. McGraw-Hill, Inc., New York, 1975.

[28] Sunde, Erling D., *Earth Conduction Effects in Transmission Systems*, Dover Publications, Inc., New York, 1968.

[29] Wagner, C. F., and Evans, R. D., *Symmetrical Components*, McGraw-Hill, New York, 1933.

[30] Weedy, B. M., *Electric Power Systems*, 2nd edition, Wiley, New York, 1972.

Problems

4-1. Consider two parallel conductors, a and b, with gmr's r'_a and r'_b separated by a center to center distance D running for a length d. Ignoring resistive and capacitive effects, derive an equivalent circuit that accounts for only the magnetic effects. Draw the circuit labeling the end terminals a_1, a_2, b_1, and b_2. As usual $d \gg D$, so ignore end effects.

4-2. Repeat problem 4-1, determining a new equivalent circuit considering only capacitive effects. Denote the conductor radii as r_a and r_b.

4-3. Repeat problem 4-1, determining a new equivalent circuit considering only resistive effects. Denote the conductor resistance per unit length as R_a and R_b.

4-4. The three effects considered in problems 4-1 through 4-3 cannot be accurately modeled separately but must be considered "mixed together." Write a brief (one page) explanation of this point, using Figure 4-5 as a conceptual aid.

4-5 The unit "circular mil" is sometimes used to measure conductor cross-sectional area. One circular mil is defined as the area of a circle of diameter 10^{-3} inch. One thousand circular mils is designated by the abbreviation "MCM"

(a) Derive an equation to convert from area in MCM to area in square metres.

(b) Determine the cross-sectional conducting area of 795 MCM conductor in square metre.

(c) Determine the ac resistance in ohm/km for 795 MCM ACSR conductor at 50°C.

4-6. A four conductor equilateral line has the following dimensions: $D = 8$ metres, $r = r' = 5$ cm, and negligible resistance. Evaluate $[Z_{012}]$ and $[Y_{012}]$ for this line at 60 Hz.

4-7. Continue problem 4-6 and evaluate Z_{c_0}, Z_{c_1}, γ_0 and γ_1 for the line.

4-8. Suppose the line of problems 4-6 and 4-7 is 100 km long. Calculate equivalent circuit element values for the circuits shown in Figure 4-7. Use equations (4-73) and (4-74).

4.9. Repeat problem 4-8 using the approximate relations (4-75d) and (4-76d). Compare answers with those of problem 4-8.

4-10. Refer to example 4-5. Recalculate Z_{c_0}, Z_{c_1}, γ_0, and γ_1 on the assumption that the resistance is negligible.

4-11. Refer to example 4-5. Recalculate Z_0, $Y_0/2$, Z_1, and $Y_1/2$ without assuming that γ_0 and γ_1 are imaginary or that Z_{c_0} and Z_{c_1} are real. Compare results to those of example 4-5.

4–12. Refer to example 4-5. Convert the equivalent circuit values into per unit given $S_{3\phi_{base}} = 100\text{MVA}$ and $V_{line_{base}} = 500\text{kV}$.

4-13. Three phase conductors each with $r = 4$ cm and $r' = 3$ cm are arranged in a plane, 20 metres between outside conductors and the third conductor centered.

(a) compute the equivalent equilateral spacing.

(b) at 60 Hz compute approximate values for X_1 in ohms/m and B_1 in siemens/m.

4-14. A 3ϕ 115 kV 200 km line has its phase conductors spaced 3 m apart in a vertical plane. Each phase conductor is 636 MCM ACSR ($r = 1.26$ cm; $r' = 1.23$ cm; $R = 0.100$ Ω/km; $I_{rated} = 770$ amperes). Calculate the thermal limit, $S_{3\phi_{rated}}$.

4-15. Approximating the line of problem 4-14 as an equilateral line, calculate the values Z_{c1} and γ_1.

4-16. For the line of problem 4-15 evaluate A, B, C, and D.

4-17. For the line of problem 4-15 calculate $P_{3\phi ss}$.

4-18. Repeat example 4-7 for the following parameters:

$d = 2\,\text{m}$ $E_s = 100/s$

$L = 1\,\text{H/m}$ $Z_s = 2\,\Omega$

$C = 1\,\text{F/m}$ $Z_r = 4\,\Omega$

4-19. Prove that $[L]$ is a unitary matrix.

4-20. Prove that the modal transformation is power invariant.

4-21. The modal transformation may also be applied to the sinusoidal steady state case. If this is the case derive a general equation to transform $[Z_{012}]$ to an equivalent $[Z_{0\alpha\beta}]$; where:

$$[Z_{012}] = \begin{bmatrix} Z_{00} & 0 & 0 \\ 0 & Z_{11} & 0 \\ 0 & 0 & Z_{22} \end{bmatrix}$$

4-22. If $Z_{11} = Z_{22}$, show that the results from problem 4-21 simplify to:

$$[Z_{o\alpha\beta}] = \begin{bmatrix} Z_{00} & 0 & 0 \\ 0 & Z_{11} & 0 \\ 0 & 0 & Z_{22} \end{bmatrix}$$

4-23. Rework example 4-9 if all values are the same except:

$Z_s(s) = 325s$ $(R_g = 0)$

$Z_r(s) = 325$

4-24. The objective of this problem is to show that equation (4-26h) can be used for X_1 for a general transposed 3ϕ line. Imagine four conductors (a, b, c, n) arranged in an arbitrary, but fixed, four position geometrical array $(1, 2, 3, 4)$ as shown below. For simplicity, assume $R_a = R_b = R_c = R_n = 0$ and $r'_a = r'_b = r'_c = r'_n = r'$.

Line Section	Position 1	Position 2	Position 3	Position 4
$0 < l < d/3$	a	b	c	n
$d/3 < l < 2d/3$	c	a	b	n
$2d/3 < l < d/3$	b	c	a	n

(a) Set up expressions for λ_a, λ_b, λ_c, and λ_n, using equation (4-13), for each line section.

(b) Calculate total flux linkages for conductors a, b, c, and n by multiplying the λ's by $d/3$ and summing the line section components together. Simplify.

(c) Calculate voltage expressions equivalent to equations (4-22). Eliminate I_n to obtain the equivalent of equations (4-23). Simplify. Write the set in matrix form as shown in equation (4-25). Identify X_s and X_m.

(d) Now show that

$$X_1 = X_s - X_m = \frac{\mu\omega}{2\pi} \ln \frac{D}{r'} \text{ ohm/m}$$

where

$$D = \sqrt[3]{D_{12}D_{23}D_{31}}$$

(e) Finally derive X_0 from:

$$X_0 = X_s + 2X_m \text{ ohm/m}$$

THE POWER
TRANSFORMER

"Exercise caution in your business affairs; for the world is full
of trickery. But let this not blind you to what virtue
there is; many persons strive for high ideals,
and everywhere life is full of heroism."

Max Ehrmann, DESIDERATA

The power transformer is a power system component of major importance, ranking with the synchronous machine and the transmission line. Because electrical power is proportional to the product of voltage and current, for a specified power level we can maintain low current levels only at the expense of high voltage, and vice versa. Differing requirements of generation, transmission, and utilization dictate different optimum values for voltage and current combinations. It therefore is imperative that we design a component capable of changing (transforming) voltage and current at high power levels, reliably and efficiently. The modern power transformer serves such a function admirably, with efficiencies approaching 100%, power ratings exceeding 1000 MW, voltage ratings exceeding 700 kV, and current ratings exceeding 23 kA. A bank of three single phase transformers is shown in Figure 5-1.

The transformer is a well-known device; many excellent texts analyze and describe its operation at practically all levels of sophistication. Our

Figure 5-1 A three-phase bank consisting of three 383 MVA 289/17 kV General Electric transformers. (Photograph courtesy of Georgia Power Company.)

need here is not for a rigorous investigation of the physical machine, but rather the development of a circuit model that predicts electrical performance with reasonable accuracy and is suitable for integration into the total power system. We should be aware that there is considerably more to be learned about transformers and interested students will want to pursue the subject in depth.

5-1 The Three Winding Ideal Transformer Equivalent Circuit

To understand the basics of transformer operation consider the three coils wrapped on a common core as shown in Figure 5-2(a). We select a three winding device for several reasons. The extension to the n-winding case is straightforward, the reduction to the simpler two winding case is likewise

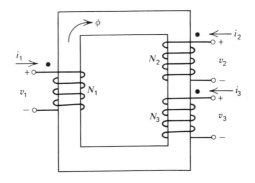

(a)   The transformer

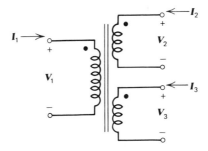

(b)   Equivalent circuit symbol

Figure 5-2 The basic three winding ideal transformer with idealizations $\mu = \infty$ and $\sigma = \infty$.

151

clear, and we are specifically interested in modeling three winding power transformers. To simplify the situation we imagine two idealizations: first, that the core permeability (μ) is infinite, and, second, that the windings are made of material of infinite conductivity (σ). As a result of the latter idealization, all windings have zero resistance. From Faraday's law we then write:

$$v_1 = N_1 \frac{d\phi}{dt} \tag{5-1a}$$

$$v_2 = N_2 \frac{d\phi}{dt} \tag{5-1b}$$

$$v_3 = N_3 \frac{d\phi}{dt} \tag{5-1c}$$

But the "ϕ" in all three equations is the same; no flux can escape the core because of its infinite permeability. We divide (5-1a) by (5-1b); (5-1b) by (5-1c); and (5-1c) by (5-1a), producing:

$$\frac{v_1}{v_2} = \frac{N_1}{N_2} \tag{5-2a}$$

$$\frac{v_2}{v_3} = \frac{N_2}{N_3} \tag{5-2b}$$

$$\frac{v_3}{v_1} = \frac{N_3}{N_1} \tag{5-2c}$$

We are interested only in sinusoidal steady state performance. Therefore, cross-multiply by the denominator voltages and transform into phasor values, producing:

$$V_1 = \frac{N_1}{N_2} V_2 \tag{5-3a}$$

$$V_2 = \frac{N_2}{N_3} V_3 \tag{5-3b}$$

$$V_3 = \frac{N_3}{N_1} V_1 \tag{5-3c}$$

The wisdom of the particular assigned positive voltage conventions selected should be apparent. Equations (5-3) have no minus signs, implying V_1, V_2 and V_3 are in phase (the turns ratios are of course positive real numbers). We are usually not provided with information as to how the coils are wound, so, in lieu of this, polarity markings (dots) are provided. Later we learn that another industrial polarity convention is used for power transformers, which amounts to the same thing. The circuit symbol to be used is shown in Figure 5-2(b). Ampere's law requires that

$$\oint \hat{H} \cdot \hat{dl} = i_{\text{enclosed}} \tag{5-4}$$

If the closed path selected is the core center line, the magnetic field intensity H must be everywhere zero, again because of infinite core permeability. Therefore, equation 5-4 becomes:

$$0 = i_{\text{enclosed}} \tag{5-5a}$$

or

$$0 = N_1 i_1 + N_2 i_2 + N_3 i_3 \tag{5-5b}$$

We transform equation (5-5b) into phasor notation:

$$N_1 I_1 + N_2 I_2 + N_3 I_3 = 0 \tag{5-6}$$

The assignment of positive current directions in Figure 5-2 is not quite so satisfactory. We were virtually forced to do it this way because unless we have knowledge of the external connections it is impossible to select "reasonable" positive directions. Our approach is symmetric (consistently define currents into the dots in all windings), and results in equations that can easily be extended to the n-winding case. Yet at least one current in (5-5b) is always negative, an annoying detail. Equations (5-3) and (5-6) are basic to understanding transformer operation. Consider equation (5-3a). This relation tells us it is possible, given V_2, to produce any V_1 we wish simply by adjusting the turns ratio N_1/N_2. Also note that V_1, V_2 and V_3 must be in phase with each other. Now consider the total input complex power S.

$$S = V_1 I_1^* + V_2 I_2^* + V_3 I_3^* \tag{5-7a}$$

$$= V_1 I_1^* + \frac{N_2}{N_1} V_1 I_2^* + \frac{N_3}{N_1} V_1 I_3^* \tag{5-7b}$$

$$= \frac{V_1}{N_1} [N_1 I_1 + N_2 I_2 + N_3 I_3]^* \tag{5-7c}$$

$$= 0 \tag{5-7d}$$

The interpretation to be made here is that the ideal transformer can absorb neither real nor reactive power. An example should clarify these properties.

Example 5-1

The transformer of Figure 5-2 has the following data: $N_1 = 1000$ turns; $N_2 = 500$ turns; $N_3 = 2000$ turns. Winding #1 is terminated in an ideal voltage source with $V_1 = 1000\underline{/0°}$. Winding #2 is terminated in a 20 Ω resistance and winding #3 in an inductance of reactance 100 Ω. Calculate the current I_1. Check the winding powers.

Solution

Let us find V_2 and V_3:

$$V_2 = \frac{N_2}{N_1} V_1$$

$$= \frac{500}{1000}(1000\underline{/0°}) = 500\underline{/0°} \tag{5-3a}$$

Similarly:

$$V_3 = \frac{N_3}{N_1} V_1$$

$$= \frac{2000}{1000}(1000\underline{/0°}) = 2000\underline{/0°} \tag{5-3c}$$

Now find the currents I_2 and I_3:

$$-I_2 = \frac{V_2}{Z_2}$$

$$= \frac{500\underline{/0°}}{20} = 25\underline{/0°}$$

$$-I_3 = \frac{V_3}{Z_3}$$

$$= \frac{2000\underline{/0°}}{j100} = 20\underline{/-90°}$$

Using equation (5-6):

$$I_1 = -\frac{N_2}{N_1}I_2 - \frac{N_3}{N_1}I_3$$

$$= \frac{500}{1000}(25\underline{/0°}) + \frac{2000}{1000}(-j20)$$

$$= 12.5 - j40$$

$$= 41.9\underline{/-72.6}$$

The output complex powers are:

$$S_2 = V_2(-I_2)^* = 12500$$

$$S_3 = V_3(-I_3)^* = +j40000$$

Totaling to $12.5 + j40$ kVA

The input complex power is:

$$S_1 = V_1 I_1^* = 41.9\underline{/+72.6°} \text{ kVA}$$

$$= 12.6 + j40 \text{ kVA}$$

Recall that when the per unit system was introduced, one of its justifications was that it drastically simplified transformer circuit models. We will deal with that point now. Suppose we arbitrarily select two base values, $V_{1_{base}}$ and $S_{1_{base}}$. We require that base values for windings 2 and 3 be

$$V_{2_{base}} = \frac{N_2}{N_1} V_{1_{base}} \tag{5-8a}$$

$$V_{3_{base}} = \frac{N_3}{N_1} V_{1_{base}} \tag{5-8b}$$

and

$$S_{1_{base}} = S_{2_{base}} = S_{3_{base}} = S_{1\emptyset_{base}} \tag{5-9}$$

The symbol "$S_{1\emptyset_{base}}$" is somewhat mysterious at this point; we intend later to apply our work to three phase transformer connections and the notation will make more sense then. By definition:

$$I_{1_{base}} = \frac{S_{1\emptyset_{base}}}{V_{1_{base}}} \tag{5-10a}$$

$$I_{2_{base}} = \frac{S_{1\emptyset_{base}}}{V_{2_{base}}} \tag{5-10b}$$

$$I_{3_{\text{base}}} = \frac{S_{1\emptyset_{\text{base}}}}{V_{3_{\text{base}}}} \tag{5-10c}$$

It follows from (5-10a), (5-8a), and (5-9) that

$$I_{2_{\text{base}}} = \frac{N_1}{N_2} I_{1_{\text{base}}} \tag{5-11a}$$

$$I_{3_{\text{base}}} = \frac{N_1}{N_3} I_{1_{\text{base}}} \tag{5-11b}$$

Recall that a per unit value is the actual value divided by its appropriate base: recall equation (5-3a)

$$V_1 = \frac{N_1}{N_2} V_2 \tag{5-3a}$$

Divide through by $V_{1_{\text{base}}}$:

$$\frac{V_1}{V_{1_{\text{base}}}} = \frac{(N_1/N_2) V_2}{V_{1_{\text{base}}}} \tag{5-12a}$$

But from (5-8a) we replace $V_{1_{\text{base}}}$ on the right side with its equivalent $(N_1/N_2) V_{2_{\text{base}}}$. We get:

$$\frac{V_1}{V_{1_{\text{base}}}} = \frac{(N_1/N_2) V_2}{(N_1/N_2) V_{2_{\text{base}}}} \tag{5-12b}$$

or

$$V_{1_{\text{pu}}} = V_{2_{\text{pu}}} \tag{5-12c}$$

where the "pu" subscript indicates per unit values. Similarly

$$\frac{V_1}{V_{1_{\text{base}}}} = \frac{(N_1/N_3) V_3}{(N_1/N_3) V_{3_{\text{base}}}} \tag{5-13a}$$

or

$$V_{1_{\text{pu}}} = V_{3_{\text{pu}}} \tag{5-13b}$$

We summarize our results as:

$$V_{1_{\text{pu}}} = V_{2_{\text{pu}}} = V_{3_{\text{pu}}} \tag{5-14}$$

Now we must deal with the currents. Divide equation (5-6) by N_1:

$$I_1 + (N_2/N_1)I_2 + (N_3/N_1)I_3 = 0 \tag{5-15a}$$

Divide through by $I_{1_{\text{base}}}$:

$$\frac{I_1}{I_{1_{\text{base}}}} + \frac{(N_2/N_1)I_2}{I_{1_{\text{base}}}} + \frac{(N_3/N_1)I_3}{I_{1_{\text{base}}}} = 0 \tag{5-15b}$$

$$\frac{I_1}{I_{1_{\text{base}}}} + \frac{(N_2/N_1)I_2}{(N_2/N_1)I_{2_{\text{base}}}} + \frac{(N_3/N_1)I_3}{(N_3/N_1)I_{3_{\text{base}}}} = 0 \tag{5-15c}$$

Simplifying to:

$$I_{1_{\text{pu}}} + I_{2_{\text{pu}}} + I_{3_{\text{pu}}} = 0 \tag{5-15d}$$

Equations (5-14) and (5-15d) inspire us to synthesize the basic equivalent circuit, shown in Figure 5-3. We find it cumbersome to carry the "pu" in the subscript past this point; no confusion should result since from now on we will consistently work in per unit. A word of caution is in order,

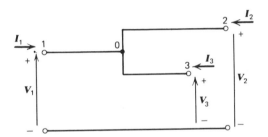

Figure 5-3 Equivalent circuit for ideal three winding transformer. All values in per unit.

however. Intuitively, we sense that the use of per unit is optional, that we could "take it or leave it," and in a sense this is correct. Unfortunately, that notion is somewhat oversimplified; some of our circuit models and corresponding equations are correct *only if the parameters and variables are in per unit.* These insights are particularly germane to the transformer. To put it another way, we cannot use the same circuit models for scaled and unscaled values. Note the internal inaccessible junction point "0." The extension to the *n*-winding case should now be obvious.

Example 5-2

Again solve for I_1 as required in example 5-1 in per unit using bases of $V_{1_{\text{base}}} = 1000$ Volts and $S_{1\emptyset_{\text{base}}} = 50$ kVA.

$$Z_{2_{base}} = \frac{V^2_{2_{base}}}{S_{1\emptyset_{base}}} = \frac{(500)^2}{50 \times 10^3} = 5 \text{ ohms}$$

$$Z_{3_{base}} = \frac{V^2_{3_{base}}}{S_{1\emptyset_{base}}} = \frac{(2000)^2}{50 \times 10^3} = 80 \text{ ohms}$$

$$Z_{2_{pu}} = \frac{20}{5} = 4 \text{ pu}$$

$$Z_{3_{pu}} = \frac{j100}{80} = j1.25 \text{ pu}$$

$$V_{1_{pu}} = \frac{1000\underline{/0°}}{1000} = 1\underline{/0°} \text{ pu}$$

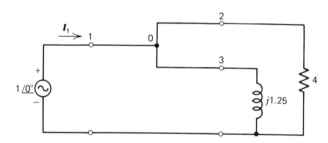

Figure 5-4 Circuit diagram for example 5-2. All values in per unit.

The corresponding circuit appears in Figure 5-4. Solving for I_1

$$I_1 = \left[\frac{1}{4} + \frac{1}{j1.25}\right] 1\underline{/0°}$$

$$= 0.25 - j0.80 \text{ pu}$$

$$= 0.838\underline{/-72.6°} \text{ pu}$$

To convert to actual values

$$I_{1_{base}} = \frac{S_{1\emptyset_{base}}}{V_{1_{base}}}$$

$$= \frac{50}{1} = 50 \text{ amperes}$$

Example 5-2

and

$$I_1 = (0.838)(50)$$

$$= 41.9 \text{ amperes}$$

which agrees with the results of example 5-1.

5-2 A Practical Three Winding Transformer Equivalent Circuit

For some power system applications the circuit of Figure 5-3 is reasonable, since the core and windings of actual transformers are constructed of materials of high μ and σ, although of course not infinite. However, for other studies, discrepancies between the performance of actual and ideal transformers are too great to be overlooked. The circuit of Figure 5-3 may be modified into that of Figure 5-5 to account for the most important discrepancies. We comment on the circuit parameters as follows:

R_1, R_2, R_3—Since the winding conductors cannot be made of material of infinite conductivity, the winding must have some resistance.

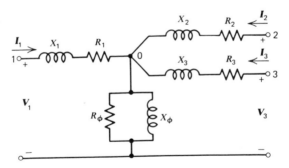

Figure 5-5 A practical three winding transformer equivalent circuit. All values in per unit.

X_1, X_2, X_3—Since the core permeability is not infinite, all of the flux created by a given winding current will not be confined to the core (most—95 to 99%—is). The part that escapes the core and seeks out parallel paths in surrounding structures and air is referred to as "leakage" flux. This ac leakage flux

159

will induce a Faraday ac voltage into each winding. Since the strength of the leakage flux (and consequently the leakage voltage) is directly proportional to the causing current, this effect may be modeled by the insertion of a linear inductive reactance as shown.

R_ϕ, X_ϕ—Also, since the core permeability is not infinite, the magnetic field intensity inside the core, while small, is not zero. Therefore, some current flow is necessary to provide this small H. The path provided in the circuit for this small magnetizing current is through X_ϕ. The core has internal power losses, referred to as "core loss," due to hysteresis and eddy current phenomena. This effect is accounted for in the resistance R_ϕ.

We should think of the circuit of Figure 5-5 as a refinement on that of Figure 5-3. The values $R_1, R_2, R_3, X_1, X_2, X_3$, are all small (less than 0.05 pu; zero for the ideal case) and R_ϕ, X_ϕ, large (greater than 10 pu; infinite for the ideal case). Again remember that the circuit of Figure 5-5 requires that all values be in per unit. Such data is available from the manufacturer or obtained from conventional tests. For more detail, consult any one of several excellent references on this subject (for example, see [3], [4], and [6] in the end-of-chapter Bibliography). Our approach is to assume such data is available, and to turn our attention to how we might use it.

Example 5-3

Refer to example 5-2 with the transformer modified to include the following per unit data:

$R_1 = R_2 = R_3 = 0.01$

$X_1 = X_2 = X_3 = 0.03$

$R_\phi = X_\phi = \infty$

Calculate I_1.

Solution

"Looking into" winding #1 we "see":

$$Z_1 = 0.01 + j0.03 + \frac{(4.01 + j0.03)(0.01 + j1.28)}{4.02 + j1.31}$$

$$= 0.38654 + j1.184 = 1.245\underline{/71.92^\circ}$$

$$I_1 = \frac{1\underline{/0°}}{1.245\underline{/71.92°}} = 0.803\underline{/-71.92°} \text{ pu.}$$

Compare the discrepancy between this result and that of example 5-2.

A comment about the terms "primary" and "secondary" is in order. In common usage the terms refer to source and load sides, respectively (i.e., energy flows from primary to secondary). However, in many applications energy can flow either way, in which case the distinction is meaningless. Also, the presence of a third winding (tertiary) confuses the issue. Here we use the terms (along with subscripts 1, 2, 3) simply to distinguish between windings. The terms "step up" and "step down" refer to what the transformer does to the voltage from source to load. Industrial standards require that for a two winding transformer the high voltage and low voltage terminals be marked as $H1$-$H2$ and $X1$-$X2$, respectively, with $H1$ and $X1$ markings having the same significance as "dots" for polarity markings.

5-3 Transformers in Three Phase Connections

We wish to use three identical transformers in three phase connections. Recall from Chapter 2 that there are only two possible symmetric and balanced connections, the wye and the delta, and we limit our investigation to those. If we use three winding transformers, we have a total of nine windings to account for. The three sets of windings may be individually connected in wye or delta in any combination. To keep matters simple, let us examine only the set of #1 windings. This is easily done by assuring that windings 2 and 3 remain open. To simplify further, define:

$$Z_\phi = \frac{R_\phi(jX_\phi)}{R_\phi + jX_\phi} \tag{5-16a}$$

and

$$Z_{\phi 1} = Z_\phi + R_1 + jX_1 \tag{5-16b}$$

Now let us connect the three windings in wye with the common point connected to the system neutral* through the impedance Z_{n1}. The results are shown in Figure 5-6. Application of KVL produces:

$$\begin{bmatrix} V_a \\ V_b \\ V_c \end{bmatrix} = \begin{bmatrix} (Z_{\phi 1}+Z_{n1}) & Z_{n1} & Z_{n1} \\ Z_{n1} & (Z_{\phi 1}+Z_{n1}) & Z_{n1} \\ Z_{n1} & Z_{n1} & (Z_{\phi 1}+Z_{n1}) \end{bmatrix} \begin{bmatrix} I_a \\ I_b \\ I_c \end{bmatrix} \tag{5-17}$$

* The system neutral is electrically connected to the earth (ground) and connections to neutral are referred to as "grounding."

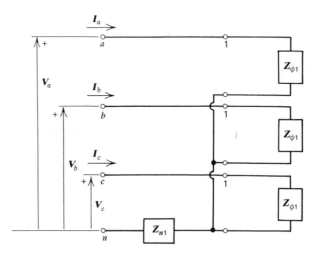

Figure 5-6 Three 3 winding transformers with windings 1 connected in wye grounded through Z_{n1} and windings 2 and 3 open.

If we next transform to sequence values through equation (2-60):

$$[Z_{012}] = [T]^{-1}[Z_{abc}][T] \tag{2-60}$$

and simplify, we get:

$$[Z_{012}]_{\text{winding 1}} = \begin{bmatrix} Z_{\phi 1} + 3Z_{n1} & 0 & 0 \\ 0 & Z_{\phi 1} & 0 \\ 0 & 0 & Z_{\phi 1} \end{bmatrix} \tag{5-18a}$$

Had we used windings 2 and 3 we would have determined

$$[Z_{012}]_{\text{winding 2}} = \begin{bmatrix} Z_{\phi 2} + 3Z_{n2} & 0 & 0 \\ 0 & Z_{\phi 2} & 0 \\ 0 & 0 & Z_{\phi 2} \end{bmatrix} \tag{5-18b}$$

$$[Z_{012}]_{\text{winding 3}} = \begin{bmatrix} Z_{\phi 3} + 3Z_{n3} & 0 & 0 \\ 0 & Z_{\phi 3} & 0 \\ 0 & 0 & Z_{\phi 3} \end{bmatrix} \tag{5-18c}$$

where

$$Z_{\phi 2} = R_2 + jX_2 + Z_\phi \tag{5-19a}$$

$$Z_{\phi 3} = R_3 + jX_3 + Z_\phi \tag{5-19b}$$

A comment on notation is appropriate. Do not confuse the "012" subscripts (for zero sequence, positive sequence, and negative sequence) with "123" subscripts (for windings 1, 2, and 3). Careful reading should eliminate any confusion.

We use equations (5-18) to synthesize the sequence networks, which are shown in Figure 5-7. The resulting networks are decoupled—that is, we may calculate sequence quantities independently in each network. If a winding

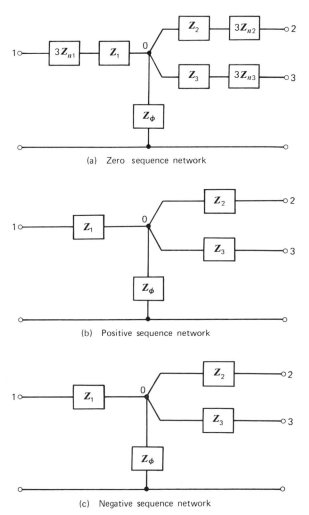

(a)   Zero   sequence   network

(b)   Positive   sequence   network

(c)   Negative   sequence   network

Figure 5-7 Transformer sequence networks corresponding to grounded wye connections.

163

set is "solidly grounded," this is jargon for stating that Z_n is zero, and therefore $3Z_n$ should be replaced with a short circuit.

If the connection is "ungrounded," this means that Z_n is infinite, and consequently that $3Z_n$ be replaced by an open circuit. Note that details of grounding only modify the zero sequence network; the positive and negative sequence networks, which are identical, are unaffected. As usual the networks of Figure 5-7 require that all values be in per unit.

Now consider the case where one winding set is connected in delta with the other two left open. The delta may be transformed to an equivalent wye. We realize that this situation is equivalent to the ungrounded wye case, and that the positive and negative sequence networks are the same* as those shown in Figures 5-7(b) and (c). The proper zero sequence network is not clear. Like the ungrounded wye, it will offer an open circuit to zero sequence current from an external circuit, but, unlike the ungrounded wye, a closed path internally around the delta is available for zero sequence current flow. (The delta-wye conversion obscured this fact.)

To examine this point, let us investigate the following situation. Look in at winding set #1 connected in wye grounded through Z_n, winding set #2 connected in delta, and the winding set #3 left open. Also, to simplify our analysis, let the transformation ratio be 1:1. See Figure 5-8. We write:

$$V_a = V_{an} = I_a Z_1 + (I_a - I_x)Z_\phi + I_n Z_n \tag{5-20a}$$

$$V_b = V_{bn} = I_b Z_1 + (I_b - I_x)Z_\phi + I_n Z_n \tag{5-20b}$$

$$V_c = V_{cn} = I_c Z_1 + (I_c - I_x)Z_\phi + I_n Z_n \tag{5-20c}$$

Also:

$$I_n = I_a + I_b + I_c \tag{5-20d}$$

Let us solve for I_x. KVL around the delta produces:

$$(I_a - I_x)Z_\phi + (I_b - I_x)Z_\phi + (I_c - I_x)Z_\phi = 3I_x Z_2 \tag{5-21a}$$

$$I_n Z_\phi = (Z_2 + Z_\phi)3I_x \tag{5-21b}$$

$$\therefore \quad I_x = \frac{Z_\phi}{3Z_2 + 3Z_\phi} I_n \tag{5-21c}$$

---

* Because of the caprices of the per unit system, even an expected factor of 3 does not appear.

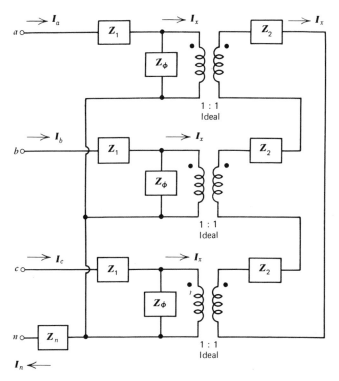

Figure 5-8 Winding set #1 connected wye; winding set #2 connected delta; winding set #3 open.

Substituting (5-21c) into (5-20), simplifying, and converting to matrix form produces:

$$\begin{bmatrix} V_a \\ V_b \\ V_c \end{bmatrix} = \begin{bmatrix} Z_s & Z_m & Z_m \\ Z_m & Z_s & Z_m \\ Z_m & Z_m & Z_s \end{bmatrix} \begin{bmatrix} I_a \\ I_b \\ I_c \end{bmatrix} \qquad (5\text{-}22\text{a})$$

where

$$Z_s = Z_1 + Z_\phi \left[\frac{3Z_2 + 2Z_\phi}{3Z_2 + 3Z_\phi} \right] + Z_n \qquad (5\text{-}22\text{b})$$

$$Z_m = Z_n - \frac{Z_\phi^2}{3Z_2 + 3Z_\phi} \qquad (5\text{-}22\text{c})$$

Transforming to sequence values using equation (2-60) produces:

$$[Z_{012}] = \begin{bmatrix} Z_{00} & 0 & 0 \\ 0 & Z_{11} & 0 \\ 0 & 0 & Z_{22} \end{bmatrix} \tag{5-23a}$$

where

$$Z_{00} = Z_s + 2Z_m \tag{5-23b}$$

$$= Z_1 + 3Z_n + \frac{Z_2 Z_\phi}{Z_2 + Z_\phi} \tag{5-23c}$$

and

$$Z_{11} = Z_s - Z_m \tag{5-23d}$$

$$= Z_1 + Z_\phi \tag{5-23e}$$

$$Z_{22} = Z_{11} \tag{5-23f}$$

The problem is then to synthesize a network that:

1. Presents an open circuit to external zero sequence current on the delta (winding 2) side.

2. Satisfies equation (5-23c) looking in from the wye (winding 1) side.

Such a network is shown in Figure 5-9 and is the correct zero sequence network for the situation discussed.

We are now in a position to generalize our findings. The circuits shown in Figure 5-7(b) and (c) are identical and are satisfactory positive and

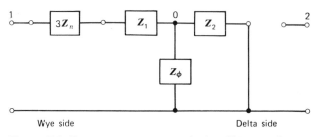

Wye side   Delta side

Figure 5-9 Zero sequence network for Y-Δ transformer connections as shown in Figure 5-8.

negative sequence networks for the three winding transformer set *indepen-dent of winding connections* (wye or delta). The general zero sequence network for the three winding transformer set is shown in Figure 5-10. The

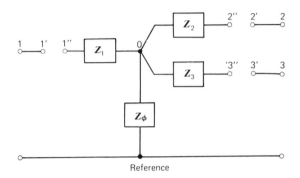

Figure 5-10 General zero sequence network for the three winding case.

terminals 1' and 1" are terminated depending on the details of how the set of #1 windings are interconnected, that is:

1. Solid ground wye—short 1' to 1"

2. Ground wye through Z_n—connect 1' to 1" through $3Z_n$

3. Ungrounded wye—leave 1' − 1" open

4. Delta—short 1" to reference

Winding sets #2 and #3 interconnections produce similar connection constraints at terminals 2'-2" and 3'-3". An example should clarify this situation.

Example 5-4

Three identical transformers as described in example 5-3 are to be used in a three phase system. They are connected at their terminals as follows:

Winding set #1—wye, grounded thru $Z_n = j0.04$
Winding set #2—wye, solid ground
Winding set #3—delta

Draw the sequence networks.

167

Solution

The sequence networks are as shown.

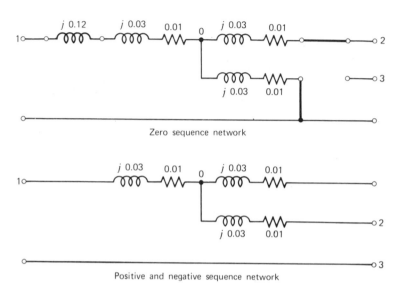

Zero sequence network

Positive and negative sequence network

5-4 Phase Shift in *Y*-Δ Connections

The positive and negative sequence networks presented in Figure 5-7(b) and (c) are misleading in one subtle, but important, detail. For *Y-Y* or Δ-Δ connections it is always possible to label the phases in such a way that there is no phase shift between corresponding primary and secondary quantities. By convention, this is always done. However for *Y*-Δ or Δ-*Y* connections, it is impossible tc label the phases so that no phase shift between corresponding quantities is introduced. To understand the problem, consider the two winding ideal transformers connected in Δ-*Y* shown in Figure 5-11. Consider the delta side to be the low voltage side. Study the phasor diagram presented in Figure 5-12. We observe that if:

$$V_{AB} = V_{LV}\underline{/0°} \qquad (5\text{-}24a)$$

It follows that:

$$V_{ab} = V_{HV}\underline{/30°} \qquad (5\text{-}24b)$$

In fact *all* high voltage quantities will lead their corresponding low voltage

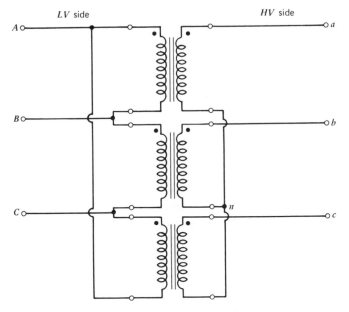

Figure 5-11 ASA standards LV-Δ:HV-Y connection.

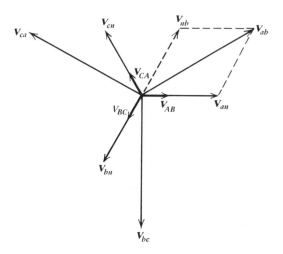

Figure 5-12 Phasor diagram for the transformer bank of Figure 5-11. Excited by positive sequence voltages.

counterparts by 30°. After some experimentation we conclude that the minimum phase shift possible is 30°. We perceive a need to standardize our Δ-Y connections so that this phase shift is predictable. Such a standard is used in industry and may be stated as follows.

> *For either wye-delta or delta-wye connections, phases shall be labeled in such a way that positive sequence quantities on the high voltage side lead their corresponding positive sequence quantities on the low voltage side by 30°. The effect on negative sequence quantities is the reverse, that is, HV values* lag *LV values by 30°.*

The connection presented in Figure 5-11 conforms to the standard for the LV-Δ:HV-Y case. The connection that satisfies the standard for the HV-Δ:LV-Y case is shown in Figure 5-13. See problems 5-8 and 5-9 for more development on these topics. From this point on in this book it will be assumed that all Y-Δ connections conform to these standards.

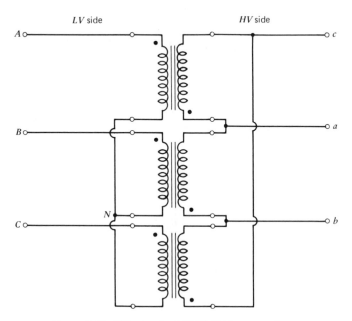

Figure 5-13 ASA standard *LV-Y:HV-Δ* connection.

The reader should realize that this 30° phase shift is *not* accounted for in the sequence networks of Figure 5-7(b) and (c). Sometimes this is of no concern to us, but in general it must be considered to realize the simplifications accrued from the per unit system. The effect only appears in the positive and negative sequence network; the zero sequence network quantities are unaffected.

5-5 The Two Winding Transformer

We have developed sequence networks for three winding transformers in three phase banks, and have gained a measure of confidence that we could

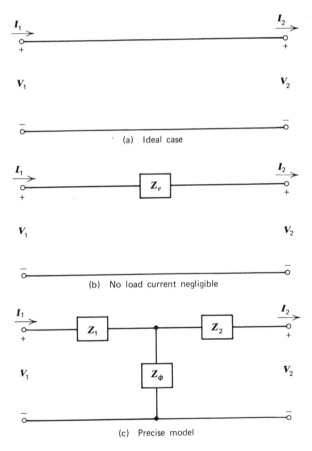

Figure 5-14 Two winding transformer equivalent circuits. All values in per unit.

171

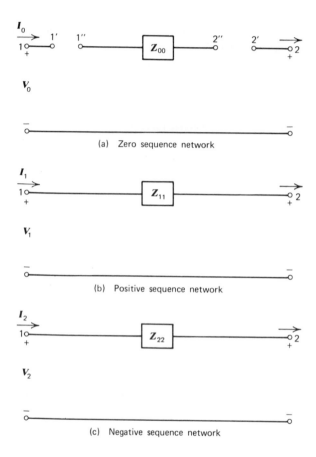

(a) Zero sequence network

(b) Positive sequence network

(c) Negative sequence network

Figure 5-15 Sequence networks for transformer as modeled in Figure 5-14(b).

deal with such devices in more complex situations. We now find it expedient to simplify our work to the important two winding case to investigate practical situations encountered in power systems. Two winding transformer circuit models, in order of increasing complexity, are shown in Figure 5-14. In particular, note the circuit of Figure 5-14(b) where

$$Z_e = Z_1 + Z_2 \tag{5-25}$$

This circuit is appropriate when $Z_\emptyset$ is large enough that magnetizing current and core loss is negligible—typically the case in power transformers. We use this model for the rest of our work in this chapter.

172

In three phase banks, the sequence networks are shown in Figure 5-15. If we are dealing with three separate physical transformers, as in our previous work:

$$Z_{00} = Z_e \qquad\qquad (5\text{-}26\text{a})$$

$$Z_{11} = Z_e \qquad\qquad (5\text{-}26\text{b})$$

$$Z_{22} = Z_e \qquad\qquad (5\text{-}26\text{c})$$

It is possible to construct a device (called a three phase transformer) that allows the phase fluxes to share common magnetic paths. Such designs allow considerable savings in core material, and corresponding economies in cost, size, and weight. The effect on our circuit model surfaces at only one point: Z_{00}, is not equal to Z_{11} or Z_{22} (Z_{11} does still equal Z_{22}). To restate this point, the difference between three interconnected $1\emptyset$ devices and one $3\emptyset$ device may be accounted for with the numerical values substituted for Z_{00}, Z_{11}, and Z_{22}. This is the reason for the notational distinction for impedances in Figure 5-15. The terminals $1'$, $1''$, $2'$, and $2''$ in the zero sequence network are to be connected as discussed in section 5-3 to account for primary and secondary wye and delta connections. Also recall that phase shifts introduced by wye and delta connections must be properly accounted for in the positive and negative sequence networks.

5-6 Autotransformers

Transformer windings, though magnetically coupled, are electrically isolated from each other. It is possible to enhance certain performance characteristics for transformers by electrically interconnecting primary and secondary windings. Such devices are called autotransformers. The benefits to be realized are lower cost, smaller size and weight, higher efficiency, and better voltage regulation. We investigate the device using a two winding ideal transformer.

Study Figure 5-16(a). Imagine the source voltage E and load Z_L adjusted to values that allow the transformer to operate at its rated conditions (i.e., V_1, I_1, V_2 are rated). Obviously:

$$E = V_1 \qquad\qquad (5\text{-}27)$$

and

$$S_{\text{source}} = S_{\text{load}} = S_{\text{rated}} \qquad\qquad (5\text{-}28\text{a})$$

where

$$S_{\text{rated}} = V_1 I_1 = V_2 I_2 \qquad\qquad (5\text{-}28\text{b})$$

The Power Transformer

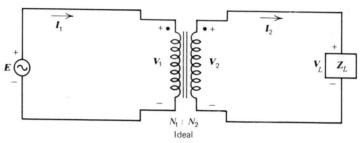

(a) A two winding transformer

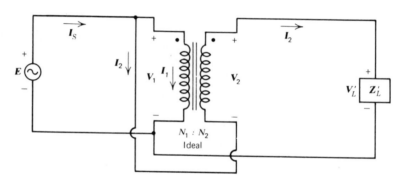

(b) The transformer of (a) Connected as an autotransformer

Figure 5-16 The two winding autotransformer.

Now consider the circuit of Figure 5-16(b) and adjust E and Z'_L until again the transformer operates at rated conditions. Now

$$S_{\text{source}} = EI_s \qquad (5\text{-}29a)$$

$$= V_1(I_1 + I_2) \qquad (5\text{-}29b)$$

$$= V_1\left[1 + \frac{N_1}{N_2}\right]I_1 \qquad (5\text{-}29c)$$

or

$$S_{\text{load}} = V'_L I_L \qquad (5\text{-}30a)$$

$$= (V_1 + V_2)I_2 \qquad (5\text{-}30b)$$

$$= V_2 I_2\left[\frac{N_1}{N_2} + 1\right] \qquad (5\text{-}30c)$$

The interpretation of equations (5-29) and (5-30) is interesting. It says that the power transferred from source to load is $(1 + N_1/N_2)$ times the transformer rating, that is:

$$S_{\text{load}} = \frac{N_2 + N_1}{N_2} S_{\text{rated}} \tag{5-31}$$

If we argue (unfairly, perhaps) that the transformer "transforms" S_{load}, we can "rerate" it at $(N_2 + N_1)/N_2\ S_{\text{rated}}$. We note that the stronger the inequality, $N_1 > N_2$, the greater the new power rating. Being greedy, we make N_2 zero, producing an *infinite* increase in transformer power rating! The only problem is that the voltage ratio (source to load) becomes $1:1$.

$$\frac{V'_L}{V_1} = \frac{V_1 + V_2}{V_1} \tag{5-32a}$$

$$= 1 + \frac{N_2}{N_1} \tag{5-32b}$$

The only possible need for a $1:1$ transformer would be for electrical isolation, a characteristic we destroyed with the autotransformer interconnection. However, if we have a need for a transformer with a voltage ratio *near* $1:1$ (say up to $1:4$) the autotransformer offers attractive benefits.

The principal advantage of the autotransformer is the increased power rating. Also, since the losses remain the same, expressed as a percentage of the new rating, they go down, and correspondingly the efficiency goes up. The machine impedances in per unit drop for similar reasons. Consider:

$$Z_{\text{base old}} = \frac{V_L^2}{S_{\text{rated}}} \tag{5-33a}$$

$$Z_{\text{base new}} = \frac{\left\{ \left[\dfrac{N_1 + N_2}{N_1} \right] \left[\dfrac{N_1}{N_2} V_L \right] \right\}^2}{\dfrac{N_2 + N_1}{N_2} S_{\text{rated}}} \tag{5-33b}$$

$$Z_{\text{base new}} = \left[\frac{N_2 + N_1}{N_2} \right] Z_{\text{base old}} \tag{5-33c}$$

Because the serial impedances (in ohms) do not change, and the base value rises, the per unit value drops, making the transformer "better" (i.e., closer to ideal).

Disadvantages include the loss of electrical isolation between primary and secondary. Also, low impedance is not necessarily good, as we shall see when we study faults on power systems. We have investigated the "step-up" case; if we interchange source and load positions we have the appropriate step-down arrangement. Autotransformers are used in three phase connections, and in voltage control applications, a topic discussed further in section 5-8.

Example 5-5

We have available a 1ϕ 480/120 volt 30 kVA two winding transformer for use on an existing 480 volt system. Can we provide 600 volt service, and if so, what should the connection be and what is the transformer's rating in its auto connection?

Solution

Yes; the proper connection is shown in Figure 5-16(b).

$$\frac{N_1}{N_2} = \frac{480}{120} = 4$$

$$S_{\text{new rating}} = \left(1 + \frac{N_1}{N_2}\right) S_{\text{rated}} \qquad (5\text{-}31)$$

$$= (1 + 4)30$$

$$= 150 \text{ kVA}$$

5-7 The Regulating Transformer

The transformer can be and is used as a control device in the power system. In such an application the transformer is referred to as a regulating transformer. The quantities we wish to control are real and reactive power flow. As we shall prove later these power flows can be controlled by adjusting the voltage phase and magnitude. An arrangement for doing this is shown in Figure 5-17. Study the diagram. The heavy lines at the top represent the power line with input (abc) at the left and output $(a'b'c')$ at the right. Suppose the input system voltages are balanced three phase, positive phase sequence, as shown in Figure 5-18. Note that the primary of the bank is delta connected.

Now examine the secondary windings numbered 1 through 6. The adjustable taps are mechanically ganged together for windings 1, 3, and 5, and separately ganged for windings 2, 4, and 6 (i.e., two independent

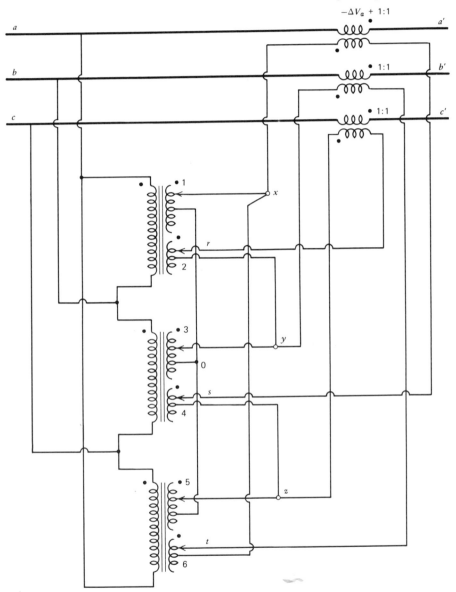

Figure 5-17 Three phase regulating transformer.

adjustments are possible). Notice that all six windings are center tapped, with adjustment to points on both sides of the center tap possible, although our discussion will assume the adjustable taps are in the position shown in the diagram. For simplicity think of all transformers as ideal.

177

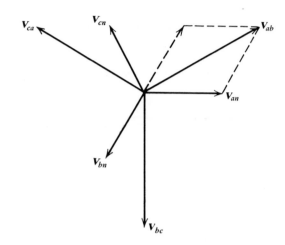

Figure 5-18 System input voltages.

To understand the device's operation, first observe:

$$V_{x0} = \frac{V_m}{\sqrt{3}} \underline{/30°}$$ (5-34a)

$$V_{y0} = \frac{V_m}{\sqrt{3}} \underline{/-90°}$$ (5-34b)

$$V_{z0} = \frac{V_m}{\sqrt{3}} \underline{/150°}$$ (5-34c)

where V_m is adjustable.
Then

$$V_{xz} = V_m \underline{/0°}$$ (5-35a)

$$V_{yx} = V_m \underline{-120°}$$ (5-35b)

$$V_{zy} = V_m \underline{/+120°}$$ (5-35c)

Now observe

$$V_{ry} = V_\phi \underline{/30°}$$ (5-36a)

$$V_{sz} = V_\phi \underline{/-90°}$$ (5-36b)

$$V_{tx} = V_\phi \underline{/150°}$$ (5-36c)

We apply KVL to obtain ΔV_a, ΔV_b, and ΔV_c:

$$\Delta V_a = V_{xz} - V_{sz} \tag{5-37a}$$

$$\Delta V_b = V_{yx} - V_{tx} \tag{5-37b}$$

$$\Delta V_c = V_{zy} - V_{ry} \tag{5-37c}$$

These "ΔV" values are transformed in series in each phase. Finally the output voltages are:

$$V_{a'n} = V_{an} + \Delta V_a \tag{5-38a}$$

$$V_{b'n} = V_{bn} + \Delta V_b \tag{5-38b}$$

$$V_{c'n} = V_{cn} + \Delta V_c \tag{5-38c}$$

Refer to Figure 5-19.

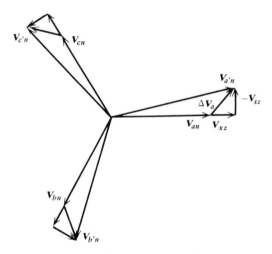

Figure 5-19 System output voltages.

Note that the xyz voltages are in phase with V_{an}, V_{bn}, V_{cn}, whereas the *rst* voltages are 90° out of phase. Adjustment of V_m therefore controls the magnitude of the system voltage, and adjustment of V_ϕ controls the phase, with the controls approximately independent. Typically, V_m and V_ϕ are variable only over a small range.

The regulating transformer is a valuable and practical means of controlling bulk power flow (up to thousands of megawatts). The control actions may be manual or automatic, complete with output sensors and feedback methods. We examine this device to gain appreciation for the possibilities for special transformer applications. There are many variations of such devices; a thorough understanding of basic transformer operation should provide us with the background to analyze any specific case we encounter.

5-8 Transformers with Off Nominal Turns Ratios Modeled in Per Unit

The use of the per unit system has simplified circuit models for transformers by eliminating the ideal transformer as a necessary circuit component. To realize this simplification it was necessary to transform current, voltage, and impedance bases according to the transformer turns ratio. There are situations where this is awkward, and sometimes impossible, to do. Consider two $3\emptyset$ transformers, the first rated $230/13.2\,\text{kV}$, and the second rated $220/12.5\,\text{kV}$, which are connected in parallel on both sides. Suppose we select $225\,\text{kV}$ as the HV voltage base. What should the LV voltage base be?

Transformer $\#1$ requires:

$$V_{LV_{base}} = \left[\frac{13.2}{230}\right](225)$$

$$= 12.91\,\text{kV}$$

Transformer $\#2$ requires:

$$V_{LV_{base}} = \left[\frac{12.5}{220}\right](225)$$

$$= 12.78\,\text{kV}$$

Which is to be used and does it make any significant difference? The answer to the second question is "yes," and to answer the first we must investigate what adjustments must be made in our models when the ratio used to transform per unit basis is *not* equal to the transformer turns ratio. Phase shift, as discussed in section 5-7 for regulating transformers, can be modeled by considering the transformer turns ratio to be complex.

For this development we use a two winding transformer modeled as discussed in section 5-5, except now we start with values *not* in per unit, as shown in Figure 5-20. We define a such that:

$$a = a\underline{/\alpha} \qquad \qquad \blacktriangleright \text{(5-39a)}$$

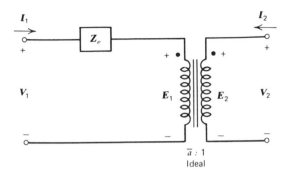

Figure 5-20 Transformer equivalent circuit: all values unscaled.

where

$$a = \frac{N_1}{N_2} = \text{Transformer turns ratio} \qquad (5\text{-}39\text{b})$$

and

$$\alpha = \text{controlled phase shift} \qquad (5\text{-}39\text{c})$$

By definition:

$$E_1 = aE_2 \qquad (5\text{-}40\text{a})$$

If we require that

$$E_1 I_1^* = E_2 I_2^* \qquad (5\text{-}40\text{b})$$

it follows that

$$I_2 = a^* I_1 \qquad (5\text{-}40\text{c})$$

Since the system "turns ratio" is typically given as a voltage ratio, we shall define it as such.

$$b = \frac{\text{Primary line voltage base}}{\text{Secondary line voltage base}} = \text{System turns ratio (real)} \qquad (5\text{-}41\text{a})$$

and select $V_{\varnothing 2_{\text{base}}}$ such that

$$V_{\varnothing 1_{\text{base}}} = b V_{\varnothing 2_{\text{base}}} \qquad (5\text{-}41\text{b})$$

Now convert to per unit by dividing through by $V_{\varnothing 1_{\text{base}}}$:

$$\frac{E_1}{V_{\varnothing 1_{\text{base}}}} = \frac{aE_2}{bV_{\varnothing 2_{\text{base}}}} \qquad (5\text{-}42\text{a})$$

or

$$E_{1_{pu}} = cE_{2_{pu}}$$
$$(5\text{-}42b)$$

where

$$c = a/b \qquad\qquad (5\text{-}42c)$$

$$= c\underline{/\alpha} \qquad\qquad (5\text{-}42d)$$

Similarly

$$I_{1_{pu}} = -\frac{1}{c^*}I_{2_{pu}} \qquad\qquad (5\text{-}43)$$

and

$$Z_{e_{pu}} = \frac{Z_e}{Z_{1_{base}}} \qquad\qquad (5\text{-}44a)$$

where

$$Z_{1_{base}} = \frac{(V_{\emptyset 1_{base}})^2}{S_{1\emptyset_{base}}} \qquad\qquad (5\text{-}44b)$$

Since everything is in per unit on the *system* bases, we drop the "pu" subscript at this point. Refer to Figure 5-21. Apparently our progress has been less than spectacular. Both our original and derived circuits contain ideal transformers, but with different turns ratios, $a:1$ and $c:1$. We do observe that when $a = b$ the ideal transformer "vanishes" from Figure 5-21.

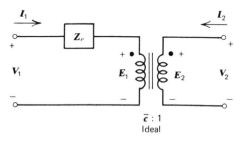

Figure 5-21 Transformer equivalent circuit: all values in per unit on system bases.

The circuit of Figure 5-21 is perfectly correct and appropriate; however, for some of our later work we will find it difficult to deal with large networks that contain ideal transformers. Because of this, we attempt to eliminate the ideal transformer from the circuit of Figure 5-21. From two port network theory, recall the short circuit admittance parameters (i.e., the y parameters):

$$I_1 = y_{11}V_1 + y_{12}V_2 \qquad\qquad (5\text{-}45a)$$

$$I_2 = y_{21}V_1 + y_{22}V_2 \tag{5-45b}$$

Calculate the y parameters for the circuit of Figure 5-21:

$$y_{11} = \frac{I_1}{V_1}\bigg|_{V_2=0} = \frac{1}{Z_e} = Y_e \tag{5-46a}$$

$$y_{12} = \frac{I_1}{V_2}\bigg|_{V_1=0} = \frac{-cV_2Y_e}{V_2} = -cY_e \tag{5-46b}$$

$$y_{21} = \frac{I_2}{V_1}\bigg|_{V_2=0} = \frac{-(V_1Y_e)c^*}{V_1} = -c^*Y_e \tag{5-46c}$$

$$y_{22} = \frac{I_2}{V_2}\bigg|_{V_1=0} = \frac{(E_1Y_e)c^*}{E_1/c} = c^2Y_e \tag{5-46d}$$

Since $y_{12} \neq y_{21}$ we cannot synthesize an equivalent network with ideal R, L, C passive elements. However, for real c, $(c = c)$, an equivalent circuit is possible. The pi network form is the simplest to obtain from the Y parameters. The general pi network appears in Figure 5-22(a). The particular pi network for our problem appears in Figure 5-22(b) for c real. Study

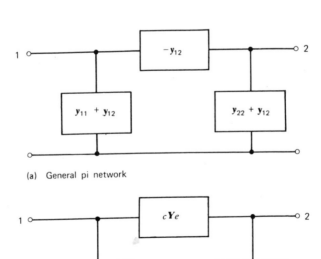

(a) General pi network

(b) Pi equivalent of circuit of Figure 5–21 for c real.

Figure 5-22 Equivalent pi networks.

183

the circuit. If $c = 1$, the shunt elements become open circuits and the series impedance reduces to Z_e. Note that for $c \neq 1$, one of the shunt elements must be negative, which may include negative resistance. We will appreciate this situation more fully in Chapter 7, when we discuss the power flow problem.

Although our development implied a single phase situation, our results are directly applicable to the three phase case. Simply interpret the pi equivalent circuit of Figure 5-22(b) as a sequence network, with Z_e understood to be Z_{00}, Z_{11}, or Z_{22} as in section 5-5. For the zero sequence network, add the $1'$–$1''$ and $2'$–$2''$ gaps as shown in Figure 5-15(a), external to the pi network. The rules for accounting for delta and wye connections remain the same.

Example 5-6

Given a $138/13.8 \, \text{kV}$ $100 \, \text{MVA}$ $3\emptyset$ transformer with $Z_e = Z_{11} = j0.05$ pu (on the transformer base). Determine the proper positive sequence equivalent circuit if the system bases are:

(a) $138/13.8 \, \text{kV}$ $100 \, \text{MVA}$

By inspection:

(b) $169/16.9 \, \text{kV}$ $200 \, \text{MVA}$

$a = b = 10$

$$Z_{e_{sys}} = \frac{Z_{e \, pu_{Trans}} Z_{base_{Trans}}}{Z_{base_{sys}}}$$

$$= \frac{j0.05 \left(\dfrac{138^2}{100} \right)}{\left(\dfrac{169^2}{200} \right)}$$

$$= j0.0667$$

(c) $169/15 \, \text{kV}$ $250 \, \text{MVA}$

$a = 10$

$$b = \frac{169}{15} = 11.27$$

$$c = a/b = 0.8873$$

(circuit diagrams, right column:)

$j.05$

$j.0667$

Example 5-6

$$Z_{\text{epu sys}} = \frac{j0.05(138^2/100)}{169^2/250}$$

$$= j0.0833$$

$$cY_e = -j10.65; \qquad \frac{1}{cY_e} = +j0.0939$$

$$(1-c)Y_e = -j1.353; \qquad \frac{1}{(1-c)Y_e} = +j0.739$$

$$(c^2-c)Y_e = +j1.200; \qquad \frac{1}{(c^2-c)Y_e} = -j0.833$$

5-9 Summary

We have completed another significant step in our creation of a circuit model of the power system. We have developed sequence networks for transformers, summarized in Figures 5-7 and 5-10. The complicated situation involving a three phase line terminated in three wye or delta connected transformer windings is simply modeled by series connections between their corresponding sequence networks. In fact, arbitrarily complicated three phase connections can be modeled with straightforward connections within and between component sequence networks. We will exploit this as we attempt to model entire power systems, but first we must deal with the next major component, the generator.

Bibliography

[1] Adkins, Bernard, *The General Theory of Electrical Machines*, Chapman and Hall, London, 1964.

[2] Elgerd, Olle I., *Electric Energy Systems Theory: An Introduction*, McGraw-Hill Inc., New York, 1971.

[3] Fitzgerald, A. E., Kingsley, Charles, and Kusko, Alexander, *Electric Machinery*, 3rd edition, McGraw-Hill, New York, 1971.

[4] Jones, C. V., *The Unified Theory of Electrical Machines*, Plenum Press, New York, 1967.

[5] Hancock, N. N., *Matrix Analysis of Electrical Machinery*, 2nd edition. Pergamon Press, Oxford, 1974.

[6] Matsch, Leander W., *Electromagnetic and Electromechanical Machines*, Intext, Scranton, 1972.

[7] Neuenswander, John R., *Modern Power Systems*, International Textbook Co., Scranton, Pa., 1971.

[8] O'Kelly, D., and Simmons S., *Introduction to Generalized Electrical Machine Theory*, McGraw-Hill, New York, 1968.

[9] Slemon, G. R., *Magnetoelectric Devices*, Wiley, New York, 1966.

[10] Stagg, Glenn W., and El-Abiad, Ahmed H., *Computer Methods in Power System Analysis*, McGraw-Hill, Inc., New York, 1968.

[11] Stevenson, Jr., William D., *Elements of Power Systems Analysis*, 3rd edition. McGraw-Hill, Inc., New York, 1975.

[12] Weedy, B. M., *Electric Power Systems*, 2nd edition. Wiley, London, 1972.

Problems

5-1. An ideal two winding transformer has $N_1 : N_2 = 3 : 1$. Suppose winding #1 is terminated in a source such that $v_1 = 10(1 - |t|)$; $-1 \le t \le 1$; $v_1 = 0$; $|t| > 1$; and winding #2 terminated in a 1 farad capacitor. Sketch v_1, v_2, i_1, i_2, and ϕ versus t for all t.

5-2. A four winding ideal transformer has windings with the following numbers of turns: $N_1 = 1000$ turns, $N_2 = 500$ turns; $N_3 = 2000$ turns, $N_4 = 3000$ turns. The windings are terminated as follows:
1 Source $V_1 = 1000 \underline{/0°}$ volts
2 Impedance $Z_2 = 10 \, \Omega$
3 Impedance $Z_3 = 40 - j30 \, \Omega$
4 Impedance $Z_4 = j100 \, \Omega$
(a) Draw a circuit diagram, defining all winding currents and voltages.
(b) Solve for all winding currents and voltages.

(c) Calculate the complex power into winding 1.
(d) Calculate the complex power out of winding 2.
(e) Calculate the complex power out of winding 3.
(f) Calculate the complex power out of winding 4.
(g) Does the complex power of (c) equal the summation of that calculated in (d), (e), and (f)? Should it?

5-3. Refer to problem 5-2. Suppose $V_{1_{base}} = 1000$ volts and $I_{1_{base}} = 100$ A. Calculate $V_{2_{base}}, V_{3_{base}}, V_{4_{base}}, I_{2_{base}}, I_{3_{base}}, I_{4_{base}}, S_{1_{base}}, S_{2_{base}}, S_{3_{base}}$, and $S_{4_{base}}$.

5-4. Refer to Problems 5-2 and 5-3. Completely rework Problem 5-2 in per unit.

5-5. Consider the three winding transformer as modeled in Figure 5-5. Transformer data follows:

$$R_1 = R_2 = R_3 = 0$$
$$X_1 = X_2 = X_3 = 0.04$$
$$X_\phi = 5$$
$$R_\phi = \infty$$

If $V_1 = 1/\underline{0°}$ and the secondary and tertiary are terminated in equal impedances $= 1 + j0.2$ solve for I_1. All values in per unit.

5-6. In problem 5-5, suppose $V_{1_{base}} = 4160$ volts and $S_{1\phi_{base}} = 500$ kVA. Compute I_1 in amperes.

5-7. Three identical transformers as described in problem 5-5 are connected as follows:

primary—solid grounded wye
secondary—delta
tertiary—delta

Draw the transformer bank sequence networks.

5-8. Consider the 3ϕ bank connected as in Figure 5-11. Apply balanced three phase negative sequence (cba) to the LV side (ABC) and prove that the connection satisfies the ASA phase shift standard for negative sequence excitation.

5-9. Prove that the connection presented in Figure 5-13 satisfies the ASA standard phase shift convention for positive and negative sequence excitation.

5-10. A 2400:240 volt 1ϕ 100 kVA transformer is to be connected for 2400:2640 volt operation.
(a) Draw a connection diagram.
(b) Compute its new kVA rating.

5-11. The transformer of problem 5-10 is 95% efficient at full load, unity *pf*. What will its full load unity *pf* efficiency be, operating as an auto transformer?·

5-12. Consider the regulating transformer shown in Figure 5-17. Assume balanced *negative* sequence voltage is applied. Work out equations similar to (5-34) through (5-38) for this case.

5-13. A 3ϕ 230 kV $\curlyvee$:23 kVΔ 300 MVA transformer has the following impedance data in per unit on its ratings:
$Z_{11} = Z_{22} = j0.06$
$Z_{00} = j0.10$

Draw the sequence networks.

5-14. Refer to problem 5-13. The transformer is to be used in a system whose base values are 240 kV : 24 kV, 100 MVA. Referring all values to the system bases, again draw the sequence networks.

5-15. Repeat problem 5-14 if the system bases are 200 kV : 24 kV, 100 MVA.

THE SYNCHRONOUS MACHINE

"Be yourself. Especially do not feign affection. Neither be
cynical about love; for in the face of all aridity and
disenchantment, it is as perennial as the grass."

Max Ehrmann, DESIDERATA

We now have circuit models for the lines and transformers that channel electrical energy from the source to load. Therefore we turn our attention to the problem of modeling the electrical source, the generator. First consider the general requirement. We must have a reliable and efficient high power energy conversion device. Recognizing that the energy is usually first encountered in the form of heat (output of a fossil fuel boiler, nuclear reactor, or solar collector) or motion (moving water or air) the problem is to design a practical thermal/mechanical/electrical energy conversion system. The turbine/generator system proves to be an optimum solution in terms of today's technology. By optimum we mean that it is the best known method, considering overall efficiency, economics, reliability, and capacity. Such a system is shown in Figure 6-1.

The scope of this book is confined to the *electrical* power system, and our detailed investigations stop at the shaft between turbine and generator. We recognize the importance of the mechanical/thermal systems, but also

Figure 6-1 Turbine–generator–exciter system. The General Electric generator is rated at 880 MW, 18 kV. (Photograph courtesy of Georgia Power Company.)

acknowledge the practical necessity of ending somewhere. There are many good references on these subjects, usually in mechanical engineering fields. When necessary we will accept and use standard mathematical models without comment.

The electrical synchronous machine is a challenging device to model. Major parts of almost all books on electrical machines are devoted to it; and entire books have been written on the subject (e.g., see [1], [2], and [7] in the end-of-chapter Bibliography). The chief difficulties are:

- Six circuits are involved: three on the rotor, three on the stator, and there is relative motion between rotor and stator.

- The presence of ferromagnetic material will introduce saturation effects.

- When rotor acceleration is considered, the turbine dynamics are relevant.

- When terminal voltage varies, the voltage control system response must be considered.

In considering the above, we must recognize that our treatment of the machine cannot be comprehensive in view of our time and space constraints. What we need is a mathematical model of the device that strikes an intelligent balance between accuracy and simplicity.

6-1 Synchronous Machine Construction

A synchronous machine has two basic parts: the stator and the rotor. The stator is a hollow laminated ferromagnetic cylindrical structure with longitudinal slots located on the inside surface (see Figure 6-2). Placed in these slots are coils that are interconnected to form three separate windings. The stator therefore serves two basic functions: it is a mounting structure for the stator windings, and it provides a low reluctance return path for the rotor magnetic field.

The rotor is a solid ferromagnetic cylindrical structure. Located on the rotor is a winding (the "field") excited from a dc source (the "exciter"). The purpose of this winding is to produce an intense magnetic field. Since the rotor is mounted so that it can rotate within the stator, this field sweeps across the stator windings, inducing a voltage by Faraday's law.

The rotor and stator are designed so that for constant rotor velocity a sinusoidal voltage is generated in each one of the stator windings. These

191

three voltages are identical in frequency and amplitude and are 120°
separated in phase. Therefore, when these three windings are interconnected in a three phase arrangement (because of other practical considerations, the connection is wye), we have formed a practical three phase source.

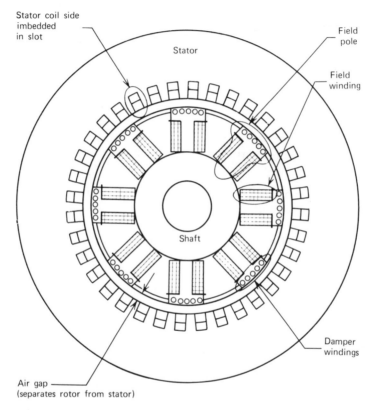

Figure 6-2 An 8 pole synchronous machine shown in cross section.

Consider the relation between rotor speed and stator voltage frequency.
If we spin a magnet inside a coil, the frequency of the induced voltage is
clearly directly related to the angular speed of the magnet. As the magnet
makes one complete "revolution" the voltage waveform goes through one
complete "cycle." In fact the rotor velocity in rev/s *is* the voltage frequency
in cycles/s (Hz). Now visualise a multi-pole rotor.

Consider a point on the stator under the center of a north rotor pole. The
time required for the next rotor north pole center to arrive at this point is

192

time required for the stator voltage to complete one cycle. For a "N" pole rotor this is:

$$\tau_e = \frac{\tau_m}{N/2} \tag{6-1}$$

where

τ_e = electrical period in seconds

τ_m = mechanical period in seconds

$$\therefore \quad f_e = \frac{N}{2} f_m \tag{6-2}$$

where

$$f_e = \frac{1}{\tau_e} \text{ in Hz} \tag{6-3a}$$

$$f_m = \frac{1}{\tau_m} \text{ in rev/s} \tag{6-3b}$$

or

$$\omega_e = \frac{N}{2} \omega_m \tag{6-4a}$$

where

$$\omega_e = 2\pi f_e \text{ in elec. rad/s} \tag{6-4b}$$

$$\omega_m = 2\pi f_m \text{ in mech. rad/s} \tag{6-4c}$$

The terminology for mechanical and electrical quantities is different; we refer to ω_m as "rotor angular velocity" and ω_e as "electrical radian frequency." In the same spirit define:

$$\theta_e = \frac{N}{2} \theta_m \tag{6-5}$$

where

θ_m = spatial angle that defines the rotor position in mechanical radians (or degrees)

θ_e = phase angle that refers to time position in the voltage waveform in electrical radians (or degrees)

Machine speeds are still commonly given in revolutions per minute (*rpm*). We observe:

$$rpm = \frac{\omega_m}{2\pi}(60) \tag{6-6}$$

Eliminating ω_m with (6-4):

$$rpm = \frac{(2/N)\omega_e(60)}{2\pi} \tag{6-7a}$$

$$= \frac{120f_e}{N} \tag{6-7b}$$

Example 6-1

We wish our generator operating frequency to be 60 Hz. Compute the rotor speed (*rpm*) if

(a) $N = 2$

(b) $N = 4$

(c) $N = 24$

Solution

(a) $rpm = \dfrac{120f_e}{N} = \dfrac{120(60)}{2} = 3600$

(b) $rpm = \dfrac{120f_e}{N} = \dfrac{120(60)}{4} = 1800$

(c) $rpm = \dfrac{120f_e}{N} = \dfrac{120(60)}{24} = 300$

There are two basic rotor designs: salient and nonsalient. The salient construction has poles that project from the rotor—that is, they exhibit a narrow air gap under the pole structure and a wider air gap between poles. Figure 6-2 shows a salient pole rotor. The salient pole design is reserved for slow speed machines. Because hydraulic turbines operate more efficiently at slow speeds, generators designed for these applications ("hydro units") generally have many salient poles.

The nonsalient design features a rotor that is circular in cross section (note that this implies a uniform air gap). This design is used in high speed machines with two or four poles. Since steam turbines operate most efficiently at these higher speeds, such generators ("steam units") are used for these applications.

6-2 The Synchronous Machine Equivalent Circuit

Consider a nonsalient two pole synchronous machine whose field winding is electrically supplied by a constant current source. Ignore magnetic saturation and assume a constant rotor velocity. Balanced three phase sinusoidal voltages will be induced in the stator windings. Since the rotor field current is constant, these voltages are independent of stator currents and are correctly modeled as ideal voltage sources. Aoso, the three stator windings are inductively coupled because of their 120° spatial displacement from each other. The symmetry of the machine forces the three self impedances to be equal, as are the mutual impedances. The appropriate equivalent circuit is shown in Figure 6-3. The neutral impedance Z_n is external to the machine.

KVL produces the following equations:

$$E_a = (R_\phi + jX_s + Z_n)I_a + (jX_m + Z_n)I_b + (jX_m + Z_n)I_c + V_a \tag{6-8a}$$

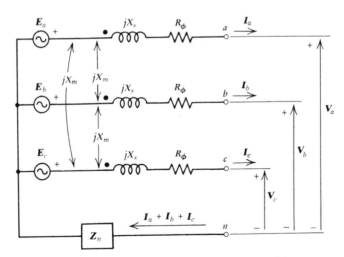

Figure 6-3 Synchronous machine equivalent circuit with constant field current.

$$E_b = (jX_m + Z_n)I_a + (R_\phi + jX_s + Z_n)I_b + (jX_m + Z_n)I_c + V_b \tag{6-8b}$$

$$E_c = (jX_m + Z_n)I_a + (jX_m + Z_n)I_b + (R_\phi + jX_s + Z_n)I_c + V_c \tag{6-8c}$$

where

$$E_a = E \tag{6-9a}$$

$$E_b = a^2 E \tag{6-9b}$$

$$E_c = aE \tag{6-9c}$$

Converting to matrix notation:

$$\begin{bmatrix} E_a \\ E_b \\ E_c \end{bmatrix} = \begin{bmatrix} Z_s & Z_m & Z_m \\ Z_m & Z_s & Z_m \\ Z_m & Z_m & Z_s \end{bmatrix} \begin{bmatrix} I_a \\ I_b \\ I_c \end{bmatrix} + \begin{bmatrix} V_a \\ V_b \\ V_c \end{bmatrix} \tag{6-10a}$$

where

$$Z_s = R_\phi + jX_s + Z_n \tag{6-10b}$$

$$Z_m = jX_m + Z_n \tag{6-10c}$$

More compactly,

$$\tilde{E}_{abc} = [Z_{abc}]\tilde{I}_{abc} + \tilde{V}_{abc} \tag{6-11}$$

The notation is the same as that of Chapter 2. Multiply by $[T]^{-1}$ (refer to equation 2-52):

$$[T]^{-1}\tilde{E}_{abc} = [T]^{-1}[Z_{abc}][T]\tilde{I}_{012} + [T]^{-1}\tilde{V}_{abc} \tag{6-12}$$

We observe that

$$[T]^{-1}\tilde{E}_{abc} = \frac{1}{3}\begin{bmatrix} 1 & 1 & 1 \\ a & a & a^2 \\ 1 & a^2 & a \end{bmatrix} \begin{bmatrix} E \\ a^2 E \\ aE \end{bmatrix} \tag{6-13a}$$

$$= \frac{E}{3}\begin{bmatrix} 0 \\ 3 \\ 0 \end{bmatrix} \tag{6-13b}$$

$$= \begin{bmatrix} 0 \\ E \\ 0 \end{bmatrix} \tag{6-13c}$$

Also recall that

$$[T]^{-1}\tilde{V}_{abc} = \tilde{V}_{012} \tag{2-54}$$

and

$$[T]^{-1}[Z_{abc}][T] = [Z_{012}] \qquad (2\text{-}60)$$

where for this case

$$[Z_{012}] = \begin{bmatrix} Z_{00} & 0 & 0 \\ 0 & Z_{11} & 0 \\ 0 & 0 & Z_{22} \end{bmatrix} \qquad (6\text{-}14a)$$

with

$$Z_{00} = Z_s + 2Z_m = R_\phi + j(X_s + 2X_m) + 3Z_n \qquad (6\text{-}14b)$$

$$Z_{11} = Z_s - Z_m = R_\phi + j(X_s - X_m) \qquad (6\text{-}14c)$$

$$Z_{22} = Z_s - Z_m = R_\phi + j(X_s - X_m) \qquad (6\text{-}14d)$$

Recognize that Z_n is external to the machine. Therefore, define the machine sequence impedances as:

$$Z_0 = Z_{00} - 3Z_n \qquad (6\text{-}15a)$$

$$Z_1 = Z_{11} \qquad (6\text{-}15b)$$

$$Z_2 = Z_{22} \qquad (6\text{-}15c)$$

Accounting for our results in (6-13c), (2-54), and (6-14a) converts equation (6-12) to:

$$\begin{bmatrix} 0 \\ E \\ 0 \end{bmatrix} = \begin{bmatrix} Z_{00} & 0 & 0 \\ 0 & Z_{11} & 0 \\ 0 & 0 & Z_{22} \end{bmatrix} \begin{bmatrix} I_0 \\ I_1 \\ I_2 \end{bmatrix} + \begin{bmatrix} V_0 \\ V_1 \\ V_2 \end{bmatrix} \qquad (6\text{-}16)$$

Synthesizing the sequence circuits from equation (6-16) is simple. The results are shown in Figure 6-4. These networks are quite satisfactory for system applications, and serve as a reasonable compromise between accuracy and simplicity. However, the values for Z_0, Z_1, and Z_2 given in equations 6-15a, b, and c are not practical, particularly the value for Z_2. The reasons for serious discrepancies are twofold:

1. Driving the field with a constant current source is not realistic; in fact, in practice the exciter behaves more like a *voltage* source. This means that unbalances in stator electrical quantities can induce voltages into the field, which in turn can cause the field current to vary. This varying field current causes a varying rotor magnetic field that makes the voltage induced in the stator windings more complex than balanced three phase.

197

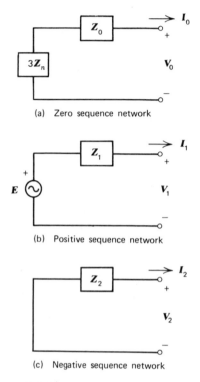

(a) Zero sequence network

(b) Positive sequence network

(c) Negative sequence network

Figure 6-4 Synchronous machine sequence networks.

2. There are other windings on the rotor. Called damper (or amortisseur) windings, their function is to damp out rotor mechanical oscillations about synchronous speed. Structurally, these windings consist of a few shorted turns placed in slots of the rotor and/or parts of the rotor structure itself. Effectively, there are two: one whose magnetic axis aligns with that of the field (i.e., the "d" winding) and the second whose magnetic axis is in quadrature with the field (i.e., the "q" axis). Currents flowing in these windings will also induce voltages into stator windings.

To summarize, the electromagnetic rotor-stator interactions have not been adequately accounted for. If these effects are considered, we find that the networks of Figure 6-4 are satisfactory is we use different values for Z_0, Z_1, and Z_2. Fortunately, the values for Z_0 and Z_2 are routinely published for synchronous generators by the manufacturer. Because these machine parameters are readily available we exert no further effort to derive appropriate values. The positive sequence impedance (Z_1) selection is a matter of engineering judgment. For sinusoidal steady state performance the value suggested in equation (6-15b) is satisfactory. Such a value is also available from the manufacturer, but is not identified as the "positive sequence" impedance. We next deal with machine parameters as supplied by the manufacturer.

6-3 Synchronous Machine Electrical Transient Performance

An important class of problems exist that require a model of the machine operating under suddenly switched conditions. The most important of these problems are described as fault studies. A "fault" is a suddenly applied short circuit. Our knowledge of circuit theory would suggest that we now must abandon our sinusoidal steady state approach in favor of a more general transient treatment. Rigorously this is correct, but experimental and field tests made over a number of years, and for a variety of conditions, show that the general electrical behaviour of the machine is still basically ac. The students' first reaction to the term "transient ac" may be to think that the author is playing without a full deck. Nonetheless, the term accurately describes the conventional analytical approach to the problem, and it is reasonable. Let us consider an experimental approach.

Imagine a synchronous machine running at constant angular speed and with its field excited by a constant voltage source. Now consider a 3ϕ short circuit suddenly applied to the stator terminals, which were previously opened. The recorded current waveforms appear in Figure 6-5. Observe that initially the currents are considerably larger than they are several cycles later (the frequency of the alternating current shown is 60 Hz). If you look closely, you will also observe that each waveform is "offset," that is, not symmetric about the time axis. This latter effect is predicted by the straightforward solution to a simple R-L circuits problem. Consider the following example.

Example 6-2

In the circuit of Figure 6-6, $e = V_m \sin(\omega t + \alpha)$, and the switch is closed at $t = 0$. Solve for $i(t)$.

199

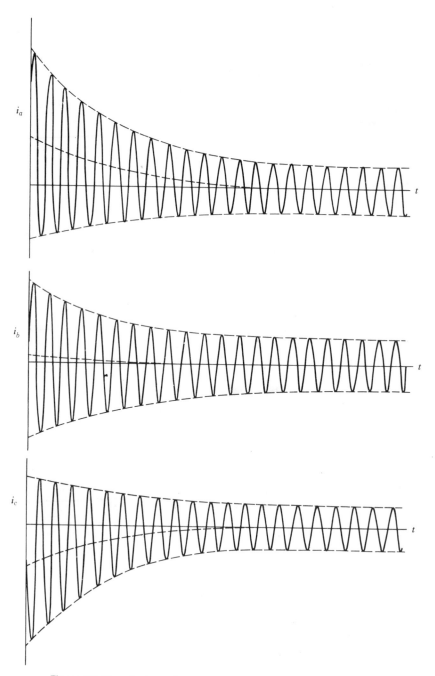

Figure 6-5 Transient synchronous machine short circuit currents.

Example 6-2

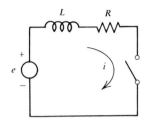

Figure 6-6 *R-L* circuit.

Solution

KVL produces

$$e = L\frac{di}{dt} + iR \tag{6-17}$$

The characteristic equation is

$$Lp + R = 0; \qquad p = \frac{d}{dt}(\ ) \tag{6-18a}$$

$$p = -R/L \tag{6-18b}$$

The source free solution is

$$i_f = K e^{-R/Lt} \tag{6-19}$$

The forced solution is:

$$i_s = \frac{V_m}{Z}\sin(\omega t + \alpha - \phi) \tag{6-20a}$$

where

$$Z = \sqrt{R^2 + (\omega L)^2} \tag{6-20b}$$

and

$$\phi = \tan^{-1}\left[\frac{\omega L}{R}\right] \tag{6-20c}$$

The total solution is:

$$i = i_f + i_s \tag{6-21a}$$

$$= K e^{-(R/L)t} + \frac{V_m}{Z}\sin(\omega t + \alpha - \phi) \tag{6-21b}$$

We observe that $i(0^-) = i(0^+) = 0$:

$$\therefore \quad K = -\frac{V_m}{Z} \sin(\alpha - \phi) \tag{6-21c}$$

Eliminating K:

$$i = \frac{V_m}{Z} [\sin(\omega t + \alpha - \phi) - \sin(\alpha - \phi)\, e^{-(R/L)t}] \tag{6-21d}$$

In the above result, the $\sin(\alpha - \phi)\, e^{-(R/L)t}$ term is the so-called "offset." The angle α relates to where in the sinusoidal cycle the switch is closed. Note that if $\alpha = \phi$ the offset term does not appear. If the switch is thrown at random, we observe unpredictable amounts of offset. Since in a practical situation we will never know when in the sinusoidal cycle a fault will occur, we can never predict how much offset we will have. Also, in the three phase case, each phase has in general a different offset.

The machine transient current i contains such an offset term, although it is more complex since the envelope of the sinusoidal term varies exponentially. Because of its unpredictability let us remove (subtract) the offset term. The resultant machine current waveform, shown in Figure 6-7,

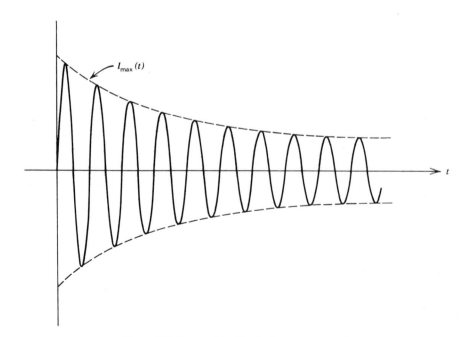

Figure 6-7 Symmetrical instantaneous current.

is referred to as the symmetrical instantaneous current. The dashed envelope shown is referred to as the symmetrical maximum current waveform. Note that over any one cycle the current is approximately sinusoidal. If we prefer to use rms values, it is reasonable to divide the maximum values of Figure 6-7 by $\sqrt{2}$. We now plot the resultant waveform in Figure 6-8a and refer to it as the symmetrical rms current, $I(t)$.

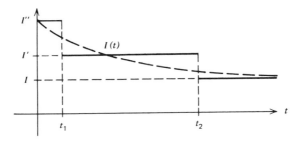

(a) Current $I(t)$ vs time with stepped approximation

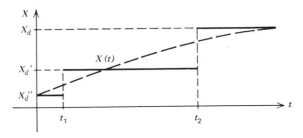

(b) Reactance $X(t)$ vs time with stepped approximation

Figure 6-8 Transient ac quantities.

If we decide to use sinusoidal steady state circuit analysis methods, we are forced into an extremely unorthodox situation. Suppose we wish to calculate $I(t)$, using the circuit of Figure 6-9. The voltage, $E(t)$, or the impedance, $Z(t)$, or a combination of both, must be *time varying*. In a practical machine $X \gg R$, so that $Z \cong X$ with insignificant error. We choose to deal with the situation by keeping E constant and allowing $X(t)$ to vary. Therefore, $X(t)$ should vary as shown in Figure 6-8b. For most practical applications it is sufficient to let $X(t)$ vary discretely (three steps) instead of continuously. This approach is considerably simpler, and can be used to obtain good engineering results.

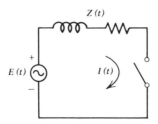

Figure 6-9 Transient ac circuit model.

Refer to Figure 6-8a. Note that the continuously varying current $I(t)$ can be approximated with the three discrete current levels I'', I', and I. They are identified as:

$I'' =$ Subtransient current; $0 \leq t \leq t_1$. $\qquad$ (6-22a)

$I' =$ Transient current; $t_1 \leq t \leq t_2$. $\qquad$ (6-22b)

$I =$ Steady state current; $t \geq t_2$. $\qquad$ (6-22c)

The related time intervals are likewise referred to as subtransient, transient, and steady state. The corresponding values of reactance are

$$X''_d = \frac{E}{I''} = \text{direct axis subtransient reactance} \qquad (6\text{-}23a)$$

$$X'_d = \frac{E}{I'} = \text{direct axis transient reactance} \qquad (6\text{-}23b)$$

$$X_d = \frac{E}{I} = \text{direct axis synchronous reactance} \qquad (6\text{-}23c)$$

where

E is the open circuit phase voltage

The terminology and symbology for X''_d, X'_d, and X_d is standard throughout the industry (refer to [4] in the end-of-chapter Bibliography). The "d" subscript refers to the direct axis; there are similar quadrature axis values available, denoted as X''_q, X'_q, and X_q. Our approach here will not permit us to properly account for saliency. For a more rigorous development see [1] in the end-of-chapter Bibliography.

For most machines t_1 extends to about 2 cycles at 60 Hz (33.3 ms), and t_2 to about 30 cycles (0.50 s). The determination of I'' and I is clear, but the appropriate intermediate value I' is not. Again, we will not develop this detail. Fortunately, all six parameters (X''_d, X'_d, X_d, X''_q, X'_q, and X_q) are routinely available for commercial machines and we write equation (6-23) more for educational than for utility reasons.

We now turn back to the results of section 6-2. Recall we have available values for Z_0 and Z_2. The difficulty was to obtain an appropriate positive sequence value. The case studied here, although of a transient nature, nonetheless involved a *balanced* three phase termi nation. As such, the circuit of Figure 6-9 is in effect a positive sequence network for the machine. Therefore we have available three values:

$$Z_1'' = R_\phi + jX_d'' \tag{6-24a}$$

$$Z_1' = R_\phi + jX_d' \tag{6-24b}$$

$$Z_1 = R_\phi + jX_d \tag{6-24c}$$

Which of these is appropriate is a matter for engineering judgment. For fault study applications, for example, Z_1'' is appropriate. As we study specific power system problems, we discuss which value is most reasonable.

6-4 The Turbine-Generator-Exciter System

Generator operation can be understood best by considering it as the basic component in a system involving the external electrical system, the prime mover, and exciter. Study such a system, illustrated in Figure 6-10. The turbine produces torque (T_m) in the direction of rotation. The torque T_m is directly controlled by the steam flow to the turbine, and therefore by the main steam valve position. This torque is opposed by an equal electromagnetic torque (T_e), physically created by the interaction between the rotor and stator magnetic fields. It is the associated power, $P_e(P_e = T_e\omega_m)$, that is converted into the electrical form. The governor senses shaft speed and attempts to correct any deviation from ω_s by adjusting the steam valve position. Here our concern is limited to the steady state constant speed balanced 3$\emptyset$ case so that $T_m = T_e$.

The internal positive sequence voltage E (refer back to Figure 6-4b) is controlled by the exciter. Note that the generator output voltage is sensed by the potential transformer (PT) and fed back to the regulator. The regulator compares this feedback voltage with a desired reference value and uses any error to control the exciter, which in turn reacts to change E in such a way that the voltage error is reduced or eliminated.

Two basic control actions are therefore apparent. The power converted from mechanical to electrical form (P_e) is controlled by the steam valve setting and the terminal voltage is controlled by the regulator setting. We wish to examine this system analytically without getting overwhelmed by

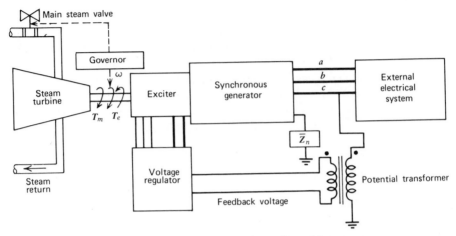

Figure 6-10 Turbine-generator-exciter system.

the intricacies of any of its particular components. Also, we postpone the more complex problem of transient performance until Chapter 12. We will think "quasi-steady state": that is, we have the option of adjusting the steam valve or regulator setting, but can only calculate steady state conditions at each setting.

We limit our interest to only the balanced three phase steady state situation. Therefore, only the positive sequence networks of the generator and system are involved. For simplicity, and practicality, circuit resistance will be ignored; also neglect other machine losses. Think of the system positive sequence network as being modeled by an ideal source as shown in Figure 6-11. Also interpret the following equations in per unit. Suppose we compute the complex power delivered to the system. First apply KVL:

$$E = jX_dI + V \tag{6-25}$$

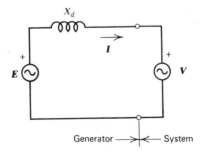

Figure 6-11 Simplified positive sequence network.

206

Now write:

$$S = VI^* \tag{6-26a}$$

$$= V\left[\frac{E - V}{jX_d}\right] \tag{6-26b}$$

Define E and V as:

$$V = V\underline{/0^\circ} = \text{Generator terminal voltage} \tag{6-27a}$$

$$E = E\underline{/\delta} = \text{Generator internal voltage} \tag{6-27b}$$

where

$$\delta = \text{power angle}\dagger \tag{6-27c}$$

so that equation (6-26b) becomes:

$$S = \frac{VE}{X_d}\underline{/90^\circ - \delta} - j\frac{V^2}{X_d} \tag{6-28a}$$

$$= \frac{VE}{X_d}\sin\delta + j\left[\frac{VE}{X_d}\cos\delta - \frac{V^2}{X_d}\right] \tag{6-28b}$$

Therefore:

$$P = Re[S] \tag{6-29a}$$

$$= \frac{VE}{X_d}\sin\delta \tag{6-29b}$$

and

$$Q = \text{Im}[S] \tag{6-30a}$$

$$= \frac{VE}{X_d}\cos\delta - \frac{V^2}{X_d} \tag{6-30b}$$

We now realize an extremely important fact. The P of equation (6-29b) is P_e, the power converted from mechanical to electrical form. Equations (6-29) are very useful in explaining generator reaction to turbine and exciter control. Let us first consider real power. Imagine adjusting *only* the main steam valve, *opening* it further. Clearly, T_m, and thus P_m, increase. Consequently, P_e and P increase. Since E, V, and X_d are constant, δ must increase. Study Figure 6-12. An increase from P_0 to P_1 causes a corresponding increase in δ from δ_0 to δ_1. Obviously there is a limit to how far we can

---

† Oversimplified when compared to the power angle as used in the literature.

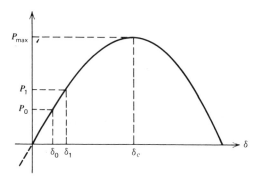

Figure 6-12 P versus δ.

increase P. That limit P_{max} is referred to as the steady state stability limit and is:

$$P_{max} = \frac{EV}{X_d} \tag{6-31a}$$

The corresponding δ_c is the critical power angle:

$$\delta_c = 90° \tag{6-31b}$$

If P_m is further increased, the· generator could no longer remain synchronized with the system. To avoid the problem, generators are normally operated at fairly small power angles ($\sim 20°$). The conclusion is clear; there is a strong interrelation between the steam valve setting, real power flow, and δ.

Now consider the second control action, namely, changing the voltage regulator setting. Suppose E increases. If we examine equation (6-30b) and assume V, X_d, and δ are constant, we observe that Q increases. Actually this is somewhat oversimplified, since δ also changes, although not very much. The general conclusion, however, that increases in the regulator settings (and therefore E) will increase Q flow out of the generator is still valid. We observe a strong interrelation between regulator setting, reactive power flow, and E. An example problem should help our understanding.

Example 6-3

The generator of Figures 6-10 and 6-11 operates at rated terminal conditions with a power factor of 0.8 lagging. The reactance $X_d = 0.7$ pu (on the generator ratings).
(a) Find P, Q, E, and δ. Draw the phasor diagram.

Example 6-3

(b) The steam valve is opened further so that P increases 20%. Reevaluate P, Q, E, and δ. Draw the new phasor diagram.

(c) The system is restored to the conditions of (a). The excitor is adjusted to raise E 20%. Reevaluate P, Q, E, and δ. Draw the new phasor diagram.

Solution

Let us use the subscript "0" to denote initial values and "1" to denote adjusted values.

(a) $V = 1\underline{/0}$

$$\theta_0 = \cos^{-1}(0.8) = 36.9°$$

$$\therefore \quad I_0 = 1\underline{/-36.9°} = 0.8 - j0.6$$

$$E_0 = jX_dI_0 + V \tag{6-25}$$

$$= j0.7(0.8 - j0.6) + 1$$

$$= 1.42 + j0.56$$

$$= 1.53\underline{/21.5°}$$

$$E_0 = 1.53$$

$$\delta_0 = 21.5°$$

$$P_0 = \frac{E_0V}{X_d}\sin\delta_0 \tag{6-29b}$$

$$= \frac{(1.53)(1.0)}{0.7}\sin(21.5°)$$

$$= 0.800$$

$$Q_0 = \frac{E_0V}{X_d}\cos\delta - \frac{V^2}{X_d} \tag{6-30b}$$

$$= \frac{(1.53)(1.0)}{0.7}\cos(21.5°) - \frac{(1.0)^2}{0.7}$$

$$= 0.60$$

Consult phasor diagram shown in Figure 6-13a.

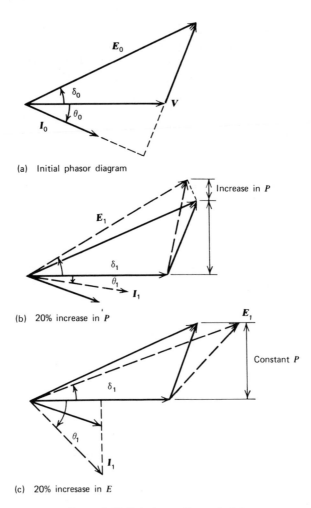

(a) Initial phasor diagram

(b) 20% increase in P

(c) 20% incresase in E

Figure 6-13 Solution to Example 6-3.

(b) $P_1 = 1.2 P_0 = 1.2(0.8) = 0.96$ (20% increase)

$E_1 = E_0 = 1.53$ (No change)

$\therefore \quad \delta_1 = \sin^{-1}\left(\dfrac{P_1 X_d}{E_1 V}\right)$

$= \sin^{-1}\left(\dfrac{0.96(0.7)}{(1.53)(1.0)}\right) = 26.1°$ (21% increase)

$$Q_1 = \frac{E_0 V}{X_d} \cos \delta_1 - \frac{V^2}{X_d} \qquad \text{(6-30b)}$$

$$= \frac{1.53(1.0)}{0.7} \cos(26.1°) - \frac{(1.0)^2}{0.7}$$

$$= 0.535 \ (11\% \text{ decrease})$$

Consult the phasor diagram shown in Figure 6-13b.

(c) $P_1 = P_0 = 0.8$ (No change)

$$E_1 = 1.2 E_0 = 1.2(1.53) = 1.84 \ (20\% \text{ increase})$$

$$\delta_1 = \sin^{-1}\left(\frac{P_1 X_d}{E_1 V}\right)$$

$$= \sin^{-1}\left(\frac{0.8(0.7)}{1.84(1.0)}\right) = 17.8° \ (17\% \text{ decrease})$$

$$Q_1 = \frac{E_1 V}{X_d} \cos \delta - \frac{V^2}{X_d} \qquad \text{(6-30b)}$$

$$= \frac{(1.84)(1.0)}{0.7} \cos(17.8°) - \frac{(1.0)^2}{0.7}$$

$$= 1.07 \ (44\% \text{ increase})$$

Consult the phasor diagram shown in Figure 6-13c.

6-5 Operating Limits on Synchronous Generators

We are interested in the synchronous generator as a power source, and therefore are concerned with limits on its power delivery capabilities. The power system is normally operated as a constant voltage system; that is, the voltage at any point in the system is held to within about ±10% of some nominal value. At fixed voltage, power is proportional to current and we turn our attention to current capacity.

For a given stator winding conductor size there must be an associated winding resistance. If a current I flows in the winding, there is an attendant $I^2 R$ power loss (we are not considering the superconducting case here). This energy must either be removed or it will raise the temperature of the conductor and its immediate environment. A major consideration in machine design is the removal of this waste heat. However, no matter how efficient such cooling systems are, there exists a current I that, if exceeded indefinitely, will cause stator winding temperatures to reach levels high enough to permanently damage the machine. Such a current is referred to as the rated current, which may be considered as an upper limit for a short time without damaging the windings (the shorter the time, the greater the excess allowed). Associated with this value, I_{rated}, along with V_{rated}, is a power rating:

$$S_{3\phi rated} = V_{line\ rated} I_{line\ rated} \sqrt{3} \tag{6-32}$$

where the above quantities are in SI units. If we define $S_{3\phi}$ as the machine output complex power, operation is constrained to:

$$|S_{3\phi}| \le S_{3\phi\ rated} \tag{6-33}$$

If we convert equation (6-33) to per unit with the machine's ratings taken as base values, we get:

$$|S| \le 1 \tag{6-34}$$

Let us interpret (6-34) geometrically in the $S(P, Q)$ plane. It is the region inside the unit circle (*abcde*) shown in Figure 6-14. Operation is constrained to within this circle due to stator winding heating.

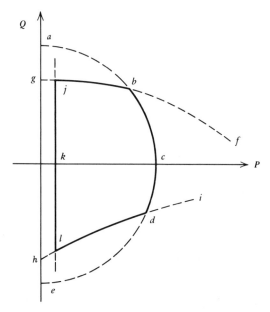

Figure 6-14 Operating characteristic for a synchronous generator.

Consider the upper region of this curve (arc ab): high Q, low P. Thinking in terms of our circuit model (Figure 6-11), I is considerably out of phase with V, and lagging. The phasor jX_dI is therefore close to in phase with V and consequently requires a large value for E. An example will illustrate.

Example 6-4

A generator is modeled as shown in Figure 6-11 and has $X_d = 1.2$ pu. Calculate the required E at rated conditions for unity power factor and zero power factor lagging. Compare.

Solution

Unity power factor: $I = 1 + j0$

$$E = jX_dI + V$$
$$= j1.2(1) + 1 = 1.56\underline{/50.2°}$$
$$\therefore \quad E = 1.56$$

Zero power factor lagging: $I = 0 - j1$

$$E = jX_dI + V$$
$$= j1.2(-j) + 1 = 2.20\underline{/0°}$$
$$\therefore \quad E = 2.20$$

A substantially larger E is required in the lagging case. Recall that E is generated by the rotor field, which in turn is created by the rotor field current, I_f. Because of $I_f^2 R_f$ heating in the rotor windings, I_f cannot be increased indefinitely. Therefore, there is a practical upper limit on E. We could of course design the machine to provide the appropriate E even for the zero power factor lagging case. However, this condition is rarely required (who needs a generator that delivers no P?), and does not justify the added expense of allowing for it. Therefore, as we move up the P, Q curve from c to b, we require increasingly larger values of E until at point b we reach E_{max}, corresponding to I_{fmax}. From this point on, E is no longer adjustable but fixed at E_{max}.

Now what is the operating constraint curve? To answer this question recall equation (6-28b) with $E = E_{max}$:

$$S = \frac{VE_{max}}{X_d} \sin \delta + j\left[\frac{VE_{max}}{X_d} \cos \delta - \frac{V^2}{X_d}\right] \tag{6-28b}$$

This is the equation of a circle in the S plane with center $(0, -V^2/X_d)$ and radius VE_{max}/X_d. A segment of this circle *jbf* appears in Figure 6-14. Operation is constrained to within the circle because of rotor field winding heating.

Now consider the lower region of the circle *abcde, cde*. Note that Q is negative implying a leading power factor situation. As we move from c to d, E decreases, which seems to be desirable. Suppose we examine a specific situation with an example.

Example 6-5

Continue with example 6-4 by calculating E for a power factor of 0.553 leading at rated terminal conditions. Comment.

Solution

$$\theta = \cos^{-1}(0.553) = 56.4°$$

$$\therefore \quad I = 1/\underline{56.4°}$$

$$E = jIX_d + V$$

$$\quad = j1.2(1/\underline{56.4°}) + 1$$

$$\quad = j0.664$$

$$\quad = 0.664/\underline{90°}$$

$$\therefore \quad E = 0.664; \qquad \delta = 90°$$

The $\delta = 90°$ value indicates that operation at this power factor is completely unacceptable. We are on the verge of instability; if any additional P_m is suddenly supplied by the prime mover, we will lose synchronism.

We realize that as we moved from c to d, E decreased, but δ increased. How far is it reasonable to allow δ to increase? If we conclude that it is prudent to keep a 10% reserve on real power delivery capability, $\delta_{max} = \sin^{-1}(0.9) = 64.2°$. For a given δ_{max} we can calculate P and Q as E varies:

$$P = \frac{EV}{X_d} \sin \delta_{max} \tag{6-35a}$$

$$Q = \frac{EV}{X_d} (\cos \delta_{max}) - \frac{V^2}{X_d} \tag{6-35b}$$

and plot the results in the P, Q plane. Such a curve is shown in Figure 6-14 as contour *hdi*. To avoid steady state instability, generator operation is constrained to the region above the curve *hdi*.

For some types of units it is undesirable to allow the real power output to go completely to zero, unless we intend to take the unit off line. An example would be a fossil fuel steam plant, where constant boiler temperatures are required to prevent slagging problems. Therefore, we require that:

$$P \geq P_{min} \tag{6-36}$$

Equation (6-36) interpreted geometrically is the region to the right of the line *jkl*, shown in Figure 6-14.

Generator operation is therefore confined to the P, Q interior region bounded by *jbcdlk* as shown in Figure 6-14. For a specific machine these curves can be accurately calculated and steady state operation limited to within the enclosed region, either by operator, or automatically, or a combination of both. We can simplify the situation by fitting the largest possible rectangle inside the operating characteristic, creating the region shown in Figure 6-15. Such a characteristic is particularly useful for

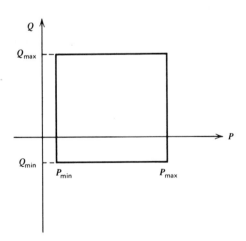

Figure 6-15 Simplified generator operating characteristic.

computer control applications. We need only provide four numbers, Q_{min}, Q_{max}, P_{min}, and P_{max}, to define the operating region.

6-6 Summary

The equivalent circuits that we shall use to model the synchronous machine are shown in Figure 6-4. These are sequence networks; the voltages and currents are symmetrical components. Phase values must be calculated from the transformation equations (2-53) and (2-55). For balanced three phase sinusoidal steady state situations, only the positive sequence network is needed. Impedance values are routinely available from the manufacturers; our interest is directed more toward the use of such values than their theoretical development.

Transient machine performance was modeled by a technique referred to as transient ac. We found that we could still use the circuits of Figure 6-4, except that we must use different values for the positive sequence impedance.

We found that real and reactive power flow was controlled by δ and E, which are in turn controlled by turbine and exciter settings. There are definite limits to the P and Q values that a generator can supply and we examined those.

It is important to take note of simplifications we have made, which can produce misleading results. To produce the intense magnetic fields required, synchronous machines are constructed of ferromagnetic materials. Such materials are subject to magnetic saturation effects. The effect on equivalent circuit parameters is that inductance is not constant. Although power losses in the machine are small, there are nonzero electrical rotor and stator, mechanical, and magnetic losses.

The most worrisome and mysterious of our approximations is the attempt to deal with transient phenomena with sinusoidal steady state (ac) concepts. When we consider the machine in general we find that the machine's state equations are considerably more complicated than the constrained situation presented in section 6-2 would suggest. If we are to solve the transient problem in more generality, the simplified models presented in this chapter are not adequate. Also remember that saliency is not properly accounted for. The literature records some brilliant engineering work that surmounts these difficulties. To do the subject justice, a formal course in the synchronous machine is recommended to the serious student.

In spite of these approximations, the circuit models presented here give good engineering results when appropriately applied to basic power systems problems. They are commonly used throughout the industry and provide good insight into normal synchronous machine behavior. We extend our work on synchronous machines as required in later chapters.

Bibliography

[1] Adkins, Bernard, *The General Theory of Electrical Machines*, Chapman and Hall, London, 1964.

[2] Concordia, Charles, *Synchronous Machines—Theory and Performance*, Wiley, New York, 1951.

[3] Elgerd, Olle I., *Electric Energy Systems Theory: An Introduction*, McGraw-Hill, Inc., New York, 1971.

[4] Fitzgerald, A. E., Kingsley, Charles, and Kusko, Alexander, *Electric Machinery*, 3rd edition. McGraw-Hill, New York, 1971.

[5] Jones, C. V., *The Unified Theory of Electrical Machines*, Plenum Press, New York, 1967.

[6] Hancock, N. N., *Matrix Analysis of Electrical Machinery*, 2nd edition. Pergamon Press, Oxford, 1974.

[7] Kimbark, E. W., *Power System Stability*, vol. 3. Synchronous Machines, Wiley, New York, 1956.

[8] Matsch, Leander W., *Electromagnetic and Electromechanical Machines*, Intext, Scranton, Pa., 1972.

[9] Slemon, G. R., *Magnetoelectric Devices*, John Wiley and Sons Inc., New York, 1966.

[10] Stagg, Glenn W., and El-Abiad, Ahmed H., *Computer Methods in Power System Analysis*, McGraw-Hill, Inc., New York, 1968.

Problems

6-1. A hydro turbine operates at peak efficiency in the speed range 235 to 215 rev/min. How many poles should an appropriate 60 Hz electrical generator have?

6-2. A steam turbine runs at 1800 rev/min and develops a maximum torque of 2653 knm. Compute the corresponding power in megawatts.

6-3. In Figure 6-3, $E_a = 1000\underline{/30°}$ volts, $R_\theta = 0$, $X_s = 20\Omega$, $X_m = -9\Omega$, and

$Z_n = j2 \ \Omega$. Draw the sequence networks shown in Figure 6-4 with all values shown.

6-4. Terminate the machine of problem 6-3 as follows: phase a, short circuit ($V_a = 0$); phases b and c, open circuit ($I_b = I_c = 0$). Calculate I_a, V_b, and V_c using equations (6-8).

6-5. In the circuit of Figure 6-6, $e = 100 \sin(377t)$, $\omega L = 10$ ohms, and $R = 1$ ohm. Sketch $i(t)$ versus t for 8 cycles ($t_0 \le t \le t_0 + 0.133$ s), switch closed at t_0, where:
(a) $t_0 = 0$
(b) $t_0 = 3.902$ ms
(c) $t_0 = 7.804$ ms

6-6. A 300 MVA 13.8 kV 3∅ wye connected 60 Hz generator is adjusted to rated voltage at open circuit. A balanced 3∅ fault is applied to the terminals at $t = 0$. After the raw data ($i(t)$) is "massaged" the symmetrical rms transient current $I(t)$ is obtained, producing $I(t) = 10^4[1 + e^{-t/\tau_1} + 6e^{-t/\tau_2}]$ amperes,
where $\tau_1 = 200$ ms; $\tau_2 = 15$ ms.
(a) Plot $I(t)$ versus t for $0 \le t \le 500$ ms.
(b) Compute X_d'' and X_d in ohms per phase.
(c) Convert X_d'' and X_d to per unit on the machine's ratings.

6-7. Consider a synchronous generator connected to a system modeled as an ideal 3∅ source. Refer to Figure 6-11. $V = 1\underline{/0°}$, $X_d = 1.0$; $E = 1.5\underline{/\delta}$. For $T_m = 0$, 0.5, and 1.0, calculate P, Q, I, and δ. Draw a complete phasor diagram for each case. All values are in per unit. $\omega_{base} = \omega_{sync}$ and $T_{base} = P_{base}/\omega_{base}$.

6-8. Repeat problem 6-7 with the following modifications. Hold T_m at 0.5 and vary E from 0.8 to 1.3 to 1.8.

6-9. A lossless synchronous machine with $X_d = 0.8$ is driven by a prime mover capable of absorbing or delivering power. The machine is terminated electrically in an ideal 3∅ source ($V = 1\underline{/0}$). If $E = 1.5\underline{/\delta}$, sketch P versus δ as $-1 \le T_m \le 1$. All values are in per unit.

6-10. A lossless synchronous machine with $X_d = 1.1$ has $V = 1\underline{/0}$. The following operating limits are to be observed: $E_{max} = 1.8$; $P_{min} = 0.1$; $S_{max} = 1.0$; and $\delta_{max} = 58°$. Accurately plot on graph paper the operating characteristic that corresponds to Figure 6-14.

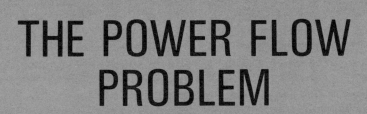

THE POWER FLOW PROBLEM

"Take kindly the counsel of the years, gracefully
surrendering the things of youth."

Max Ehrmann, DESIDERATA

Power system electrical performance in the normal balanced three phase sinusoidal steady state condition is of primary importance to the power system engineer. Given a specified load condition, the basic questions to be answered are:

- What are the line and transformer loads throughout the system?

- What are the voltages throughout the system?

We are particularly interested in answering these questions as we evaluate proposed changes to an existing system such as:

- New generation sites

- Projected load growth

- New transmission line locations

Thinking of the power system as modeled by an electrical network it would seem that the straightforward approach would be to use either conventional loop or nodal analysis methods to solve for voltages and currents of interest. This direct approach is not possible because the loads are known as complex powers, not impedances; also the generators cannot be modelled as "voltage sources" in the circuit analysis sense, but behave more like "power sources." The problem is basically that of solving $2n$ nonlinear algebraic equations in $2n$ unknowns for an n bus system and therefore requires numerical analysis techniques for its solution.

Since the system is in its balanced 3ϕ ac mode, only the positive sequence network is required. We work in per unit throughout this chapter and all equations should be correspondingly interpreted.

7-1 A Statement of the Power Flow Problem

Consider the ith bus of an n bus power system, as depicted in Figure 7-1. We realize:

$$S_{Gi} = S_{Li} + S_{Ti} \qquad (7\text{-}1)$$

where

$S_{Gi} = 3\phi$ complex generated power flowing *into* the ith bus

$S_{Li} = 3\phi$ complex load power flowing *out of* the ith bus

$S_{Ti} = 3\phi$ complex transmitted power flowing *out of* the ith bus

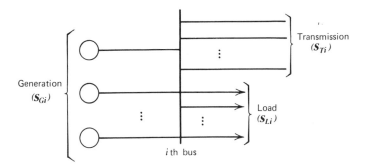

Figure 7-1 Complex powers at *i*th bus.

Since

$$S_{Gi} = P_{Gi} + jQ_{Gi} \qquad\qquad\qquad (7\text{-}2a)$$

$$S_{Li} = P_{Li} + jQ_{Li} \qquad\qquad\qquad (7\text{-}2b)$$

$$S_{Ti} = P_{Ti} + jQ_{Ti} \qquad\qquad\qquad (7\text{-}2c)$$

it follows that:

$$P_{Gi} = P_{Li} + P_{Ti} \qquad\qquad\qquad (7\text{-}3a)$$

and independently that:

$$Q_{Gi} = Q_{Li} + Q_{Ti} \qquad\qquad\qquad (7\text{-}3b)$$

Mind your P's and Q's!—they represent real and reactive powers. These total to six variables per bus, related by two independent equations (7-3a and b). For an n bus system we have $2n$ equations involving $6n$ variables. Power flow calculations are made at specified load conditions. Therefore P_{Li} and Q_{Li} are given, or known. This leaves us with four variables per bus: P_{Gi}, Q_{Gi}, P_{Ti}, and Q_{Ti}. The generation terms are straightforward and will cause us no problems; however, the transmission terms are a different matter. Let us concentrate on the complex transmitted power S_{Ti}.

Consider an n bus power system arranged in the manner suggested by Figure 7-2. Note that generation and loads are located externally to the boxed-in transmission network. The network in the box is passive, and can be treated with n port network theory. We can use either the impedance or admittance approach; the admittance method proves more suited to our problem. We write:

$$\tilde{I}_{\text{bus}} = [Y_{\text{bus}}]\tilde{V}_{\text{bus}} \qquad\qquad\qquad (7\text{-}4)$$

221

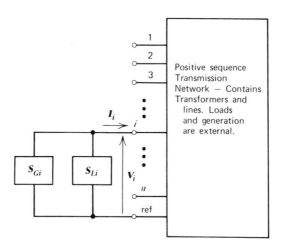

Figure 7-2 General *n* bus power system arranged for power flow analysis.

where

$\tilde{I}_{\text{bus}}$ is an $n \times 1$ vector with general entry I_i

$\tilde{V}_{\text{bus}}$ is an $n \times 1$ vector with general entry $V_i = V_i\underline{/\delta_i}$

$[Y_{\text{bus}}]$ is an $n \times n$ matrix with entries

$Y_{ii} = Y_{ii}\underline{/\gamma_{ii}} = $ short circuit driving point admittance at port *i*.

$Y_{ij} = Y_{ij}\underline{/\gamma_{ij}} = $ short circuit transfer admittance between ports *i* and *j*.

The vector $\tilde{V}_{\text{bus}}$ is the easiest to understand; its entries are simply the bus voltages. The vector $\tilde{I}_{\text{bus}}$ is a type of "source" current vector that injects a current that accounts for generation less load at each of the *n* system busses. The matrix $[Y_{\text{bus}}]$ contains all the transformer and transmission line circuit information. To understand its structure, let us write the *i*th entry of equation (7-4)

$$I_i = Y_{i1}V_1 + Y_{i2}V_2 + \ldots + Y_{ii}V_i + \ldots + Y_{in}V_n \qquad (7\text{-}5)$$

How might we find one of these Y's; say, Y_{i2}, for example? Suppose $V_1 = V_3 = V_4 = \ldots = V_n = 0$ (all voltages except V_2). Then:

$$I_i = 0 + Y_{i2}V_2 + 0 + \ldots + 0 \qquad (7\text{-}6a)$$

or

$$Y_{i2} = \frac{I_i}{V_2}\Bigg]_{\text{All } V=0 \text{ except } V_2} \qquad (7\text{-}6b)$$

If we generalize on this, we write:

$$Y_{ij} = \frac{I_i}{V_j}\bigg]_{\text{All } V=0 \text{ except } V_j} \tag{7-7}$$

The i, j values are any one of the n bus numbers and may be selected independently of one another.

To force the V's to zero, we terminate all ports in a short circuit, except the jth one, which we terminate in a voltage source of strength V_j. Next solve for I_i. To compute Y_{ij} substitute into (7-7). The Y's are called "short circuit parameters" for obvious reasons. For reciprocal networks Y_{ij} equals Y_{ji}, a condition satisfied by the power system.

If the n busses represent *all* the nodes of the transmission network, an even simpler formulation of $[Y_{\text{bus}}]$ is possible. The Y's are simply the coefficients of the voltages that result from application of the conventional node voltage method of circuit analysis:

$$Y_{ii} = \text{sum of admittances connected to the } i\text{th bus} \tag{7-8a}$$

$$Y_{ij} = \text{negative of the total admittance directly connected between the } i\text{th and } j\text{th busses} \tag{7-8b}$$

Example 7-1

Consider the simple transmission network shown in Figure 7-3. Formulate $[Y_{\text{bus}}]$ by (a) treating the network as two port. (b) using conventional nodal analysis.

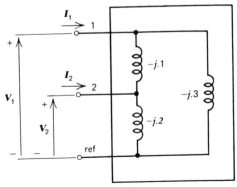

Figure 7-3 Circuit for Example 7-1.

Solution

(a) The "port" approach:

$$Y_{11} = \frac{I_1}{V_1}\bigg]_{V_2=0} = \frac{(-j0.1-j0.3)V_1}{V_1} = -j0.4$$

$$Y_{12} = \frac{I_1}{V_2}\bigg]_{V_1=0} = \frac{-(-j0.1)V_2}{V_2} = +j0.1$$

$$Y_{21} = Y_{12} = +j0.1$$

$$Y_{22} = \frac{I_2}{V_2}\bigg]_{V_1=0} = \frac{(-j0.1-j0.2)V_2}{V_2} = -j0.3$$

$$\therefore \quad [Y_{bus}] = \begin{bmatrix} -j0.4 & +j0.1 \\ +j0.1 & -j0.3 \end{bmatrix}$$

(b) The "nodal analysis" approach:

$$I_1 = (-j0.1-j0.3)V_1 - (-j0.1)V_2$$

$$I_2 = -(-j0.1)V_1 + (-j0.1-j0.2)V_2$$

$$\therefore \quad [Y_{bus}] = \begin{bmatrix} -j0.4 & +j0.1 \\ +j0.1 & -j0.3 \end{bmatrix}$$

The nodal analysis approach is amenable to computer formulation. We start with the $[Y_{bus}]$ array initially set to zero. Now consider an element of admittance y connected between busses i and j. Four entries in $[Y_{bus}]$ are affected: Y_{ii}, Y_{ij}, Y_{ji}, and Y_{jj}. We modify these entries as follows:

$$Y_{ii_{new}} = Y_{ii_{old}} + y \tag{7-9a}$$

$$Y_{ij_{new}} = Y_{ij_{old}} - y \tag{7-9b}$$

$$Y_{ji_{new}} = Y_{ji_{old}} - y \tag{7-9c}$$

$$Y_{jj_{new}} = Y_{jj_{old}} + y \tag{7-9d}$$

The matrix $[Y_{bus}]$ is modified for each element added. If the element is connected from i to reference, only the entry Y_{ii} is affected. An example will demonstrate the algorithm.

Example 7-2

Construct $[Y_{bus}]$ for the circuit of example 7-1 using equations 7-9.

Solution

We start with

$$[Y_{bus}] = \begin{bmatrix} 0 & 0 \\ 0 & 0 \end{bmatrix}$$

Add the $-j0.1$ element:

$$[Y_{bus}] = \begin{bmatrix} -j0.1 & +j0.1 \\ +j0.1 & -j0.1 \end{bmatrix}$$

Add the $-j0.2$ element:

$$[Y_{bus}] = \begin{bmatrix} -j0.1 & +j0.1 \\ +j0.1 & -j0.3 \end{bmatrix}$$

Add the $-j0.3$ element:

$$[Y_{bus}] = \begin{bmatrix} -j0.4 & +j0.1 \\ +j0.1 & -j0.3 \end{bmatrix}$$

Equation (7-9) will deal with all power system devices encountered, except for two cases: the regulating transformer and transmission lines with mutual coupling. We examine the regulating transformer in section 7-2; we will consider mutually coupled lines now. A mutually coupled transmission line is shown in Figure 7-4. Shunt elements are omitted for simplicity; they are handled in a straightforward manner. We write:

$$\begin{bmatrix} V_i \\ V_k \end{bmatrix} - \begin{bmatrix} V_j \\ V_l \end{bmatrix} = \begin{bmatrix} z_{s1} & z_m \\ z_m & z_{s2} \end{bmatrix} \begin{bmatrix} I_i \\ I_k \end{bmatrix} \qquad (7\text{-}10)$$

Solving for the currents

$$\begin{bmatrix} I_i \\ I_k \end{bmatrix} = \begin{bmatrix} y_{s1} & y_m \\ y_m & y_{s2} \end{bmatrix} \left\{ \begin{bmatrix} V_i \\ V_k \end{bmatrix} - \begin{bmatrix} V_j \\ V_l \end{bmatrix} \right\} \qquad (7\text{-}11)$$

where

$$\begin{bmatrix} y_{s1} & y_m \\ y_m & y_{s2} \end{bmatrix} = \begin{bmatrix} z_{s1} & z_m \\ z_m & z_{s2} \end{bmatrix}^{-1} \qquad (7\text{-}12)$$

Viewing the situation from the j, l end:

$$\begin{bmatrix} I_j \\ I_l \end{bmatrix} = \begin{bmatrix} y_{s1} & y_m \\ y_m & y_{s2} \end{bmatrix} \left\{ \begin{bmatrix} V_j \\ V_l \end{bmatrix} - \begin{bmatrix} V_i \\ V_k \end{bmatrix} \right\} \qquad (7\text{-}13)$$

225

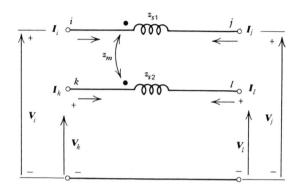

Figure 7-4 Equivalent circuit for mutually coupled lines.

The proper modifications to the sixteen affected entries follow:

$$Y_{ii_{new}} = Y_{ii_{old}} + y_{s1} \tag{7-14a}$$

$$Y_{jj_{new}} = Y_{jj_{old}} + y_{s1} \tag{7-14b}$$

$$Y_{ij_{new}} = Y_{ji_{new}} = Y_{ij_{old}} - y_{s1} \tag{7-14c}$$

$$Y_{kk_{new}} = Y_{kk_{old}} + y_{s2} \tag{7-14d}$$

$$Y_{ll_{new}} = Y_{ll_{old}} + y_{s2} \tag{7-14e}$$

$$Y_{kl_{new}} = Y_{lk_{new}} = Y_{kl_{old}} - y_{s2} \tag{7-14f}$$

$$Y_{ik_{new}} = Y_{ki_{new}} = Y_{ik_{old}} + y_m \tag{7-14g}$$

$$Y_{il_{new}} = Y_{li_{new}} = Y_{il_{old}} - y_m \tag{7-14h}$$

$$Y_{jl_{new}} = Y_{lj_{new}} = Y_{jl_{old}} + y_m \tag{7-14i}$$

$$Y_{jk_{new}} = Y_{kj_{new}} = Y_{jk_{old}} - y_m \tag{7-14j}$$

We return to consideration of solving for S_{Ti}. Equation (7-5) may be written as:

$$I_i = \sum_{j=1}^{n} Y_{ij} V_j \tag{7-15a}$$

$$= \sum_{j=1}^{n} Y_{ij} V_j \underline{/\delta_j + \gamma_{ij}} \tag{7-15b}$$

From Figure 7-2 it is clear that

$$S_{Ti} = V_i I_i^* \tag{7-16}$$

Substituting (7-15b) into (7-16) and simplifying we get:

$$S_{Ti} = V_i \left[\sum_{j=1}^{n} Y_{ij} V_j \underline{/\delta_j + \gamma_{ij}} \right]^* \tag{7-17a}$$

$$= \sum_{j=1}^{n} V_i V_j Y_{ij} \underline{/\delta_i - \delta_j - \gamma_{ij}} \tag{7-17b}$$

We may break S_{Ti} down into rectangular components:

$$P_{Ti} = \sum_{j=1}^{n} V_i V_j Y_{ij} \cos(\delta_i - \delta_j - \gamma_{ij}) \tag{7-18a}$$

$$Q_{Ti} = \sum_{j=1}^{n} V_i V_j Y_{ij} \sin(\delta_i - \delta_j - \gamma_{ij}) \tag{7-18b}$$

We see that we may replace the $2n$ variables (P_{Ti}, Q_{Ti}) through equations (7-18a) and (7-18b) with the $2n$ variables (V_1, δ_1), which we do. Equations (7-18) illustrate the complexity of the power flow problem. They show that the transmitted real and reactive powers at a given bus in general will be a function of the voltage magnitude and phase at *all* the other busses in the system. Using equations (7-18) we rewrite (7-3) as:

$$P_{Gi} - P_{Li} = \sum_{j=1}^{n} V_i V_j Y_{ij} \cos(\delta_i - \delta_j - \gamma_{ij}) \tag{7-19a}$$

$$Q_{Gi} - Q_{Li} = \sum_{j=1}^{n} V_i V_j Y_{ij} \sin(\delta_i - \delta_j - \gamma_{ij}) \tag{7-19b}$$

or

$$S_{Gi} - S_{Li} = \sum_{j=1}^{n} V_i V_j^* Y_{ij}^* \tag{7-20}$$

Let us review our position. At each bus we are concerned with six variables: P_{Gi}, Q_{Gi}, P_{Li}, Q_{Li}, V_i, and δ_i. Two are specified (P_{Li} and Q_{Li}) leaving P_{Gi}, Q_{Gi}, V_i, and δ_i. Equations 7-19a and b must be satisfied, forcing two more variables to be specified. As far as the mathematics is concerned, *any* two variables may be specified; however, when the physical system is considered we realize that we may specify only variables over which we have physical control. Our choices are influenced to some degree by what devices are connected to a particular bus. It turns out that we have essentially four options that are summarized in Table 7-1. We identify these options by defining "bus types."

Table 7-1. Bus types for power flow analysis.

Bus Type	Code	Knowns	Unknowns	Approximate Number
Reference	0	$V_i = 1.0$ $\delta_i = 0$	P_{Gi} Q_{Gi}	1
Load	1	P_{Gi} Q_{Gi}	V_i δ_i	85%
Generator	2	P_{Gi} V_i	δ_i Q_{Gi}	10%
Voltage controlled	3	P_{Gi} Q_{Gi} V_i	δ_i c	5%

The first type (0) is referred to as the reference, swing, or slack bus. It is essentially a generator bus with no constraints. We recall that in any ac circuit, one phasor quantity may be selected as phase reference; we exercise this option at the reference bus, setting the phase angle of its voltage to zero. We also set the voltage to 1.0 per unit, insuring that the voltages throughout the system will be close to unity. For convenience we will always number the reference bus "1" so that:

$$V_1 = 1\underline{/0°} \tag{7-21}$$

Physically the reference bus must be a generator bus or tie bus (i.e., terminates lines to other systems) so that a wide range of P_{Gi} and Q_{Gi} values are possible. Only one is chosen per system.

The type one bus would identify any bus for which P_{Gi} and Q_{Gi} are known; this would include any bus with no generation since P_{Gi} and Q_{Gi} would be zero. For this reason such busses are referred to as "load." Type one busses are by far the most common, typically comprising more than 80% of all busses. At such busses we must solve for the unknown V_i and δ_i variables.

Types two and three busses are frequently grouped together and identified as "voltage controlled" busses; we separate them because of physical differences and slightly different calculation strategies at these bus types. The type two bus always has generation connected. Recall from Chapter 6 that the two main direct control actions possible in the turbine-governor-exciter-generator system allow us to control P_{Gi} and V_i. Because we can "set" these values we take them as specified or known. The generator operating characteristics require that we stay within the region outlined in Figure 6-15. Constraining P_{Gi} is simple; we simply set $P_{Gi_{min}} \leq P_{Gi} \leq P_{Gi_{max}}$.

228

Unfortunately, honoring the limits on Q_{Gi} is not so straightforward—it is one of our unknowns. This means that as we calculate Q_{Gi} we must check to see if $Q_{Gi_{min}} \leq Q_{Gi} \leq Q_{Gi_{max}}$. If Q_{Gi} is not within limits, we set it at the appropriate limit ($Q_{Gi_{max}}$ if $Q_{Gi} > Q_{Gi_{max}}$ and $Q_{Gi_{min}}$ if $Q_{Gi} < Q_{Gi_{min}}$) and release our constraint that V_i is fixed. That is, V_i and Q_{Gi} exchange roles as knowns and unknowns. This essentially changes the type of the bus in question from 2 to 1. We continue to check Q_{Gi} and whenever it falls within acceptable limits we allow the bus type to again be 2. This complicated situation should become clearer as we spell out our calculation strategy for each bus type.

The type three bus also has voltage control capability, but uses an adjustable tap-adjustable phase transformer instead of a generator with turbine and exciter control. We realize $P_{Gi} = Q_{Gi} = 0$ but still wish to control the voltage at such busses.

Our statement of the general power flow problem follows:

Iterate to solve for the unknown variables listed in Table 7-1 taken as given the known variables listed in Table 7-1 for a system of known $[Y_{bus}]$ with known operating limits on generators and regulating transformers at a specified load condition. Now solve for the power flows on all lines.

We shall now proceed to solve it.

7-2 Power Flow Solution by the Gauss–Seidel Method

A general nonlinear algebraic equation set may have no formal method for direct solution; that is, it may be impossible to arrange the equations so that the unknowns can be equated to a finite number of functional operations on known values. In such cases we turn to numerical iterative techniques. These methods are designed to progressively compute more accurate estimates of the unknowns until results are obtained to any desired accuracy in a finite number of iterations. When this is possible, the solution is said to "converge." The key idea with regard to such methods is that it is possible to write a general expression that instructs us how to compute the $k+1$ estimates from the k estimates.

Solution of the power flow problem requires iterative techniques. Two methods commonly used are the Gauss-Seidel Method and the Newton–Raphson Method. Virtually all commercial power flow programs are based on one or both of these methods. We first investigate the Gauss–Seidel Method. Consider the following example.

Example 7-3

Solve the equation $x + \sin x - 2 = 0$ by Gauss–Seidel iteration, starting from $x = 0$.

Solution

We solve for x:

$$x = 2 - \sin x$$

Starting from $x^0 = 0$ we calculate:

$$x^1 = 2 - \sin(0) = 2$$

which we use for our next approximation for x. Use the superscript i to represent the ith approximation for x. We calculate:

$$x^2 = 2 - \sin(2)$$
$$= 1.09$$

We repeat the process until x ceases to change significantly. Results for 14 trials are shown.

Approximation	x
0	0
1	2
2	1.09
3	1.113
4	1.103
5	1.1075
6	1.1054
7	1.10634
8	1.10593
9	1.10612
10	1.10603
11	1.10607
12	1.10606
13	1.10606

Now consider a system of n equations in n unknowns. Manipulating until we solve for each unknown x, we get equations of the form:

$$x_i = f_i(x_1 x_2 \ldots x_i \ldots x_n) \tag{7-22a}$$

where

$$1 \le i \le n \tag{7-22b}$$

We start with initial known approximate values for x's, written as x_1^0, $x_2^0, \ldots x_n^0$. Our first approximation for x_1 is:

$$x_1^1 = f_1(x_1^0 x_2^0 \ldots x_n^0) \tag{7-23a}$$

For x_2:

$$x_2^1 = f_2(x_1^1 x_2^0 x_3^0 \ldots x_n^0) \tag{7-23b}$$

For x_i:

$$x_i^1 = f_i(x_1^1 x_2^1 \ldots x_{i-1}^1 x_i^0 x_{i+1}^0 \ldots x_n^0) \tag{7-23c}$$

In general the kth approximation for x_i is computed from:

$$x_i^k = f_i(x_1^k x_2^k \ldots x_{i-1}^k x_i^{k-1} x_{i+1}^{k-1} \ldots x_n^{k-1}) \tag{7-24}$$

We examine the change in each variable

$$\Delta x_i = x_i^k - x_i^{k-1} \tag{7-25a}$$

When

$$\Delta x_i < \varepsilon; \text{ all } i \tag{7-25b}$$

we claim convergence. From experience with particular problems, the number of iterations required for convergence can be reduced if we change the old value by something more than Δx_i; we use

$$x_i^k = x_i^{k-1} + \sigma \Delta x_i \tag{7-26a}$$

$$\sigma \ge 1 \tag{7-26b}$$

The factor σ is called the acceleration factor and is chosen from experience with a particular problem; $\sigma = 1.6$ is a good value for use in the power flow problem.

Our calculation strategy is different at our four bus types because of the different mix of knowns and unknowns. We discuss each in order. Understand the subscript "i" to represent the bus at which we are presently calculating ($1 \le i \le n$).

TYPE 0.

No iterative calculations are required at the reference bus; we simply skip it and proceed to the next bus.

TYPE 1.

Recall equation (7-20)

$$S_{Gi} - S_{Li} = \sum_{j=1}^{n} V_i V_j^* Y_{ij}^* \qquad (7\text{-}20)$$

Taking its conjugate:

$$S_{Gi}^* - S_{Li}^* = \sum_{j=1}^{n} V_i^* V_j Y_{ij} \qquad (7\text{-}27a)$$

$$= V_i^* V_i Y_{ii} + \sum_{\substack{j=1 \\ j \neq i}}^{n} V_i^* V_j Y_{ij} \qquad (7\text{-}27b)$$

Rearrange and divide through by $V_i^* Y_{ii}$:

$$V_i = \frac{S_{Gi}^* - S_{Li}^*}{V_i^* Y_{ii}} - \frac{1}{Y_{ii}} \sum_{\substack{j=1 \\ j \neq i}}^{n} V_j Y_{ij} \qquad (7\text{-}28)$$

At a given point in any iteration at a load bus (i), S_{Gi}^*, S_{Li}^*, and $[Y_{bus}]$ are known, and we have estimates for all voltages. We use (7-28) to compute the next approximation for V_i. Recall that it is $V_i = V_i \underline{/\delta_i}$ that is unknown and that we wish to calculate. Sometimes a second pass through (7-28) is made, using the previously calculated V_i on the right side in V_i^*. This is the most common bus type.

TYPE 2.

Recall that at a generator bus we wish to hold the voltage magnitude constant at $V_{i_{spec}}$, which must be stored. Unfortunately, this may not be possible, depending on the value of Q_{Gi}. Therefore, in a given iteration V_i is not necessarily $V_{i_{spec}}$, depending on what happened in the last iteration. First we save the present value and name it V_{i_0}.

$$\therefore \quad V_{i_0} = V_{i_0} \underline{/\delta_{i_0}} = V_i \qquad (7\text{-}29)$$

We then calculate Q_{Gi} from (7-19b):

$$Q_{Gi} = Q_{Li} + \sum_{j=1}^{n} V_{i_{spec}} V_j Y_{ij} \sin(\delta_{i_0} - \delta_j - \gamma_{ij}) \qquad (7\text{-}19b)$$

Next check Q_{Gi} to see if it is within limits; specifically, is $Q_{Gi_{max}} \geq Q_{Gi} \geq Q_{Gi_{min}}$? If so, use equation (7-28) to calculate the next approximation for V_i

using $V_{i_{spec}}$ and δ_{i_0} for the magnitude and phase for V_i on the right side. Keep the new δ_i, but reset V_i to $V_{i_{spec}}$ and continue to the next bus.

If $Q_{Gi} > Q_{Gi_{max}}$, set Q_{Gi} to $Q_{Gi_{max}}$, set V_i back to V_{i_0}, and default to Type 1 computational logic for the rest of the iteration. Similarly, if $Q_{Gi} < Q_{Gi_{min}}$, set Q_{Gi} to $Q_{Gi_{min}}$, set V_i back to V_{i_0}, and default to Type 1 computational logic for the rest of the iteration.

TYPE 3.

Some power systems utilize transformers capable of tap changing and phase shifting operations. Such devices were discussed in Chapter 5 (see sections 5-7 and 5-8). The device circuit performance is described with equations 5-45) and (5-46). Recall that the complex turns ratio c is:

$$c = c \underline{/\alpha} \tag{5-42d}$$

The value c is the turns ratio and α the phase shift. These two parameters are independently adjustable, usually over specified ranges. For our purposes consider:

$$0.90 \le c \le 1.10 \tag{7-30a}$$

$$-10° \le \alpha \le +10° \tag{7-30b}$$

Because both variables are physically adjustable by tap changing under load, adjustment is made in fixed steps of Δc and $\Delta \alpha$. For our work:

$$\Delta c = 0.025 \tag{7-31a}$$

$$\Delta \alpha = 2.5° \tag{7-31b}$$

Consider such a transformer to be connected between busses i and j, with bus i the bus at which voltage is to be controlled. We intend to control voltage (V_i) by adjusting c and real power flow through the transformer (P_{ij}) by adjusting α. Recall the equations:

$$I_i = y_{ii}V_i + y_{ij}V_j \tag{5-45a}$$

$$I_j = y_{ji}V_i + y_{jj}V_j \tag{5-45b}$$

where:

$$y_{ii} = Y_e \tag{5-46a}$$

$$y_{ij} = -cY_e \tag{5-46b}$$

$$y_{ji} = -c^*Y_e \tag{5-46c}$$

$$y_{jj} = c^2Y_e \tag{5-46d}$$

We start the calculation by comparing V_i to $V_{i_{spec}}$; specifically, is $|V_i - V_{i_{spec}}| \leq 0.0125 V_{i_{spec}}$?$(0.0125 = \Delta c/2.)$ If we find low voltage, we increase c by Δc (one step); similarly, high voltage requires a decrease in c by Δc. More than one step could be required to bring the voltage within tolerance; however, changing by only one step will minimize overshoot problems, and we can rely on the next iteration to make further needed adjustments. Next we deal with P_{ij}. This power is somewhat ambiguous since the transformer is not completely lossless. It is sufficient for our purposes to use:

$$P_{ij} \cong \text{Re}\,[\,V_i I_i^*\,] \tag{7-32a}$$

Using equations (5-45 and 46) and simplifying, we get:

$$P_{ij} = +V_i^2 Y_e \cos(\gamma_e) - V_i V_j c Y_e \cos(\delta i - \delta_j - \gamma_e - \alpha) \tag{7-32b}$$

We need to estimate the change in $P_{ij}(\Delta P_{ij})$ produced by a change in $\alpha(\Delta \alpha)$.

$$\frac{\partial P_{ij}}{\partial \alpha} = -V_i V_j c Y_e \sin(\delta_i - \delta_j - \gamma_e - \alpha) \tag{7-33}$$

Now approximate ΔP_{ij} as

$$\Delta P_{ij} = \frac{\partial P_{ij}}{\partial \alpha} \Delta \alpha \tag{7-34}$$

We check to see if $|P_{ij} - \dot{P}_{ij_{spec}}| \leq \Delta P_{ij}/2$. If so, no change in α is required. If not, change α one step $\Delta \alpha$ in the appropriate direction (an increase in α will cause more real power flow from i to j). We now have an updated version of c, and are ready to update the appropriate entries in $[Y_{bus}]$:

$$Y_{ii_{new}} = Y_{ii_{old}} + y_{ii_{new}} - y_{ii_{old}} \tag{7-35a}$$

$$Y_{ij_{new}} = Y_{ij_{old}} + y_{ij_{new}} - y_{ij_{old}} \tag{7-35b}$$

$$Y_{ji_{new}} = Y_{ji_{old}} + y_{ji_{new}} - y_{ji_{old}} \tag{7-35c}$$

$$Y_{jj_{new}} = Y_{jj_{old}} + y_{jj_{new}} - y_{jj_{old}} \tag{7-35d}$$

The "old" values are those available from the previous iteration; the y "new" values are calculated from equation (5-46), using our updated c value. We now calculate V_i from equation (7-28) and proceed to the next bus. In each iteration c and α are checked and kept within the limits defined in equation (7-30).

Having dealt with all four bus types, we can understand the basic calculation strategy. Starting at bus 1, and proceeding consecutively over all the busses, we first determine the type and calculate accordingly. A

complete cycle over all the busses is termed one iteration. We continue to
iterate until the solution converges, or until we reach the maximum allowable number of iterations, and conclude that the solution will not converge.
The number of iterations required depends on the problem and starting
values used, but might typically range from 4 or 5 to over 50. Assuming that
convergence is achieved, the next order of business is to calculate the line
flows. Using the circuit model of Figure 4-7(b) for a line connected from bus i
to bus j:

$$I_{ij} = \frac{Y_1 V_i}{2} + \frac{V_i - V_j}{Z_1} \qquad (7\text{-}36a)$$

from which

$$S_{ij} = V_i I_{ij}^* \qquad (7\text{-}36b)$$

$$= P_{ij} + jQ_{ij} \qquad (7\text{-}36c)$$

If a transformer interconnects bus i to bus j we use the circuit of Figure
5-14b to calculate I_{ij}:

$$I_{ij} = \frac{V_i - V_j}{Z_e} \qquad (7\text{-}37)$$

Now use equation (7-36b) to calculate the power flows. If a regulating
transformer interconnects bus i to bus j, we use equation (6-45a) to calculate
I_{ij} and equation (7-36b) to calculate the power flows.

At the reference bus we calculate $S_{T_i}(i = 1)$ from equation (7-17b) and
finally S_{Gi}:

$$S_{Gi} = S_{Li} + S_{Ti} \qquad (7\text{-}1)$$

At generator busses (Type 2) we calculate Q_{Ti} from (7-18b) and finally
Q_{Gi} from:

$$Q_{Gi} = Q_{Li} + Q_{Ti} \qquad (7\text{-}3b)$$

The results of a solved power flow problem are typically displayed by
bus, and include the following information:

Bus name and number
Bus type
Bus voltage magnitude (V_i)
Bus voltage phase angle (δ_i)
Real load (P_{Li})
Reactive load (Q_{Li})
Real generation (P_{Gi})

235

Reactive generation (Q_{Gi})
Line (and transformer) flows (P_{ij}, Q_{ij})
P and Q mismatch after convergence

Sometimes "line charging" values (Q delivered to the shunt capacitance elements in the line models) are separated from the "line flow" values; also, shunt reactive elements used for voltage control are usually separated from the load. Other useful information such as number of iterations required for convergence and net system power mismatch is usually printed out.

7-3 Power Flow Solution by the Newton–Raphson Method

For very large systems the Gauss-Seidel method may require an excessive number of iterations to converge, if it converges at all. Negative resistances also tend to prevent convergence by the method; recall that Type 3 busses may have negative resistance involved, which are present in our model of the regulating transformer. The Newton–Raphson method is used for large programs.

To understand the method consider the equation

$$f(x) = 0 \tag{7-38}$$

Suppose we wish to solve for a root, x. We expand $f(x)$ by Taylor's Series about a point x^0.

$$f(x) = f(x^0) + \frac{1}{1!} \frac{df(x^0)}{dx}(x - x^0) + \frac{1}{2!} \frac{df^2(x^0)}{dx^2}(x - x^0)^2 + \ldots \tag{7-39}$$

Our first estimate of the root is computed by truncating the series after the first derivative. We manipulate to:

$$x^1 = x^0 - \frac{f(x^0)}{\dfrac{df(x^0)}{dx}} \tag{7-40}$$

Use the above equation as a recursion formula to compute successive approximations for x from previous results.

$$x^{k+1} = x^k - \frac{f(x^k)}{\dfrac{df(x^k)}{dx}} \tag{7-41a}$$

Streamline the notation as follows:

$$f(x^k) = f^k \tag{7-42a}$$

$$\frac{df(x^k)}{dx} = f_x^k \tag{7-42b}$$

Then

$$x^{k+1} = x^k - \frac{f^k}{f_x^k} \tag{7-41b}$$

An example should prove helpful.

Example 7-4

Repeat example 7-3 using the Newton–Raphson approach.

Solution

Define $f(x) = x + \sin x - 2 = 0$

Then $f_x = 1 + \cos x$

The recursion equation is:

$$
\begin{aligned}
x^{k+1} &= x^k - \frac{f^k}{f_x^k} \\
&= x^k - \frac{x^k + \sin x^k - 2}{1 + \cos x^k}
\end{aligned}
$$

Approximation$^{(k)}$	x^k
0	0
1	1
2	1.103
3	1.10606

Compare the results of example 7-2 to those of example 7-3. Note that the Newton–Raphson method converges at a faster rate than the Gauss–Seidel method, but that the calculations in each iteration are more involved because calculation of and division by the derivative is required.

We extend the method to two equations in two variables

$$f(x, y) = 0 \qquad \text{(7-43a)}$$

$$g(x, y) = 0 \qquad \text{(7-43b)}$$

Expansion of the functions $f(x, y)$ and $g(x, y)$ by Taylor's Series about a point x^k, y^k produces

$$f(x, y) = f^k + \frac{1}{1!} \frac{\partial f(x^k, y^k)}{\partial x}(x - x^k) + \frac{1}{1!} \frac{\partial f(x^k, y^k)}{\partial y}(y - y^k) + \ldots \qquad \text{(7-44a)}$$

$$g(x, y) = g(x^k, y^k) + \frac{1}{1!} \frac{\partial g(x^k, y^k)}{\partial x}(x - x^k) + \frac{1}{1!} \frac{\partial g(x^k, y^k)}{\partial y}(y - y^k) + \ldots$$
$$\text{(7-44b)}$$

Simplify the notation as follows:

$$f(x^k, y^k) = f^k \qquad \text{(7-45a)}$$

$$g(x^k, y^k) = g^k \qquad \text{(7-45b)}$$

$$\frac{\partial f(x^k, y^k)}{\partial x} = f_x^k \qquad \text{(7-45c)}$$

$$\frac{\partial f(x^k, y^k)}{\partial y} = f_y^k \qquad \text{(7-45d)}$$

$$\frac{\partial g(x^k, y^k)}{\partial x} = g_x^k \qquad \text{(7-45e)}$$

$$\frac{\partial g(x^k, y^k)}{\partial y} = g_y^k \qquad \text{(7-45f)}$$

$$\Delta x = x - x^k \qquad \text{(7-45g)}$$

$$\Delta y = y - y^k \qquad \text{(7-45h)}$$

$$\Delta f = f - f^k \qquad \text{(7-45i)}$$

$$\Delta g = g - g^k \qquad \text{(7-45j)}$$

so that equation (7-44) becomes

$$\Delta f = f_x^k \Delta x + f_y^k \Delta y \qquad \text{(7-46a)}$$

$$\Delta g = g_x^k \Delta x + g_y^k \Delta y \qquad \text{(7-46b)}$$

In matrix form:

$$\begin{bmatrix} \Delta f \\ \Delta g \end{bmatrix} = \begin{bmatrix} f_x^k & f_y^k \\ g_x^k & g_y^k \end{bmatrix} \begin{bmatrix} \Delta x \\ \Delta y \end{bmatrix} \qquad \text{(7-47a)}$$

The coefficient matrix in equation (7-47a) is called the Jacobian and is written as:

$$[J^k] = \begin{bmatrix} f_x^k & f_y^k \\ g_x^k & g_y^k \end{bmatrix}$$

(7-48)

so that (7-47a) becomes:

$$\begin{bmatrix} \Delta f \\ \Delta g \end{bmatrix} = [J^k] \begin{bmatrix} \Delta x \\ \Delta y \end{bmatrix}$$

(7-47b)

We will use equation (7-47b) to iterate as follows. In the kth iteration we have values for x and y (denoted as x^k and y^k). Therefore, we can evaluate f, g, and J to produce f^k, g^k, and J^k.

Since we desire that:

$$f = 0$$

(7-43a)

$$g = 0$$

(7-43b)

we calculate

$$\Delta f^k = 0 - f^k$$

(7-49a)

$$\Delta g^k = 0 - g^k$$

(7-49b)

The unknowns are Δx^k and Δy^k. We calculate from (7-47b):

$$\begin{bmatrix} \Delta x^k \\ \Delta y^k \end{bmatrix} = [J^k]^{-1} \begin{bmatrix} \Delta f^k \\ \Delta g^k \end{bmatrix}$$

(7-50)

We set the predicted values for x and y to:

$$x^{k+1} = x^k + \Delta x^k$$

(7-51a)

$$y^{k+1} = y^k + \Delta y^k$$

(7-51b)

We now are prepared to begin the next $(k+1)$ iteration. We continue the process until convergence is obtained or until we have completed the maximum allowable number of iterations.

Let us extend the method to $2n$ equations in $2n$ unknowns. The equations are:

$$f_i(\tilde{x}, \tilde{y}) = 0; \quad i = 1, 2, \ldots n$$

(7-52a)

$$g_i(\tilde{x}, \tilde{y}) = 0; \quad i = 1, 2, \ldots n$$

(7-52b)

where the unknown $\tilde{x}$ and $\tilde{y}$ vectors are:

$$\tilde{x} = \begin{bmatrix} x_1 \\ x_2 \\ \vdots \\ x_n \end{bmatrix} \qquad (7\text{-}53a)$$

$$\tilde{y} = \begin{bmatrix} y_1 \\ y_2 \\ \vdots \\ y_n \end{bmatrix} \qquad (7\text{-}53b)$$

The equation corresponding to equation (7-46) extended to $2n$ variables follows.

$$\begin{bmatrix} \Delta f_1^k \\ \Delta f_2^k \\ \vdots \\ \Delta f_n^k \\ \hline \Delta g_1^k \\ \Delta g_2^k \\ \vdots \\ \Delta g_n^k \end{bmatrix} = \begin{bmatrix} \dfrac{\partial f_1^k}{\partial x_1} & \cdots & \dfrac{\partial f_1^k}{\partial x_n} & \dfrac{\partial f_1^k}{\partial y_1} & \cdots & \dfrac{\partial f_1^k}{\partial y_n} \\ \vdots & & \vdots & \vdots & & \vdots \\ \dfrac{\partial f_n^k}{\partial x_1} & \cdots & \dfrac{\partial f_n^k}{\partial x_n} & \dfrac{\partial f_n^k}{\partial y_1} & \cdots & \dfrac{\partial f_n^k}{\partial y_n} \\ \hline \dfrac{\partial g_1^k}{\partial x_1} & \cdots & \dfrac{\partial g_1^k}{\partial x_n} & \dfrac{\partial g_1^k}{\partial y_1} & \cdots & \dfrac{\partial g_1^k}{\partial y_n} \\ \vdots & & \vdots & \vdots & & \vdots \\ \dfrac{\partial g_n^k}{\partial x_1} & \cdots & \dfrac{\partial g_n^k}{\partial x_n} & \dfrac{\partial g_n^k}{\partial y_1} & \cdots & \dfrac{\partial g_n^k}{\partial y_n} \end{bmatrix} \begin{bmatrix} \Delta x_1^k \\ \Delta x_2^k \\ \vdots \\ \Delta x_n^k \\ \hline \Delta y_1^k \\ \Delta y_1^k \\ \vdots \\ \Delta y_n^k \end{bmatrix} \qquad (7\text{-}54a)$$

or

$$\begin{bmatrix} \tilde{\Delta} f^k \\ \tilde{\Delta} g^k \end{bmatrix} = \begin{bmatrix} J_{fx}^k & J_{fy}^k \\ J_{gx}^k & J_{gy}^k \end{bmatrix} \begin{bmatrix} \tilde{\Delta} x^k \\ \tilde{\Delta} y^k \end{bmatrix} \qquad (7\text{-}54b)$$

where the submatrices J_{fx}^k, J_{fy}^k, J_{gx}^k, and J_{gy}^k have definitions that are obvious if we compare (7-54a) with (7-54b). We continue to write:

$$\begin{bmatrix} \tilde{\Delta} f^k \\ \tilde{\Delta} g^k \end{bmatrix} = [J^k] \begin{bmatrix} \tilde{\Delta} x^k \\ \tilde{\Delta} y^k \end{bmatrix} \qquad (7\text{-}54c)$$

Solve for $\tilde{\Delta} x$ and $\tilde{\Delta} y$ from (9-54c):

$$\begin{bmatrix} \tilde{\Delta} x^k \\ \tilde{\Delta} y^k \end{bmatrix} = [J^k]^{-1} \begin{bmatrix} \tilde{\Delta} f^k \\ \tilde{\Delta} g^k \end{bmatrix} \qquad (7\text{-}55)$$

We are now prepared to apply the method to the power flow problem. We formulate the basic equations from (7-3):

$$f_i = -P_{G_i} + P_{T_i} + P_{L_i} = 0 \qquad (7\text{-}56a)$$

$$g_i = -Q_{G_i} + Q_{T_i} + Q_{L_i} = 0$$

240

where

$$P_{T_i} = \sum_{j=1}^{n} V_i V_j Y_{ij} \cos(\delta_i - \delta_j - \gamma_{ij}) \qquad (7\text{-}18a)$$

$$Q_{T_i} = \sum_{j=1}^{n} V_i V_j Y_{ij} \sin(\delta_i - \delta_j - \gamma_{ij}) \qquad (7\text{-}18b)$$

We define:

$$\tilde{x} = \tilde{\boldsymbol{\delta}} \qquad (7\text{-}57a)$$

$$\tilde{y} = \tilde{\boldsymbol{V}} \qquad (7\text{-}57b)$$

To form the Jacobian (J) the following eight general partial derivatives are required.

$$\frac{\partial f_i}{\partial x_j} = \frac{\partial P_{Ti}}{\partial \delta_j} = V_i V_j Y_{ij} \sin(\delta_i - \delta_j - \gamma_{ij}) \qquad (7.58a)$$

$$\frac{\partial f_i}{\partial x_i} = \frac{\partial P_{Ti}}{\partial \delta_i} = - \sum_{\substack{j=1 \\ j \neq i}}^{n} V_i V_j Y_{ij} \sin(\delta_i - \delta_j - \gamma_{ij}) \qquad (7.58b)$$

$$\frac{\partial f_i}{\partial y_j} = \frac{\partial P_{Ti}}{\partial V_j} = V_i Y_{ij} \cos(\delta_i - \delta_j - \gamma_{ij}) \qquad (7.58c)$$

$$\frac{\partial f_i}{\partial y_i} = \frac{\partial P_{Ti}}{\partial V_i} = V_i Y_{ij} \cos(\gamma_{ii}) + \sum_{j=1}^{n} V_j Y_{ij} \cos(\delta_i - \delta_j - \gamma_{ij}) \qquad (7.58d)$$

$$\frac{\partial g_i}{\partial x_j} = \frac{\partial Q_{Ti}}{\partial \delta_j} = -V_i V_j Y_{ij} \cos(\delta_i - \delta_j - \gamma_{ij}) \qquad (7.58e)$$

$$\frac{\partial g_i}{\partial x_i} = \frac{\partial Q_{Ti}}{\partial \delta_i} = \sum_{\substack{j=1 \\ j \neq i}}^{n} V_i V_j Y_{ij} \cos(\delta_i - \delta_j - \gamma_{ij}) \qquad (7.58f)$$

$$\frac{\partial g_i}{\partial y_j} = \frac{\partial Q_{Ti}}{\partial V_j} = V_i Y_{ij} \sin(\delta_i - \delta_j - \gamma_{ij}) \qquad (7.58g)$$

$$\frac{\partial g_i}{\partial y_i} = \frac{\partial Q_{Ti}}{\partial V_i} = V_i Y_{ii} \sin(-\gamma_{ii}) + \sum_{j=1}^{n} V_j Y_{ij} \sin(\delta_i - \delta_j - \gamma_{ij}) \qquad (7.58h)$$

The basic Jacobian matrix J is now formulated. Interpretations and modifications for specific bus types are necessary and we discuss each type separately. Understand the subscript i to indicate the bus under discussion.

241

TYPE 0.

No iterative calculations are required at the reference bus. The Jacobian matrix may be reduced by two orders; we delete the appropriate row and column, eliminate $\Delta\delta_i$, ΔV_i, ΔP_i, ΔQ_i as variables and renumber the remaining busses.

TYPE 1.

The results computed from equation (7-55) can be directly applied. We update δ_i and V_i by:

$$\delta_i^{k+1} = \delta_i^k + \Delta\delta_i^k \tag{7-59a}$$

$$V_i^{k+1} = V_i^k + \Delta V_i^k \tag{7-59b}$$

TYPE 2.

There is no need of the g_i function at a generator bus since the voltage is known. We therefore remove that relation from the equation set represented by (7-54), reducing the size of the Jacobian by one order. We calculate:

$$\delta_i^{k+1} = \delta_i^k + \Delta\delta_i^k \tag{7-59a}$$

At the end of the iteration we compute Q_{Gi} from (7-19b) using $V_{i_{\text{spec}}}$ for V_i. Now check $Q_{Gi_{\min}} \leq Q_{Gi} \leq Q_{Gi_{\max}}$. If Q_{Gi} is within limits, no change in V_i is allowed; V_i is kept at $V_{i_{\text{spec}}}$. If not, Q_{Gi} is set at the appropriate limit and the function g_i is returned to our equation set; in other words, we treat the bus as type one for the next iteration, allowing V_i to vary. We must check Q_{Gi} each iteration since it is possible that as other bus values change, it may wander in and out of limits.

TYPE 3.

Several techniques for dealing with type 3 busses are possible. One approach is to treat them as type one busses when solving (7-55). We then compare V_i with $V_{i_{\text{spec}}}$. If the voltage is too low (or high) we adjust the tap setting c as explained in section 7-2. Now calculate the real power flow P_{ij} from equation (7-26b). If it is out of tolerance, adjust α as explained in section 7-2. With the adjusted $c = c/\alpha$, we calculate new y parameters from equations (5-46). Finally, we update the appropriate value in $[Y_{\text{bus}}]$ using equations (7-35) and we are ready for the next iteration.

An alternative approach is to make the complex turns ratio a variable and derive an appropriate equation to enter into the Jacobian. This latter method is superior and will result in faster convergence. The reader should consult [5] in the end-of-chapter Bibliography for specific details of implementation.

We should summarize the use of the Newton–Raphson Method. Starting with initial estimates for $\tilde{\delta}$ and $\tilde{V}$ we can calculate corresponding values for $\tilde{\Delta}f$ and $\tilde{\Delta}g$ and the Jacobian matrix J. Subject to the modifications caused by the four bus types, we invert J and calculate $\tilde{\Delta}\delta$ and $\tilde{\Delta}V$ from equation (7-55). Again, subject to the constraints required at the different bus types, we update $\tilde{\delta}$ and $\tilde{V}$ using $\tilde{\Delta}\delta$ and $\tilde{\Delta}V$. We repeat the process until convergence is obtained, or until the maximum allowable number of iterations is reached. Assuming that convergence is achieved, the rest of the calculation proceeds exactly in the same manner as discussed for the Gauss–Seidel method (refer back to section 7-2).

For educational reasons the Newton–Raphson method has been presented in terms of complex variables interpreted in polar form ($V_i = V_i \underline{/\delta_i}$; $Y_{ij} = Y_{ij} \underline{/\gamma_{ij}}$). Experience shows that machine computation proceeds faster if the equations are developed in rectangular form ($V_i = E_i + jF_i$; $Y_{ij} = G_i + jB_i$). For this reason large commercial computer programs are developed on this basis. Also, the power flow may be developed from an impedance viewpoint. See [6] and [8] in the end-of-chapter Bibliography for further development of this important and interesting problem.

7-4 An Example Application

Up to this point we have been concerned with understanding the power flow problem and programming its solution. The problem has been discussed in sufficient detail to allow students to write their own power flow programs. We now consider the question of how such a program can be used.

A power flow program is an all-purpose analytical tool that can provide answers to a variety of questions concerning the balanced $3\emptyset$ electrical performance of a power system. Basically, we want to know:

- Are the voltages within tolerances at all busses?

- Are all transformer and line loadings within device ratings?

The above questions must be answered for a variety of load and generation configurations. Also, the optimum placement of new lines must be studied. Power flows between interconnected systems are of interest. To demonstrate the kind of situation suitable for power flow analysis, consider example 7-5.

Example 7-5

Consider the system shown in Figure 7-5. The input data for the system follows in Figures 7-6 and 7-7. For the most part the values are self-explanatory, but a few clarifying comments should be helpful. All values are in per unit on a common base.

R, X = Serial impedance values
G, B = Shunt admittance values
SRAT = Apparent power rating

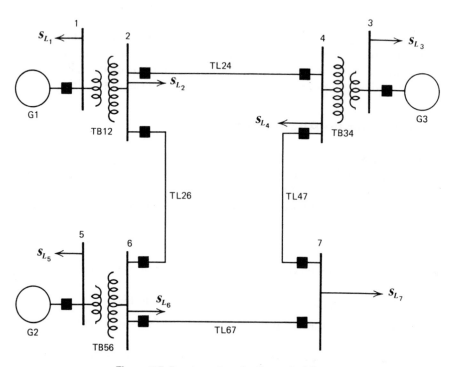

Figure 7-5 Power system for Example 7-5.

Example 7-5

* * * * * * INPUT DATA * * * * * *

7 BUSSES,   4 LINES,   3 TRANSFORMERS

* * * LINE DATA * * *

BUS TO BUS		R	X	G	B	SRAT
2	6	0.02300	0.13800	0.00000	0.27100	2.000
4	7	0.02300	0.13800	0.00000	0.27100	2.000
2	4	0.01500	0.09200	0.00000	0.18100	2.000
6	7	0.01500	0.09200	0.00000	0.18100	2.000

* * * TRANSFORMER DATA * * *

BUS TO BUS		R	X	G	B	SRAT
1	2	0.00120	0.01500	0.00000	0.00000	4.000
3	4	0.00100	0.01200	0.00000	0.00000	5.000
5	6	0.00200	0.02400	0.00000	0.00000	2.500

Figure 7-6 Input data for Example 7-5, Case I.

* * * * * BUS DATA * * * * * *

BUS	TYPE	VMAG	VANG	PG	PL	QL	QMAX	QMIN
1	3	1.000	0.000	0.000	0.000	0.000	*****	*****
2	1	1.000	0.000	0.000	2.000	0.300	*****	*****
3	2	1.050	0.000	5.000	0.600	0.080	4.000	-3.000
4	1	1.000	0.000	0.000	2.000	0.200	*****	*****
5	2	1.050	0.000	2.000	0.500	0.050	1.400	-1.000
6	1	0.950	0.000	0.000	1.000	0.300	*****	*****
7	1	0.850	0.000	0.000	4.000	1.000	*****	*****

****

Figure 7-7 Bus input data for Example 7-5, Cases I and II.

In Figure 7-7 reading from left to right, the bus data values are:

BUS = Bus number
TYPE = Code number: 1 = load; 2 = generator; 3 = reference
VMAG = Specified voltage magnitude for reference and generator
    buses; starting value for load busses.
VANG = Starting angle in degrees for all busses.

PG = Generated power.
PL = Load real power
QL = Load reactive power
QMAX, QMIN = Reactive power limits for generators

Suppose that the given loads represent maximum expected values, and the generation is set to full capacity. We define satisfactory system operation to mean that no transformers or lines are overloaded, and that voltage at all points in the system is within ±5% of nominal (i.e. $0.95 \le V_i \le 1.05$). Let our problem then be:

(a) Show that system performance is satisfactory (or unsatisfactory) at the maximum load condition.
(b) If system operation is unsatisfactory investigate the effects of adding a fifth line TL27 with parameters as given in Figure 7-10.·

Solution

The system as given is designated as Case I. The power flow program is run and results are presented in Figures 7-8 and 7-9. We observe:

1. The voltage at bus 7 is low (0.864).
2. Line TL47 is overloaded.
3. Line TL67 is overloaded.

POWER FLOW CONVERGED, 43 ITERATIONS REQUIRED

*************** FINAL VALUES *****************

* * * * * BUS DATA * * * * * *

BUS	TYPE	VMAG	VANG	PG	QG	PL	QL	DELP
1	3	1.000	0.000	3.415	0.001	0.000	0.000	-0.000
2	1	0.997	-2.944	0.000	-0.000	2.000	0.300	0.000
3	2	1.050	-0.855	5.000	1.840	0.600	0.080	0.000
4	1	1.027	-3.568	0.000	-0.000	2.000	0.200	-0.000
5	2	1.007	-11.820	2.000	1.400	0.500	0.050	0.000
6	1	0.972	-13.769	0.000	-0.000	1.000	0.300	-0.000
7	1	0.864	-23.996	0.000	0.000	4.000	1.000	-0.000

*** LOW VOLTAGE AT BUS 7 ***

Figure 7-8 Final bus data for Example 7-5, Case I.

246

```
* * * * * * POWER FLOWS * * * * *


    * * * LINE DATA * * *

BUS TO BUS         P           Q          S

  2   6        -1.333      -0.052     1.334
  6   2        -1.292       0.037     1.293

  4   7         2.445      -1.068     2.668      ***OVERLOADED***
  7   4        -2.282      -0.338     2.307

  2   4         0.067      -0.422     0.427
  4   2        -0.065       0.247     0.255

  6   7        -1.784       0.917     2.006      ***OVERLOADED***
  7   6        -1.718      -0.662     1.841

     * * * TRANSFORMER DATA * * *

BUS TO BUS         P           Q          S


  1   2        -3.415       0.001     3.415
  2   1        -3.401       0.174     3.405

  3   4        -4.400       1.760     4.738
  4   3        -4.379      -1.515     4.634

  5   6         1.500       1.350     2.018
  6   5        -1.492      -1.253     1.948

   * * * * * SYSTEM TOTALS * * * * * *


   GENERATION: PG = 10.415    QG =    3.240

     LOAD: PL = 10.100    QL =    1.930
```

Figure 7-9 Final-line data for Example 7-5, Case I.

To correct the situation we are to investigate the addition of a fifth transmission line to run from bus 2 to bus 7 (TL27). The system so modified is designated as Case II. The bus input data is identical to Case I. Results are presented in Figures 7-10, 7-11, and 7-12. We observe:

1. No lines are overloaded.
2. The voltage at Bus 7 has risen but it is still low (0.946).

Connecting capacitance at a bus generally has the effect of raising the voltage at that bus. Let us try this technique to raise the voltage at bus 7. We observe from Figure 7-7 that $Q_L = 1.0$ at bus 7. Positive Q is indicative of an inductive load. Addition of capacitance will add negative Q, resulting in a net reduction of Q_L. Let us change Q_L to 0.2, implying an addition of "0.8

* * * * * INPUT DATA * * * * * *

7 BUSSES, 5 LINES, 3 TRANSFORMERS
* * * LINE DATA * * *

BUS TO BUS		R	X	G	B	SRAT
2	7	0.02700	0.16600	0.00000	0.32600	2.000
2	6	0.02300	0.13800	0.00000	0.27100	2.000
4	7	0.02300	0.13800	0.00000	0.27100	2.000
2	4	0.01500	0.09200	0.00000	0.18100	2.000
6	7	0.01500	0.09200	0.00000	0.18100	2.000

* * * TRANSFORMER DATA * * *

BUS TO BUS		R	X	G	B	SRAT
1	2	0.00120	0.01500	0.00000	0.00000	4.000
3	4	0.00100	0.01200	0.00000	0.00000	5.000
5	6	0.00200	0.02400	0.00000	0.00000	2.500

Figure 7-10 Input line data for Example 7-5, Cases II and III.

POWER FLOW CONVERGED, 23 ITERATIONS REQUIRED

***************** FINAL VALUES *****************

* * * * * BUS DATA * * * * * *

BUS	TYPE	VMAG	VANG	PG	QG	PL	QL	DELP
1	3	1.000	0.000	3.298	-0.187	0.000	0.000	-0.000
2	1	1.000	-2.848	0.000	-0.000	2.000	0.300	-0.000
3	2	1.050	2.523	5.000	1.143	0.600	0.080	-0.000
4	1	1.035	-0.206	0.000	-0.000	2.000	0.200	0.000
5	2	1.050	-6.658	2.000	1.194	0.500	0.050	-0.000
6	1	1.021	-8.459	0.000	-0.000	1.000	0.300	0.000
7	1	0.946	-14.210	0.000	-0.000	4.000	1.000	0.000

*** LOW VOLTAGE AT BUS 7 ***

Figure 7-11 Final bus data for Example 7-5, Case II.

```
* * * * * POWER FLOWS * * * * *

* * * LINE DATA * * *
```

BUS TO BUS	P	Q	S
2 7	-1.163	0.083	1.166
7 2	-1.125	-0.157	1.136
2 6	0.685	-0.369	0.778
6 2	-0.673	0.165	0.693
4 7	1.813	0.427	1.862
7 4	-1.735	-0.228	1.750
2 4	-0.563	-0.364	0.671
4 2	0.569	0.213	0.608
6 7	-1.166	0.602	1.313
7 6	-1.140	-0.615	1.295

```
* * * TRANSFORMER DATA * * *
```

BUS TO BUS	P	Q	S
1 2	3.298	-0.187	3.303
2 1	-3.285	0.351	3.304
3 4	4.400	1.063	4.527
4 3	-4.382	-0.840	4.462
5 6	1.500	1.144	1.887
6 5	-1.494	-1.067	1.836

```
* * * * * SYSTEM TOTALS * * * * * *

GENERATION: PG = 10.298    QG = 2.150

LOAD: PL = 10.100    QL = 1.930
```

Figure 7-12 Final line data for Example 7-5, Case II.

worth" of capacitors (i.e., if $S_{base} = 100$ MVA, we are installing a capacitor bank rated at 80 Mvar). Otherwise the system is unchanged; let us designate it as Case III. Again a power flow case is run and the results are presented in Figures 7.13 and 7-14. We observe:

1. All voltages are within tolerance.
2. No lines or transformers are overloaded.

∴ The proposed modifications will result in satisfactory system performance. The value $Q_c = -0.8$ was selected arbitrarily; if the voltage at bus 7 was not satisfactory, we could easily determine a more appropriate value with additional runs. A knowledgeable power system engineer soon

```
POWER FLOW CONVERGED, 28 ITERATIONS REQUIRED

*************** FINAL VALUES *****************

      * * * * * BUS DATA * * * * * *
```

BUS	TYPE	VMAG	VANG	PG	QG	PL	QL	DELP
1	3	1.000	0.000	3.282	-0.455	0.000	0.000	-0.000
2	1	1.004	-2.841	0.000	-0.000	2.000	0.300	-0.000
3	2	1.050	2.500	5.000	0.833	0.600	0.080	-0.000
4	1	1.038	-0.236	0.000	-0.000	2.000	0.200	0.000
5	2	1.050	-6.470	2.000	0.822	0.500	0.050	-0.000
6	1	1.030	-8.296	0.000	-0.000	1.000	0.300	0.000
7	1	0.990	-13.992	0.000	-0.000	4.000	0.200	0.000

Figure 7-13 Final bus data for Example 7-5, Case III.

develops a feel for the particular system and from experience can estimate such values quite accurately. We see that the power flow program can be a valuable analytical tool to build and verify such experience.

7-5 Reducing Storage Requirements for [Y_{bus}]

The algorithm for formulating [Y_{bus}] presented in Section 7-1 was correct and simple; however, its storage requirements are prohibitive when applied to a large system. The storage of a complete $n \times n$ [Y_{bus}] would require storing n^2 complex entries; for a 1000 bus problem, 10^6 complex entries would need to be stored, requiring a memory size of roughly 8000K. We observe that [Y_{bus}] is a sparse matrix, meaning that most of its entries are zero. In a practical system a given bus will average from 1.5 to 2.5 terminating lines. This means that a given row (or column) of [Y_{bus}] will average about three (the diagonal plus about two off-diagonal) nonzero entries. If only these values are stored, our 1000 bus example would require a memory size of about 40K, allowing for double entires of y_{ij} terms (y_{ij} and y_{ji} terms are both stored).

A program for storing Y_{bus} in a vector is outlined in Table 7-2. The following notation is defined:

$YA(K) =$ Original Kth branch admittance, terminated at bus $IA(K)$ and $JA(K)$. Entered in random order. NE of these.

★ ★ ★ ★ ★ POWER FLOWS ★ ★ ★ ★ ★

★ ★ ★ LINE DATA ★ ★ ★

BUS	TO BUS	P	Q	S
2 7	7 2	-1.160 -1.124	-0.155 0.053	1.170 1.125
2 6	6 2	0.668 -0.656	-0.402 0.193	0.780 0.684
4 7	7 4	-1.817 -1.745	0.128 0.026	1.822 1.745
2 4	4 2	-0.559 0.565	-0.362 0.209	0.666 0.602
6 7	7 6	-1.151 -1.131	0.217 -0.278	1.172 1.165

★ ★ ★ TRANSFORMER DATA ★ ★ ★

BUS	TO BUS	P	Q	S
1 2	2 1	3.282 -3.269	-0.455 0.620	3.313 3.327
3 4	4 3	4.400 -4.382	0.753 -0.536	4.464 4.415
5 6	6 5	-1.500 -1.495	0.772 -0.710	1.687 1.655

★ ★ ★ ★ SYSTEM TOTALS ★ ★ ★ ★ ★ ★

GENERATION: PG = 10.282   QG = 1.200

LOAD: PL = 10.100   QL = 1.130

Figure 7-14 Final line data for Example 7-5, Case III.

$YB(K)$ = Ordered Kth branch admittance, terminated at busses $IB(K)$ and $JB(K)$. Ordered first by consecutive IB number, and second, for a given IB, by consecutive JB number. Two NE of these.

NB = No. of busses, including the reference, which is numbered last.

NE = No. of elements.

$Y(L)$ = The vector Y_{bus}; the first NB entries are the y_{ii} values, where i = bus number. The next $2(NE)$ entries are the y_{ij} values, doubly entered, in the same order as explained for $YB(K)$.

251

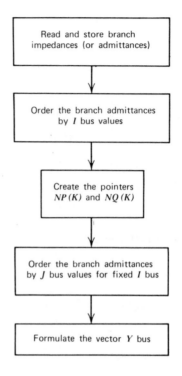

$NP(K) =$ The vector location for the first of $NQ(K)$ y_{ij} values for the Kth bus.

Once the branch admittances are ordered as described above, the construction of the vector Y_{bus} is simple:

$$Y(L) = \text{Summation of } YB(K) \text{ from } K = N1 \text{ to } K = N2 \text{ where:} \qquad (7\text{-}60a)$$

$$N1 = NP(L) \qquad (7\text{-}60b)$$

$$N2 = NP(L) + NQ(L) - 1 \qquad (7\text{-}60c)$$

$$Y(L) = -YB(L) \qquad NB \leq L \leq NB + 2(NE) \qquad (7\text{-}60d)$$

An example will clarify the situation

Example 7-6

Evaluate the vector Y_{bus} for the circuit of Figure 7-15.

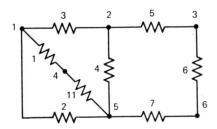

Figure 7-15 Circuit for Example 7-5, all values admittances.

Solution

The method presented in this section was programmed. Results are presented in Figures 7-16 and 7-17. We observe:

$$NB = 6$$

$$NE = 8$$

The branch admittances are first listed in the order in which they are read, which was random. For comparison the Y matrix was formulated by conventional methods (equation 7-9) and is displayed in standard form in Figure 7-16. The branch admittances are then ordered as explained for $YB(K)$ and listed.

Now refer to Figure 7-17. The vector formulation of $[Y]_{bus}$ has been completed and is displayed. The locater parameters NP and NQ are also shown. For example consider

$$NP(2) = 10 \text{ and}$$

$$NQ(2) = 3$$

This means the y_{2j} ($j \neq 2$) entries start at position 10 in the Y vector and there are 3 of them (-3, -5, and -4). The y_{22} entry is stored in position 2. The j index is also stored.

Note that the vector Y has 22 entries compared to Y_{bus}, which has 36. The savings are not impressive until one realizes that Y will increase in proportion to n while Y_{bus} increases at a rate proportional to n^2. The double entries (y_{ij} and y_{ji}) were made in Y for convenience of access. A single entry scheme would save even more storage.

7-6 Summary

The power flow problem is primarily concerned with the calculation of energy flows over the transmission lines of a system. The mathematical

NO. OF BUSSES=   6   NO. OF ELEMENTS=   8

BRANCH ADMITTANCES

IA	JA	YA
1	2	3.0
1	5	2.0
4	5	11.0
1	4	1.0
3	6	6.0
2	5	4.0
5	6	7.0
2	3	5.0

CONVENTIONAL Y MATRIX

6.0	−3.0	0.0	−1.0	−2.0	0.0
−3.0	12.0	−5.0	0.0	−4.0	0.0
0.0	−5.0	11.0	0.0	0.0	−6.0
−1.0	0.0	0.0	12.0	−11.0	0.0
−2.0	−4.0	0.0	−11.0	24.0	−7.0
0.0	0.0	−6.0	0.0	−7.0	13.0

ORDERED BRANCH ADMITTANCES

IB	JB	YB
1	2	3.0
1	4	1.0
1	5	2.0
2	1	3.0
2	3	5.0
2	5	4.0
3	2	5.0
3	6	6.0
4	1	1.0
4	5	11.0
5	1	2.0
5	2	4.0
5	4	11.0
5	6	7.0
6	3	6.0
6	5	7.0

Figure 7-16 Data for Example 7-6.

constraints imposed on variables at a given bus must be harmonious with control actions that are physically possible. Another matter of concern is the voltage profile of the system, requiring that all bus voltages be within a specified tolerance of a nominal value.

The construction of $[Y_{bus}]$ is a basic consideration, requiring proper attention to computer storage constraints. Tap changing transformer operation may be modeled by making changes on appropriate entries in $[Y_{bus}]$.

The problem is solved using either Gauss–Seidel or Newton–Raphson numerical analysis methods. Comparing the two, Gauss–Seidel is simple to

Y MATRIX IN VECTOR FORM

K	Y(K)	I(K)	J(K)
1	6.0	1	1
2	12.0	2	2
3	11.0	3	3
4	12.0	4	4
5	24.0	5	5
6	13.0	6	6
7	-3.0	1	2
8	-1.0	1	4
9	-2.0	1	5
10	-3.0	2	1
11	-5.0	2	3
12	-4.0	2	5
13	-5.0	3	2
14	-6.0	3	6
15	-1.0	4	1
16	-11.0	4	5
17	-2.0	5	1
18	-4.0	5	2
19	-11.0	5	4
20	-7.0	5	6
21	-6.0	6	3
22	-7.0	6	5

LOCATER PARAMETERS NP AND NQ

BUS	NP	NQ
1	7	3
2	10	3
3	13	2
4	15	2
5	17	4
6	21	2

Figure 7-17 Vector formulation for Y bus.

program, performs a single iteration quickly, but requires many iterations to converge, the number increasing with system size. A Newson–Raphson iteration is quite complex, requiring a large matrix inversion, but fewer are required and the number required for convergence is essentially independent of system size.

The power flow calculation is a basic and useful tool of the power system engineer. It is used to evaluate the impact of the addition or removal of lines and generators. For the nonpower specialty student it serves as an example of a problem of large dimension involving nonlinear algebraic equations that must be solved on a computer.

Bibliography

[1] Brown, H. E., Carter, G. K., and Person, C. E., "Power flow solution by impedance matrix method," *Trans. AIEE*, vol. 82, Part 3 (1963), p. 1.

[2] Brown, Homer E., *Solution of Large Networks by Matrix Methods*, John Wiley and Sons, Inc., 1975.

[3] Brown, R. J., and Tinney, W. F., "Digital solution for large power networks," *Trans. AIEE*, vol. 76, Part 3 (1957), pp. 347–355.

[4] Elgerd. Olle I., *Electric Energy Systems Theory: An Introduction*, McGraw-Hill Inc., New York, 1971.

[5] Peterson, N. M., and Meyer, W. S., "Automatic adjustment of transformer and phase shifter taps in the Newton power flow," *IEEE Trans. PAS*, vol. 90 (January-February 1971), pp. 103–108.

[6] Stagg, Glenn W., and El-Abiad, Ahmed H., *Computer Methods in Power System Analysis*, McGraw-Hill, Inc., New York, 1968.

[7] Stevenson, Jr., William D., *Elements of Power Systems Analysis*, 3rd edition. McGraw-Hill, Inc., New York, 1975.

[8] Tinney, W. F., and Hart, C. E., "Power flow solutions by Newton's method," *Trans. IEEE PAS*, vol. 86 (November 1967), p. 1449.

[9] Sato, N., and Tinney, W. F., "Technique for exploiting the sparsity of the network admittance matrix," *Trans. IEEE PAS*, vol. 82 (December 1963), p. 944.

[10] Tinney, W. F., and Walker, J. W., "Direct solution of sparse network equations by optimally ordered triangular factorization," *Proc. IEEE*, vol. 55 (November 1967), pp. 1801–1809.

[11] Ward, J. B., and Hale, H. W., "Digital computer solution of power flow problems," *Trans. AIEE*, vol. 75, Part 3 (June 1956), pp. 398–404.

[12] Van Ness, J. E., and Griffin, J. H., "Elimination methods for load flow studies," *Trans. AIEE*, vol. 80, Part 3 (1961), p. 299.

Problems

7-1. Consider $f(x) = x^2 - x - 2 = 0$
   (a) Solve for x algebraically.
   (b) Applying the Gauss-Seidel method, start from $x^0 = 1$ using:

$$x^{i+1} = \sqrt{x^i + 2}$$

   to compute x to four places.
   (c) Repeat (b) for $x^0 = 3$
   (d) Repeat (b) for $x^0 = -2$

(e) Repeat (b) for $x^0 = -3$

(f) Repeat (b) using:

$$x^{i+1} = -\sqrt{x^i + 2}$$

starting from $x^0 = 0$.

7-2 Repeat problem 7-1 using the Newton–Raphson method for starting values of:

(a) $x^0 = -2$

(b) $x^0 = +3$

(c) $x^0 = +1/2$

7-3 In the given dc circuit

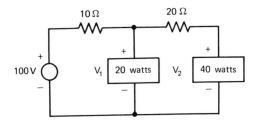

(a) Show that:

$$V_1 = \sqrt{100 V_1 - 200 - 400 \, V_1/V_2}$$

$$V_2 = \sqrt{V_1 V_2 - 800}$$

(b) Solve for V_1 and V_2 to three places starting from $V_1^0 = V_2^0 = 100$, by Gauss–Seidel methods.

7-4. Repeat problem 7-3 using Newton–Raphson methods.

7-5. Consult example 7-5. Use the results from Case III Figure 7-13 to compute the line flows 2-7 and 7-2 tabulated in Figure 7-14.

7-6. Consult example 7-5. For Case III results refer to Figure 7-14. Verify all system totals. Explain the discrepancy between real and reactive generation and load.

7-7. Write $[Y_{bus}]$ for the circuit of Figure 7-15 by inspection. Check with the result published in Figure 7-16.

7-8. The $[Y_{bus}]$ in problem 7-7 is 6×6. If Bus 6 is considered the reference, $[Y_{bus}]$ is considered as 5×5. How do we form $[Y_{bus}]_{5 \times 5}$ from $[Y_{bus}]_{6 \times 6}$?

7-9. Consult the input impedance data for example 7-5 as listed in Figure

7-6. List the branch admittances along with their terminating busses. (This problem, along with the next two, can be done by computer.)

7-10. From problem 7-9 formulate $[Y_{bus}]$, using equation (7-9).

7-11. From problem 7-10, write $[Y_{bus}]$ in vector form as explained in section 7-5.

BALANCED AND UNBALANCED FAULTS

"Nurture strength of spirit to shield you in sudden misfortune.
But do not distress yourself with dark imaginings.
Many fears are born of fatigue and loneliness."

Max Ehrmann, DESIDERATA

The normal operating mode of a power system is balanced three phase ac. A number of undesirable but unavoidable incidents can temporarily disrupt this condition. If the insulation of the system should fail at any point, or if a conducting object should come in contact with a bare power conductor, a "short circuit" or "fault" is said to occur. The causes of faults are many: they include lightning, wind damage, trees falling across lines, vehicles colliding with towers or poles, birds shorting out lines, aircraft colliding with lines, vandalism, small animals entering switchgear, and line breaks due to excessive ice loading. Power system faults may be categorized as one of four types in order of frequency of occurrence: single line to ground, line to line, double line to ground, and balanced three phase. The first three types constitute severe unbalanced operating conditions.

It is important to determine the values of system voltages and currents during faulted conditions so that protective devices may be set to detect and minimize the harmful effects of such contingencies. Therefore, it is necessary to analyze the power system operating in unbalanced modes. The time constants of the associated transients are such that sinusoidal steady state methods may still be used. The method of symmetrical components is admirably suited to unbalanced system analysis.

Our objective in this chapter is to understand how symmetrical components may be applied specifically to the four general fault types mentioned, and how the method can be extended to any unbalanced three phase system problem. Although the three phase fault is the least common and does not represent an unbalanced condition, we begin with it because of its simplicity. The approach used is followed consistently for all four fault types.

8-1 Simplifications in the System Model

We shall use the zero, positive, and negative sequence equivalent circuits for the power system. Certain simplifications are possible that will not substantially affect the accuracy of our results. These include the following.

- Shunt elements in the transformer model that account for magnetizing current and core loss are neglected.

- Shunt capacitance in the line model is ignored.

- Sinusoidal steady state circuit analysis techniques are used. The so-called dc offset is accounted for by using correction factors.

• We set all internal system voltage sources to $1\underline{/0^\circ}$. We select unity by arguing that the system voltage is at its nominal value prior to the application of a fault, which is reasonable. The selection of zero phase for one source is arbitrary and convenient. Assuming that all sources are in phase and of the same magnitude is equivalent to neglecting prefault load current. When desirable we shall account for load current using superposition.

In addition to the above, for hand calculations and educational purposes, we shall neglect series resistance (this approximation will not be necessary for computer solution). Also, the only difference in the positive and negative sequence networks is introduced by the machine impedances. If we select the subtransient reactance X''_d for the positive sequence reactance, the difference is slight (in fact, the two are identical for nonsalient machines). The simplication is important, as we shall see in Chapter 9, since we reduce computer storage requirements by roughly one-third.

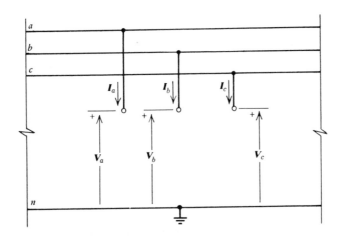

Figure 8-1 General three phase access port.

Our basic approach to the problem is to consider the general situation suggested in Figure 8-1. The general terminals brought out are for purposes of external connections that will simulate faults. Note carefully the positive assignments of phase quantities. Particularly note that the currents flow *out of* the system. Recall that we do not have a *phase* circuit model for the power system.

We can construct general *sequence* equivalent circuits for the system, and such circuits are indicated in Figure 8-2. The ports indicated

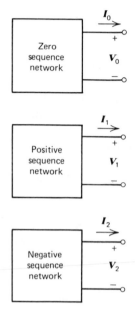

Figure 8-2 General sequence networks.

correspond to the general three phase entry port of Figure 8-1. Note the positive sense of sequence values are compatible with those for phase values.

An example system is introduced in example 8-1 that is used throughout this chapter. Consider the following.

Example 8-1

Consider the system shown in Figure 8-3. The system data is as follows:

System data

ITEM	MVA RATING	VOLTAGE RATING	X_1	X_2	X_0
G1	100	25 kV	0.2	0.2	0.05
G2	100	13.8 kV	0.2	0.2	0.05
T1	100	25/230 kV	0.05	0.05	0.05
T2	100	13.8/230 kV	0.05	0.05	0.05
TL12	100	230 kV	0.1	0.1	0.3
TL23	100	230 kV	0.1	0.1	0.3
TL13	100	230 kV	0.1	0.1	0.3

Example 8-1

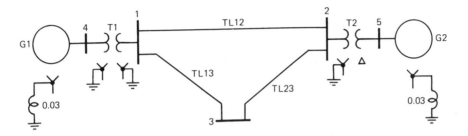

Figure 8-3 Example system for Chapter 8.

Pu Bases: 100 Mva

$\underline{25\,kV}$ $\quad I = \dfrac{100}{0.025\sqrt{3}} = 2310$ A

$\underline{230\,kV}$ $\quad I = \dfrac{100}{0.23\sqrt{3}} = 251$ A

$\underline{13.8\,kV}$ $\quad I = \dfrac{100}{0.0138\sqrt{3}} = 4184$ A

(a) Draw the sequence networks.
(b) Reduce the networks in (a) to their Thevenin equivalents "looking in" at bus 3.

Solution

(a) The sequence networks are shown in Figure 8-4.
(b) We start with the zero sequence network. Using series combinations:

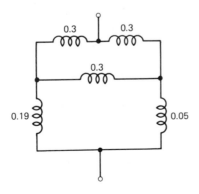

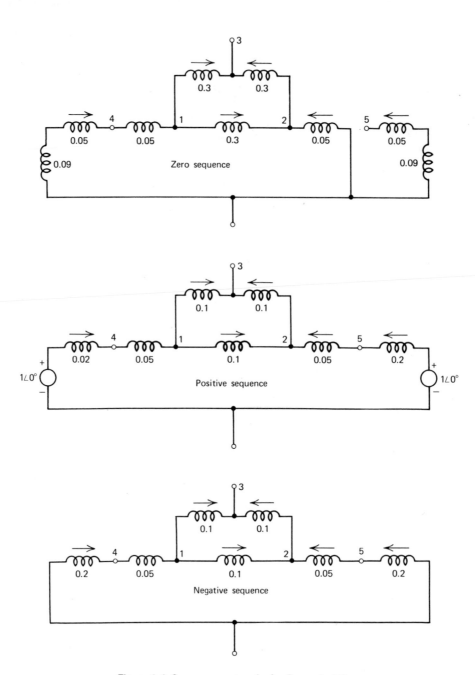

Figure 8-4 Sequence networks for Example 8-1.

Using a delta-wye transformation:

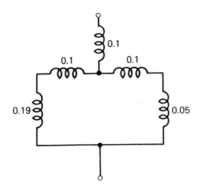

Series/parallel combinations produce:

Now consider the positive sequence network. The Thevenin voltage is $1\underline{/0°}$. The Thevenin impedance may be computed by repeating the steps used for zero sequence. Thus:

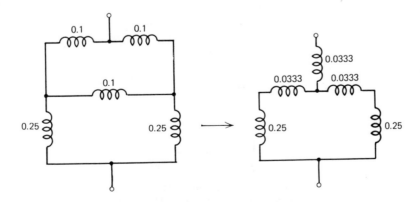

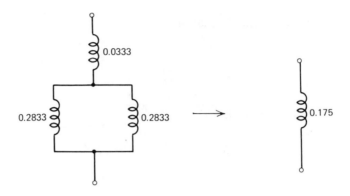

The negative sequence impedance is identical. Results are presented in Figure 8-5.

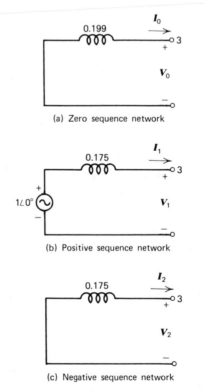

Figure 8-5 Sequence networks for Example 8-1. Thevenin equivalents looking into bus 3.

8-2 The Balanced Three Phase Fault

Imagine our general three phase access port terminated in a fault impedance Z_f as shown in Figure 8-6. Typically, Z_f is set to zero for fault study

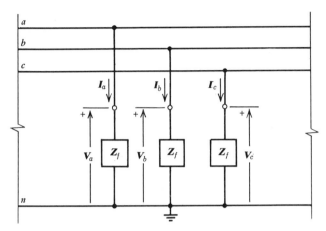

Figure 8-6 General balanced 3ϕ fault.

purposes; we include it in our development only for the sake of generality. The terminal conditions are such that we may write:

$$V_a = I_a Z_f \tag{8-1a}$$

$$V_b = I_b Z_f \tag{8-1b}$$

$$V_c = I_c Z_f \tag{8-1c}$$

In matrix form:

$$\begin{bmatrix} V_a \\ V_b \\ V_c \end{bmatrix} = \begin{bmatrix} Z_f & 0 & 0 \\ 0 & Z_f & 0 \\ 0 & 0 & Z_f \end{bmatrix} \begin{bmatrix} I_a \\ I_b \\ I_c \end{bmatrix} \tag{8-1d}$$

Using equation (2-60):

$$[Z_{012}] = [T]^{-1} \begin{bmatrix} Z_f & 0 & 0 \\ 0 & Z_f & 0 \\ 0 & 0 & Z_f \end{bmatrix} [T] \tag{8-2a}$$

$$= \begin{bmatrix} Z_f & 0 & 0 \\ 0 & Z_f & 0 \\ 0 & 0 & Z_f \end{bmatrix} \tag{8-2b}$$

It follows that:

$$V_0 = Z_f I_0 \qquad\qquad (8\text{-}3a)$$

$$V_1 = Z_f I_1 \qquad\qquad (8\text{-}3b)$$

$$V_2 = Z_f I_2 \qquad\qquad (8\text{-}3c)$$

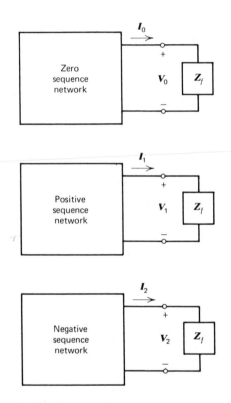

Figure 8-7 Sequence network terminations for a balanced 3ϕ fault.

The corresponding network connections are given in Figure 8-7. Because the zero and negative sequence networks are passive, only the positive sequence network is nontrivial.

$$V_0 = V_2 = 0 \qquad\qquad (8\text{-}4a)$$

$$I_0 = I_2 = 0 \qquad\qquad (8\text{-}4b)$$

We apply these general results to the system of example 8-1.

Example 8-2

Example 8-2

A balanced three phase fault occurs at bus 3 in the system of example 8-1. Compute the fault currents and voltages.

Solution

The sequence networks are as shown in Figure 8-8. Obviously:

$$V_0 = I_0 = V_2 = I_2 = 0$$

We calculate

$$I_1 = \frac{1\underline{/0^\circ}}{j0.175} = -j5.71$$

Also

$$V_1 = 0$$

To compute the phase values we use equations (2-53) and (2-55):

$$\begin{bmatrix} V_a \\ V_b \\ V_c \end{bmatrix} = [T] \begin{bmatrix} 0 \\ 0 \\ 0 \end{bmatrix} = \begin{bmatrix} 0 \\ 0 \\ 0 \end{bmatrix}$$

$$\begin{bmatrix} I_a \\ I_b \\ I_c \end{bmatrix} = \begin{bmatrix} 1 & 1 & 1 \\ 1 & a^2 & a \\ 1 & a & a^2 \end{bmatrix} \begin{bmatrix} 0 \\ -j5.71 \\ 0 \end{bmatrix} = \begin{bmatrix} 5.71\underline{/-90^\circ} \\ 5.71\underline{/150^\circ} \\ 5.71\underline{/30^\circ} \end{bmatrix}$$

It is useful to reflect on just what currents we have calculated. Refer back to Figure 8-6. We calculated the *fault* values; there are several other "I_a's" throughout the system. If we wish to compute the currents in, for example, TL13, the reduced simplified networks of Figure 8-8 are not adequate—we would need the more detailed networks of Figure 8-4.

8-3 The Single Line to Ground Fault

Imagine our general three phase access port terminated as shown in Figure 8-9. The terminal conditions are such that we may write:

$$I_b = 0 \tag{8-5a}$$

$$I_c = 0 \tag{8-5b}$$

$$V_a = I_a Z_f \tag{8-5c}$$

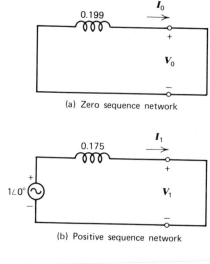

(a) Zero sequence network

(b) Positive sequence network

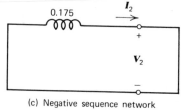

(c) Negative sequence network

Figure 8-8 Sequence networks for Example 8-2. Balanced 3ϕ fault applied at bus 3.

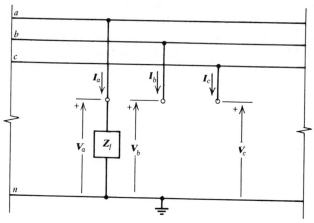

Figure 8-9 General single line to ground fault.

Obviously, from (8-5a) and (8-5b)

$$I_b = I_c \qquad (8\text{-}6)$$

We use equation (2-55) to conclude that equation (8-6) requires:

$$I_0 + a^2 I_1 + a I_2 = I_0 + a I_1 + a^2 I_2 \qquad (8\text{-}7a)$$

We manipulate (8-7a) to:

$$(a^2 - a)I_1 = (a^2 - a)I_2 \qquad (8\text{-}7b)$$

or

$$I_1 = I_2 \qquad (8\text{-}7c)$$

Also

$$I_b = I_0 + a^2 I_1 + a I_2 = 0 \qquad (8\text{-}8a)$$
$$I_0 + (a^2 + a)I_1 = 0 \qquad (8\text{-}8b)$$
$$I_0 = -(a^2 + a)I_1 \qquad (8\text{-}8c)$$
$$I_0 = I_1 \qquad (8\text{-}8d)$$

Furthermore equation (8-5c) requires that:

$$V_a = Z_f I_a \qquad (8\text{-}5c)$$
$$\therefore \quad V_0 + V_1 + V_2 = Z_f(I_0 + I_1 + I_2) \qquad (8\text{-}9a)$$
$$V_0 + V_1 + V_2 = 3 Z_f I_1 \qquad (8\text{-}9b)$$

In general, equations (8-7c), (8-8d), and (8-9b) must be simultaneously satisfied. These conditions can be met by interconnecting the sequence networks as shown in Figure 8-10. Study the circuit to be sure that you see this is true.

Example 8-3

A single line to ground fault occurs at bus 3 in the system of example 8-1. Compute the fault currents and voltages.

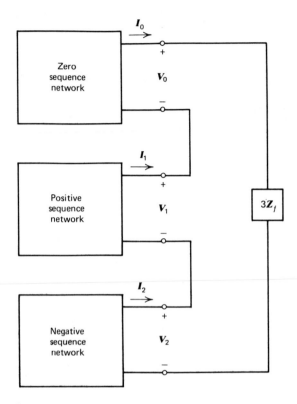

Figure 8-10 Sequence network terminations for a single line to ground fault.

Solution

The sequence networks are interconnected as shown in Figure 8-11. We write:

$$I_0 = I_1 = I_2 = \frac{1\underline{/0°}}{j0.199 + j0.175 + j0.175}$$

$$= -j1.82$$

From (2-55):

$$\begin{bmatrix} I_a \\ I_b \\ I_c \end{bmatrix} = \begin{bmatrix} 1 & 1 & 1 \\ 1 & a^2 & a \\ 1 & a & a^2 \end{bmatrix} \begin{bmatrix} -j1.82 \\ -j1.82 \\ -j1.82 \end{bmatrix} = \begin{bmatrix} -j5.46 \\ 0 \\ 0 \end{bmatrix}$$

272

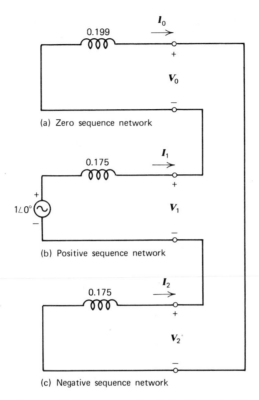

(a) Zero sequence network

(b) Positive sequence network

(c) Negative sequence network

Figure 8-11 Sequence networks for Example 8-3 single line to ground fault applied to bus 3.

The sequence voltages are:

$$V_0 = j0.199(-j1.82) = -0.362$$

$$V_1 = 1 - j0.175(-j1.82) = 0.681$$

$$V_2 = -j0.175(-j1.82) = -0.319$$

From (2-51) the phase voltages are:

$$\begin{bmatrix} V_a \\ V_b \\ V_c \end{bmatrix} = \begin{bmatrix} 1 & 1 & 1 \\ 1 & a^2 & a \\ 1 & a & a^2 \end{bmatrix} \begin{bmatrix} -0.362 \\ 0.681 \\ -0.319 \end{bmatrix} = \begin{bmatrix} 0 \\ 1.022\underline{/238°} \\ 1.022\underline{/122°} \end{bmatrix}$$

Observe that V_a, I_b, and I_c compute to zero as required. (Why?)

273

8-4 The Line to Line Fault

Imagine our general three phase access port terminated as shown in Figure 8-12. The terminal conditions are such that we may write:

$$I_a = 0 \tag{8-10a}$$

$$I_b = -I_c \tag{8-10b}$$

$$V_b = Z_f I_b + V_c \tag{8-10c}$$

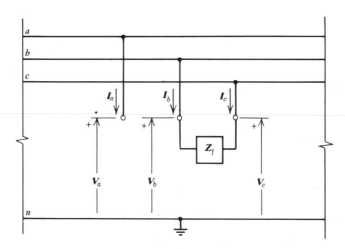

Figure 8-12 General line to line fault.

From (8-10a) it follows that:

$$I_0 + I_1 + I_2 = 0 \tag{8-11}$$

Equation (8-10b), transformed to sequence values, produces:

$$I_0 + a^2 I_1 + a I_2 = -(I_0 + a I_1 + a^2 I_2) \tag{8-12a}$$

Manipulating (8-12a):

$$2I_0 + (a^2 + a)(I_1 + I_2) = 0 \tag{8-12b}$$

Using (8-11) in (8-12b):

$$3I_0 = 0 \tag{8-13a}$$

or

$$I_0 = 0 \tag{8-13b}$$

274

Equation (8-11) then simplifies to:

$$I_1 = -I_2 \tag{8-14}$$

We transform (8-10c) to:

$$V_0 + a^2 V_1 + a V_2 = Z_f(I_0 + a^2 I_1 + a I_2) + V_0 + a V_1 + a^2 V_2 \tag{8-15a}$$

Simplifying and manipulating

$$(a^2 - a)V_1 = (a^2 - a)I_1 Z_f + (a^2 - a)V_2 \tag{8-15b}$$

or

$$V_1 = Z_f I_1 + V_2 \tag{8-15c}$$

In general, equations (8-13b), (8-14), and (8-15c) must be simultaneously satisfied. What interconnections between sequence networks can satisfy these constraints? The solution appears in Figure 8-13. Again we apply our general results to a particular problem.

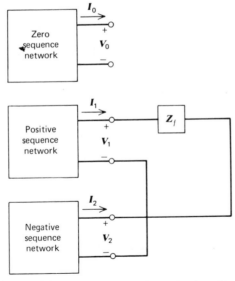

Figure 8-13 Sequence network terminations for a line to line fault.

Example 8-4

A line to line fault occurs at bus 3 in the system of example 8-1. Compute the fault currents and voltages.

Solution

The sequence networks are interconnected as shown in Figure 8-14. We write:

$$I_1 = -I_2 = \frac{1/0°}{j0.175 + j0.175}$$

$$= -j2.86$$

$$I_0 = 0$$

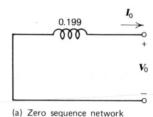

(a) Zero sequence network

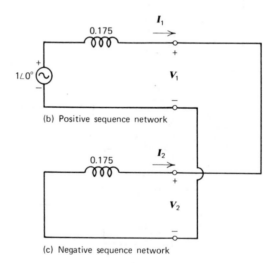

(b) Positive sequence network

(c) Negative sequence network

Figure 8-14 Sequence networks for Example 10-4. Line to line fault applied to bus 3.

The phase currents are:

$$\begin{bmatrix} I_a \\ I_b \\ I_c \end{bmatrix} = \begin{bmatrix} 1 & 1 & 1 \\ 1 & a^2 & a \\ 1 & a & a^2 \end{bmatrix} \begin{bmatrix} 0 \\ -j2.86 \\ +j2.86 \end{bmatrix} = \begin{bmatrix} 0 \\ -4.95 \\ +4.95 \end{bmatrix}$$

Also:

$$V_1 = V_2 = I_1(j0.175)$$
$$= 0.5$$

$$V_0 = 0$$

The phase voltages are:

$$\begin{bmatrix} V_a \\ V_b \\ V_c \end{bmatrix} = \begin{bmatrix} 1 & 1 & 1 \\ 1 & a^2 & a \\ 1 & a & a^2 \end{bmatrix} \begin{bmatrix} 0 \\ 0.5 \\ 0.5 \end{bmatrix} = \begin{bmatrix} 1.0 \\ -0.5 \\ -0.5 \end{bmatrix}$$

8-5 The Double Line to Ground Fault

Imagine our general three phase access port terminated as shown in Figure 8-15. The terminal conditions are such that we may write:

$$I_a = 0 \tag{8-16a}$$

$$V_b = V_c \tag{8-16b}$$

$$V_b = (I_b + I_c)Z_f \tag{8-16c}$$

It follows from (8-16a) that:

$$I_0 + I_1 + I_2 = 0 \tag{8-17}$$

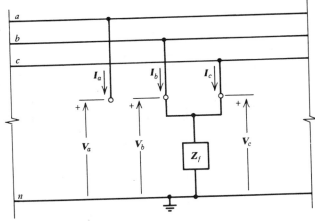

Figure 8-15 General double line to ground fault.

From equation (8-16b):

$$V_0 + a^2V_1 + aV_2 = V_0 + aV_1 + a^2V_2 \qquad (8\text{-}18a)$$

Simplifying:

$$V_1 = V_2 \qquad (8\text{-}18b)$$

From equation (8-16c):

$$V_0 + a^2V_1 + aV_2 = (I_0 + a^2I_1 + aI_2 + I_0 + aI_1 + a^2I_2)Z_f \qquad (8\text{-}19a)$$

Simplifying:

$$V_0 - V_1 = (2I_0 + (a^2 + a)(I_1 + I_2))Z_f \qquad (8\text{-}19b)$$

$$V_0 - V_1 = 3Z_fI_0 \qquad (8\text{-}19c)$$

For the general double line to ground fault, equations (8-17), (8-18b), and (8-19c) must be simultaneously satisfied. How are we to interconnect

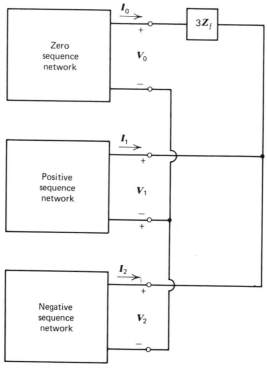

Figure 8-16 Sequence network terminations for a double line to ground fault.

the sequence networks this time? The solution appears in Figure 8-16. For a specific application consider example 8-5.

Example 8-5

A double line to ground fault occurs at bus 3 in the system of example 8-1. Compute the fault currents and voltages.

Solution

The sequence networks are as shown in Figure 8-17.

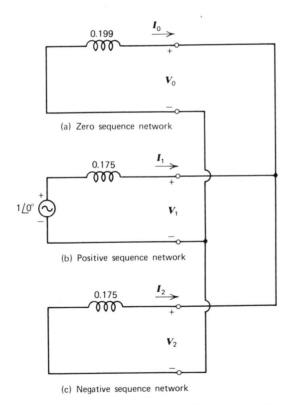

(a) Zero sequence network

(b) Positive sequence network

(c) Negative sequence network

Figure 8-17 Sequence networks for Example 8-5. Double line to ground fault applied to bus 3.

$$I_1 = \frac{1\underline{/0}}{j0.175 + \dfrac{(j0.175)(j0.199)}{j0.175 + j0.199}}$$

$$= -j3.73$$

$$I_0 = \frac{0.175}{0.175 + 0.199}(+j3.73) = +j1.75$$

$$I_2 = \frac{0.199}{0.175 + 0.199}(j3.73) = +j1.99$$

The phase currents are:

$$\begin{bmatrix} I_a \\ I_b \\ I_c \end{bmatrix} = \begin{bmatrix} 1 & 1 & 1 \\ 1 & a^2 & a \\ 1 & a & a^2 \end{bmatrix} \begin{bmatrix} j1.75 \\ -j3.73 \\ j1.99 \end{bmatrix} = \begin{bmatrix} 0 \\ 5.60\underline{/152.1°} \\ 5.60\underline{/27.9°} \end{bmatrix}$$

The sequence voltages:

$$V_0 = V_1 = V_2 = -(j1.75)(j0.199)$$

$$= 0.348$$

The phase voltages are:

$$\begin{bmatrix} V_a \\ V_b \\ V_c \end{bmatrix} = \begin{bmatrix} 1 & 1 & 1 \\ 1 & a^2 & a \\ 1 & a & a^2 \end{bmatrix} \begin{bmatrix} 0.348 \\ 0.348 \\ 0.348 \end{bmatrix} = \begin{bmatrix} 1.044 \\ 0 \\ 0 \end{bmatrix}$$

8-6 The Calculation of Interior Voltages and Currents in Faulted Systems

Thevenin equivalent circuits are equivalent to the original network only at some particular port of interest. When a given complex network is replaced by its simpler Thevenin equivalent, in general it is not possible to calculate interior currents and voltages, since the original network topology was destroyed in the Thevenin reduction process. For this information we must return to the original network. In Chapter 9 we present a systematic, general way of calculating such voltages and currents. Here we investigate the problem by considering the following example.

Example 8-6

Example 8-6

A single line to ground fault occurs at Bus 3 in the system of example 8-1. Compute the currents and voltages at the terminals of generators G1 and G2.

Solution

In this case the circuits of Figure 8-5 are inadequate. We return to the original circuits shown in Figure 8-4, properly interconnected for a single line to ground fault, as shown in Figure 8-18. We divide the problem into two parts, considering Generator G1 first.

Generator G1 (Bus 4)
Study Figure 8-18, locating I_f, I_0, I_1, I_2, V_0, V_1, and V_2. From example 8-3:

$$I_f = -j1.82$$

By symmetry:

$$I_1 = I_2 = -\tfrac{1}{2}I_f$$

$$= -j0.91$$

The current I_0 is more difficult. Transform the delta (0.3) into a wye (0.1), and use the current divider. Thus:

$$I_0 = \frac{0.15}{0.29 + 0.15}(-j1.82) = -j0.62$$

The phase currents are:

$$\begin{bmatrix} I_a \\ I_b \\ I_c \end{bmatrix} = \begin{bmatrix} 1 & 1 & 1 \\ 1 & a^2 & a \\ 1 & a & a^2 \end{bmatrix} \begin{bmatrix} -j0.62 \\ -j0.91 \\ -j0.91 \end{bmatrix} = \begin{bmatrix} 2.44\underline{/-90^\circ} \\ 0.29\underline{/+90^\circ} \\ 0.29\underline{/+90^\circ} \end{bmatrix}$$

The sequence voltages are:

$$V_0 = -(-j0.62)(j0.14) = -0.087$$

$$V_1 = 1 - j0.2(-j0.91) = 0.818$$

$$V_2 = -j0.2(-j0.91) = -0.182$$

The phase voltages are:

$$\begin{bmatrix} V_a \\ V_b \\ V_c \end{bmatrix} = \begin{bmatrix} 1 & 1 & 1 \\ 1 & a^2 & a \\ 1 & a & a^2 \end{bmatrix} \begin{bmatrix} -0.087 \\ 0.818 \\ -0.182 \end{bmatrix} = \begin{bmatrix} 0.549\underline{/0^\circ} \\ 0.956\underline{/245^\circ} \\ 0.956\underline{/115^\circ} \end{bmatrix}$$

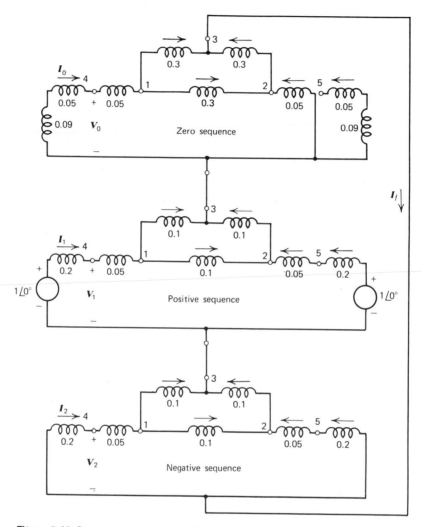

Figure 8-18 Sequence networks for Example 8-1 interconnected for a single line to ground fault.

Generator G2 (Bus 5)

Study Figure 8-19, locating I_f, I_0, I_1, I_2, V_0, V_1, and V_2. Again, from example 8-3:

$$I_f = -j1.82$$

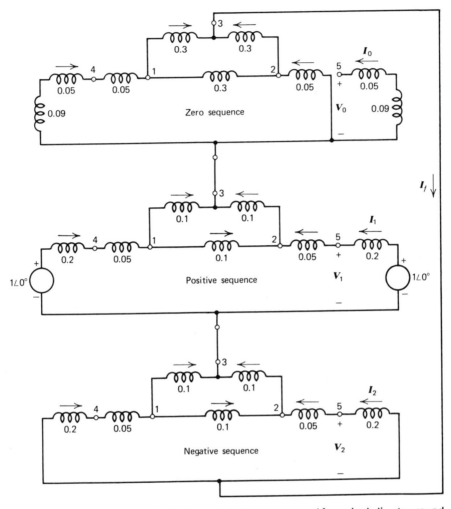

Figure 8-19 Sequence networks for Example 8-1 interconnected for a single line to ground fault.

Again, by symmetry:

$$I_1 = I_2 = \tfrac{1}{2}I_f$$

$$= -j0.91$$

By inspection

$$I_0 = 0$$

283

A consideration now arises that we first encountered in our study of transformers. Wye-delta transformer connections produce 30° phase shifts in sequence quantities. Refer back to section 5-4. Remember that the HV quantities are to be shifted 30° ahead of the corresponding LV quantities for positive sequence, and vice versa for negative sequence; therefore

$$\therefore \quad I_1 = 0.91\underline{/-90° - 30°} = 0.91\underline{/-120°}$$

$$I_2 = 0.91\underline{/-90° + 30°} = 0.91\underline{/-60°}$$

since bus 5 is the LV side. We now are prepared to calculate the phase currents:

$$\begin{bmatrix} I_a \\ I_b \\ I_c \end{bmatrix} = \begin{bmatrix} 1 & 1 & 1 \\ 1 & a^2 & a \\ 1 & a & a^2 \end{bmatrix} \begin{bmatrix} 0 \\ 0.91\underline{/-120°} \\ 0.91\underline{/-60°} \end{bmatrix} = \begin{bmatrix} 1.58\underline{/-90°} \\ 1.58\underline{/+90°} \\ 0 \end{bmatrix}$$

The positive and negative sequence voltages are the same as on the G1 side:

$$V_1 = 0.818$$

$$V_2 = -0.182$$

Again, a phase shift is required.

$$V_1 = 0.818\underline{/0° - 30°} = 0.818\underline{/-30°}$$

$$V_2 = 0.182\underline{/180° + 30°} = 0.182\underline{/210°}$$

The zero sequence voltage is, obviously:

$$V_0 = 0$$

We are now prepared to calculate phase voltages:

$$\begin{bmatrix} V_a \\ V_b \\ V_c \end{bmatrix} = \begin{bmatrix} 1 & 1 & 1 \\ 1 & a^2 & a \\ 1 & a & a^2 \end{bmatrix} \begin{bmatrix} 0 \\ 0.818\underline{/-30°} \\ 0.182\underline{/210°} \end{bmatrix} = \begin{bmatrix} 0.744\underline{/-42.2°} \\ 0.744\underline{/222.2°} \\ 1.00\underline{/90°} \end{bmatrix}$$

In general, interior currents and voltages may be computed by conventional circuit analysis methods, taking care to properly account for wye-delta transformer connections. For systems of practical size, hand calculations are unreasonable and we need a computer-oriented approach to the problem.

8-7 Consideration of Prefault Load Current

Our previous work has ignored the effects of prefault system loading. Usually this is reasonable because fault currents are typically much greater than load currents. However, there are situations where it is important to consider loads. It would seem that the straightforward approach would be to connect load impedances into the sequence networks at the appropriate locations, and simply deal with a more elaborate network. This is possible, but usually not done for two reasons:

- All loads are not correctly modeled as simple impedances.
- If results for several different loads are to be considered, load impedances must be recalculated and a new fault study made for each case.

The prefault line loadings are typically available as output data from a power flow study (see Chapter 7). These line loadings cause the internal generator voltages to assume magnitude and phase values other than unity and zero. It is possible to calculate proper values for a given load condition and use these values in a fault study. However, the same results may be obtained by superposition. An example will demonstrate.

Example 8-7

For the circuit of Figure 8-20 calculate the fault currents I, I_1, and I_2 (switch closed) by:
(a) Calculating correct values for E_1 and E_2, and directly calculating the currents.
(b) Using superposition.
The prefault currents are $I_1 = -I_2 = 1\underline{/0^\circ}$.

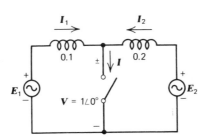

Figure 8-20 Circuit for Example 8-7.

Solution

(a) $E_1 = 1\underline{/0} + (1\underline{/0})(j0.1)$

$\quad\quad = 1 + j0.1$

$\quad E_2 = 1\underline{/0} - (1\underline{/0})(j0.2)$

$\quad\quad = 1 - j0.2$

Switch closed:

$$I_1 = \frac{E_1}{j0.1} = -j10 + 1$$

$$I_2 = \frac{E_2}{j0.2} = -j5 - 1$$

$$I = I_1 + I_2$$

$$\quad = -j10 + 1 - j5 - 1$$

$$\quad = -j15$$

(b) By superposition.

We start by ignoring prefault current, concluding:

$$E_1 = E_2 = 1\underline{/0}$$

$$I_1 = \frac{1\underline{/0}}{j0.1} = -j10$$

$$I_2 = \frac{1\underline{/0}}{j0.2} = -j5$$

$$I = I_1 + I_2 = -j15$$

Now we "superimpose" the load currents:

$$I_1 = I_{1\ fault} + I_{1\ load}$$

$$\quad = -j10 + 1$$

$$I_2 = I_{2\ fault} + I_{2\ load}$$

$$\quad = -j5 + (-1)$$

$$\quad = -j5 - 1$$

$$I = I_{\text{fault}} + I_{\text{load}}$$
$$= -j15 + 0$$

Note that the results are identical.

8-8 Further Considerations

We have implied that generators are the only sources in the system. Actually all rotating machines are capable of contributing to fault current, at least momentarily. Synchronous motors will continue to rotate due to inertia and function as generators in a faulted situation. The impedance used for such machines is usually the transient reactance X'_d or the subtransient reactance X''_d, depending on protective equipment speed of response.

Induction motors are also momentarily capable of contributing to fault current; usually they are modeled as sources in series with X''_d, or neglected entirely, again considering protective equipment speed of response. Frequently, motors smaller than 50 HP are neglected. Connecting systems are modeled with their Thevenin equivalents.

Although we have used ac circuit techniques to deal with faults, the problem is fundamentally transient since it involves sudden switching actions. We learned in Chapter 6 (section 6-3) of the so-called "dc offset" term. Should offset terms be considered? The answer is determined by deciding to what use our results will be put. For example, consider that we are calculating fault currents to determine the interrupting capacity of a circuit breaker. Suppose the breaker's speed of response is such that an 8 cycle (133 ms) delay is observed from the instant a breaker is commanded to open until the contacts actually part. Depending on the circuit L/R time constant, the dc offset may well have decayed to the vanishing point by this time.

A second point is how much offset is it reasonable to expect? The offset will vary depending at what point in the cycle the fault occurs. To consider the effect of offset consult the circuit of Figure 8-21. Consider the circuit values to be in SI units (not per unit). Our objective is to predict the largest rms current that the circuit breaker would be expected to interrupt.

$$I_{60\,\text{Hz}} = \frac{E}{\sqrt{R^2 + X^2}} \tag{8-20a}$$

and:

$$I_{\text{max}} = I_{60\,\text{Hz}}\sqrt{2} \tag{8-20b}$$

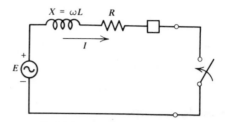

Figure 8-21 Circuit for Example 8-8.

The maximum dc offset possible would be:

$$I_{dc_{max}} = I_{max} \tag{8-21a}$$

$$= \sqrt{2} I_{60\,Hz} \tag{8-21b}$$

The dc offset will exponentially decay with time constant τ where:

$$\tau = \frac{L}{R} \tag{8-22a}$$

$$= \frac{X}{\omega R} \tag{8-22b}$$

Note that equation (8-22b) is valid for both SI and per unit values of X and R, since Z_{base} would cancel. The transient dc offset current would be $I_{dc}(t)$:

$$\therefore \quad I_{dc}(t) = I_{dc_{max}} e^{-t/\tau} \tag{8-23a}$$

$$= \sqrt{2} I_{60\,Hz} e^{-t/\tau} \tag{8-23b}$$

The "transient rms" current $I(t)$, accounting for both the 60 Hz and dc terms, would be:

$$I(t) = \sqrt{I^2{}_{60\,Hz} + I^2{}_{dc}(t)} \tag{8-24a}$$

$$= I_{60\,Hz}\sqrt{1 + 2\, e^{-2t/\tau}} \tag{8-24b}$$

Define a multiplying factor k_i such that $I_{60\,Hz}$ is to be multiplied by k_i to estimate the interrupting capacity of a breaker that operates in time T_{op}. Therefore

$$k_i = \frac{I(T_{op})}{I_{60\,Hz}} \tag{8-25a}$$

$$= \sqrt{1 + 2\, e^{-2T_{op}/\tau}} \tag{8-25b}$$

Observe that the maximum possible value for k_i is $\sqrt{3}$.

288

Example 8-8

Example 8-8

In the circuit of Figure 8-21, $E = 2400$ volts, $X = 2$ ohms, $R = 0.1$ ohms, and $f = 60$ Hz. Compute k_i and determine the interrupting capacity for the circuit breaker if it is designed to operate in 2 cycles. The fault is applied at $t = 0$.

Solution

$$I_{60\,Hz} \cong \frac{2400}{2} = 1200 \text{ amperes}$$

$$T_{op} = \frac{2}{60} = 0.0333 \text{ seconds}$$

$$\tau = \frac{X}{\omega R}$$

$$= \frac{2}{377(.1)} = 0.053 \text{ seconds}$$

$$k_i = \sqrt{1 + 2\,e^{-2T_{op}/\tau}}$$

$$= \sqrt{1 + 2\,e^{-0.0667/0.053}}$$

$$= 1.252$$

$$I = k_i I_{60\,Hz}$$

$$\therefore \quad = 1.252(1200)$$

$$= 1503 \text{ amperes}$$

8-9 Summary

We have observed that the application of a fault at a particular point in a power system is modeled by appropriate interconnections between sequence networks. Such interconnections were derived in sections 8-2 through 8-5 for the four basic fault types. The approach for any arbitrary unbalanced termination should now be clear: write equations describing the termination in phase quantities, transform to sequence quantities, manipulate to a form that facilitates synthesis of interconnections between sequence networks, and make the required interconnections.

This process is not necessarily simple; in fact it may prove impossible to derive appropriate interconnections for some cases. For example, consider a single line to ground fault, with the fault applied to phase b. We discover that in terms of sequence quantities certain factors of "a" and "a^2" appear that cannot be eliminated. Fortunately, such cases are not really required, since the single line to ground fault on phase a is essentially general (we can always relabel the phases).

The steps in performing a fault study on a power system are summarized as follows:

- A single line diagram of the system is required.

- Sequence impedances for all components are necessary. These include values for generators, motors, transformers, lines, and connected external systems.

- Points (busses) at which faults are to be applied are identified in all three sequence networks.

- The fault is produced by making appropriate interconnections between sequence networks at the desired bus.

- Required sequence voltages and currents are calculated by conventional ac circuit analysis methods.

- Proper 30° phase shifts are made in sequence quantities to account for wye-delta transformer connections.

- Phase quantities are computed from sequence quantities using equations (2-53) and (2-55).

- When necessary, the resulting symmetrical values are adjusted for dc offset to produce asymmetrical values.

The results are used to size, set, and coordinate system protection equipment, such as circuit breakers, fuses, instrument transformers, and relays. We continue our investigation into application of these results in Chapters 10 and 11.

It is clear that for systems of even modest size it is not practical to calculate the necessary quantities by hand. Observe that if the sequence networks contain only one type of impedance (inductance), it is possible to formulate an equivalent dc network in which dc current values match the magnitudes of corresponding ac currents. This dc network requires only resistors and a single dc source, making it simple and economical to physically construct. Such devices, called "dc Network Analyzers," are commercially available and can be used to model a given sytem, wherein the desired analogous sequence currents and voltages are measured instead of calculated. Fault analysis may also be performed by digital computer and this topic is the subject of Chapter 9.

Bibliography

[1] Anderson, Paul M., *Analysis of Faulted Power Systems*, Iowa State Press, Ames, Iowa, 1973.

[2] Brown, Homer E., *Solution of Large Networks by Matrix Methods*, John Wiley and Sons, Inc., 1975.

[3] IEEE., *Recommended Practice for Protection and Coordination of Industrial and Commercial Power Systems*, Wiley-Interscience, New York, 1975.

[4] Manufacturer's Publication, *Short-Circuit Current Calculations for Industrial and Commercial Power Systems*, General Electric, Publication GET-3550.

[5] Manufacturer's Publication, *Short-Circuit Currents in Low and Medium Voltage A-C Power Systems*, General Electric, Publication GET-1470D.

[6] Neuenswander, John R., *Modern Power Systems*, International Textbook Co., Scranton, 1971.

[7] Stagg, Glenn W., and El-Abiad, Ahmed H., *Computer Methods in Power System Analysis*, McGraw-Hill, Inc., New York, 1968.

[8] Stevenson, Jr., William D., *Elements of Power Systems Analysis*, 3rd edition. McGraw-Hill, Inc., New York, 1975.

[9] Weedy, B. M., *Electric Power Systems*, 2nd edition. John Wiley and Sons Ltd., London, 1972.

Problems

8-1. Consider the power system terminated as shown in Figure P8-1. Derive the appropriate interconnections between sequence networks.

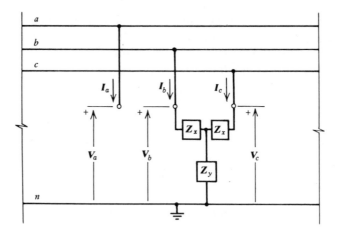

Figure P8-1 Circuit for Problem 8-1.

8-2. Consider the system of example 8-1. Derive Thevenin equivalent sequence networks "looking in" at bus X.(Instructor will select $X = 1$, 2, 4, or 5). Bus X will be the faulted bus in problem 8-3 through problem 8-14.

8-3. For a $3\emptyset$ fault at the bus selected in problem 8-2, calculate the fault phase voltages and currents (six values).

8-4. Repeat problem 8-3 for a single line to ground fault.

8-5. Repeat problem 8-3 for a line to line fault.

8-6. Repeat problem 8-3 for a double line to ground fault.

8-7. Calculate the phase voltages at bus 3 for the fault described in problem 8-3.

8-8. Repeat problem 8-7 for the fault described in problem 8-4.

8-9. Repeat problem 8-7 for the fault described in problem 8-5.

8-10. Repeat problem 8-7 for the fault described in problem 8-6.

8-11. Calculate the line currents in the line TL12 for the fault described in problem 8-3.

8-12. Repeat problem 8-11 for the fault described in problem 8-4.

8-13. Repeat problem 8-11 for the fault described in problem 8-5.

8-14. Repeat problem 8-11 for the fault described in problem 8-6.

8-15. Suppose the power system of example 8-1 had prefault loading such that the current $I_1 = 0.5\underline{/-30°}$ in line TL13 ($I_0 = I_2 = 0$). Calculate the line currents (I_a, I_b, and I_c) for a 3Ø fault at bus 3 accounting for this prefault loading.

8-16. In the circuit of Figure 8-21, for $R = 1\,\Omega$; $X = 15\,\Omega$; $E = 10$ kV:
   (a) Calculate the "worst" (maximum) possible $I(t)$.
   (b) Estimate the fastest breaker speed of response if dc offset may be neglected when $I(t) \leq 1.1 I_{60\ Hz}$.

FAULT ANALYSIS
BY COMPUTER
METHODS

"Beyond a wholesome discipline, be gentle with yourself. You
are a child of the universe no less than the trees and the
stars; you have a right to be here. And whether
or not it is clear to you, no doubt the
universe is unfolding as it should."
Max Ehrmann, DESIDERATA

In Chapter 8 we observed that fault studies required the solution of large networks. As in the power flow problem, computer methods are suitable for the solution of such networks. We observed in Chapter 7 that the network equations could be formulated in either of two formats: impedance or admittance. The latter was used for the power flow problem, collecting the system's passive data into the matrix $[Y_{\text{bus}}]$.

We shall use the impedance formulation for application to the fault analysis problem. Comparing fault analysis to power flow analysis, the problem is more complicated in one sense since all three sequence networks are involved. However, it is simpler in another sense because iteration is not required. Since three sequence networks are involved we shall use the notation $[Z_0]$, $[Z_1]$, and $[Z_2]$ for the zero, positive, and negative sequence impedance matrices. Our first order of business is the development of a programmable algorithm for construction of these arrays.

9-1 Formulation of the Impedance Matrix

As in Chapter 7, we need a clear understanding of the viewpoint to be taken for a general formulation of the problem. Consider Figure 9-1. Note particularly the sign conventions on voltage and current. The network to be represented is passive, n-port, and linear. We describe the network with:

$$V_1 = z_{11}I_1 + z_{12}I_2 + \ldots + z_{1n}I_n$$
$$V_2 = z_{21}I_1 + z_{22}I_2 + \ldots + z_{2n}I_n$$
$$\vdots \qquad\qquad \vdots$$
$$V_n = z_{n1}I_1 + z_{n2}I_2 + \cdots + z_{nn}I_n \tag{9-1a}$$

or in matrix notation:

$$\tilde{V} = [Z]\tilde{I} \tag{9-1b}$$

where

$\tilde{V}$ is $n \times 1$

$\tilde{I}$ is $n \times 1$

$[Z]$ is $n \times n$

The general entry z_{ij} may be calculated from:

$$z_{ij} = \frac{V_i}{I_j}\bigg]_{\substack{I_1 = I_2 = \ldots = I_n = 0 \\ I_j \neq 0}} \tag{9-2}$$

296

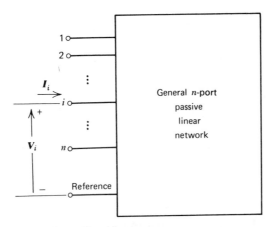

Figure 9-1 General n-port network.

Equation (9-2) suggests a method for evaluating z_{ij} for a given network. Terminate the n ports in open circuits, except for the jth port, which is terminated in a current source of strength I_j. Now solve for V_i and divide the result by I_j, which will cancel. The result is z_{ij}. An example will illustrate the technique.

Example 9-1

Consider the resistive two port network shown in Figure 9-2a. Evaluate $[Z]$.

Solution

We plan to use equation (9-2). First terminate the network as shown in Figure 9-2b.

$$z_{11} = \frac{V_1}{I_1}\bigg]_{I_2=0} = \frac{5I_1}{I_1} = 5$$

$$z_{21} = \frac{V_2}{I_1}\bigg]_{I_2=0} = \frac{3I_1}{I_1} = 3$$

Now terminate the network as shown in Figure 9-2c.

$$z_{12} = \frac{V_1}{I_2}\bigg]_{I_1=0} = \frac{3I_2}{I_2} = 3$$

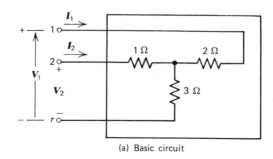

(a) Basic circuit

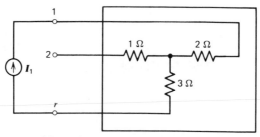

(b) Modification for calculating z_{11} and z_{21}

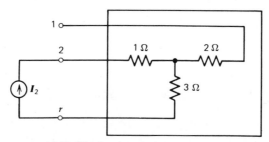

(c) Modification for calculating z_{12} and z_{22}

Figure 9-2 Circuit for Example 9-1.

$$z_{22} = \frac{V_2}{I_2}\bigg|_{I_1=0} = \frac{4I_2}{I_2} = 4$$

Collecting our results:

$$[Z] = \begin{bmatrix} 5 & 3 \\ 3 & 4 \end{bmatrix} \tag{9-3}$$

Observe that $z_{12} = z_{21}$ in example 9-1. It happens that in general $z_{ij} = z_{ji}$ for power system networks since they are reciprocal. The matrix $[Z]$ is

298

sometimes referred to as the "open circuit" impedance matrix because computation of the general entries z_{ij} utilized open circuits at all ports except the jth port.

Although application of equation (9-2) is a correct and general technique for the formulation of $[Z]$, it is not suitable for our purposes. The difficulty is that solving for V_i is not much simpler than solving the original problem. We need a method that produces $[Z]$ without requiring the solution of a conventional circuits problem (given V, find I or vice versa).

9-2 A Programmable Method for Formulating [Z]

A practical formulation of $[Z]$ must have two properties:

- The technique must be a programmable step by step method, working from branch impedance values.

- Any modification of the network must not require a complete rebuilding of [Z].

Such a technique is now described in terms of modifying an existing $[Z]$. (See [6] in the end-of-chapter Bibliography.) This method can also completely build $[Z]$ from scratch. Let us consider the basic problem of adding a branch (Z_b) to an existing impedance matrix ($[Z]_{old}$) to produce a revised impedance matrix ($[Z]_{new}$). The reference node (bus) is different from other nodes because the voltages are defined with respect to reference and the reference voltage is trivially zero. Therefore, there are three types of nodes (busses): an "old" bus, existing before Z_b is added; a "new" bus, created by the addition of Z_b; and the reference bus. We realize, then, that Z_b may be added five nontrivial ways:

1. Add Z_b from a new bus to reference.

2. Add Z_b from a new bus to an old bus.

3. Add Z_b from an old bus to reference.

4. Connect Z_b between two old busses.

5. Connect Z_b between two new busses.

If we are selective in the order in which we add elements, we can avoid the fifth situation and still be general. Let us discuss the four remaining modifications in order. We will use "i, j" to indicate old busses, "r" to

299

indicate the reference bus, and "k" to indicate the new bus. The matrix $[Z]$ is understood to be $n \times n$ before modification, and symbolized as $[Z]_{\text{old}}$. Realize that $[Z]$ is diagonally symmetric $(z_{ij} = z_{ji})$—a property we shall exploit. The sign convention to be used for voltages and currents will be as defined in Figure 9-1.

TYPE 1 MODIFICATION

Addition of a branch Z_b from a new bus (k) to reference. Refer to Figure 9-3. We write:

$$V_k = Z_b I_k \tag{9-4}$$

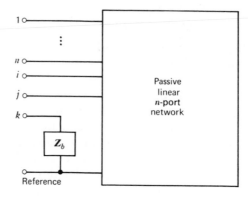

Figure 9-3 Type 1 modification.

and add the equation to the set. The modification to $[Z]_{\text{old}}$ is to add a kth row and column where:

$$z_{ik} = z_{ki} = 0 \qquad i = 1, 2, \ldots n \tag{9-5a}$$

$$z_{kk} = Z_b \tag{9-5b}$$

$$\therefore \quad [Z]_{\text{new}} = \begin{bmatrix} & & & & 0 \\ & [Z]_{\text{old}} & & & \vdots \\ & & & & \vdots \\ & & & & 0 \\ \hline 0 \ldots 0 & & & Z_b \end{bmatrix} \tag{9-5c}$$

TYPE 2 MODIFICATION

Addition of a branch Z_b from a new bus k to an old bus j. Consult Figure 9-4. We write:

$$V_k = Z_b I_k + V_j \tag{9-6a}$$

$$= Z_b I_k + z_{j1} I_1 + z_{j2} I_2 + \ldots + z_{jj}(I_j + I_k) + \ldots + z_{jn} I_n \tag{9-6b}$$

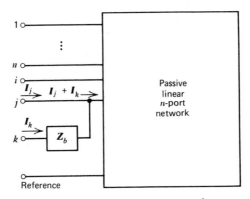

Figure 9-4 Type 2 modification.

Rearranging:

$$V_k = z_{j1} I_1 + z_{j2} I_2 + \ldots + z_{jj} I_j + \ldots + (z_{jj} + Z_b) I_k \tag{9-6c}$$

Consequently, the modified $[Z]$ is:

$$[Z]_{\text{new}} = \begin{bmatrix} & & & \vdots & z_{1j} \\ & & & \vdots & z_{2j} \\ & [Z]_{\text{old}} & & \vdots & \vdots \\ & & & \vdots & \\ & & & \vdots & z_{nj} \\ \hline z_{j1} & z_{j2} \ldots & z_{jn} & \vdots & (z_{jj} + Z_b) \end{bmatrix} \tag{9-6d}$$

TYPE 3 MODIFICATION

Addition of a branch Z_b from an old bus (j) to reference. Consult Figure 9-5. Consider the Type 2 Modification and connect the new bus to reference.

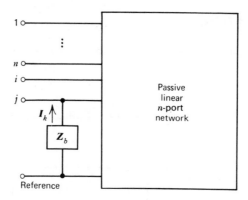

Figure 9-5 Type 3 modification.

Then $V_k = 0$ so that:

$$
\begin{bmatrix} V_1 \\ V_2 \\ \vdots \\ V_n \\ \hline 0 \end{bmatrix} = \left[\begin{array}{c|c} & z_{1j} \\ [Z]_{\text{old}} & \vdots \\ & z_{nj} \\ \hline z_{j1} \dots z_{jn} & (z_{jj} + Z_b) \end{array} \right] \begin{bmatrix} I_1 \\ \vdots \\ I_n \\ \hline I_k \end{bmatrix}
$$

(9-7a)

We eliminate I_k to produce:

$$
[Z]_{\text{new}} = [Z]_{\text{old}} - \frac{1}{z_{jj} + Z_b} \begin{bmatrix} z_{1j} \\ \vdots \\ z_{nj} \end{bmatrix} [z_{j1} \dots z_{jn}]
$$

(9-7b)

See problem 9-5 for clarification.

TYPE 4 MODIFICATION

Addition of a branch Z_b from an old bus (i) to an old bus (j). Consider Figure 9-6.

$$
V_1 = z_{11}I_1 + z_{12}I_2 + \dots + z_{1i}(I_i + I_k) + z_{1j}(I_j - I_k) + \dots + z_{1n}I_n
$$

(9-8a)

Rearranging:

$$
V_1 = z_{11}I_1 + z_{12}I_2 + \dots + (z_{1i} - z_{1j})I_k
$$

(9-8b)

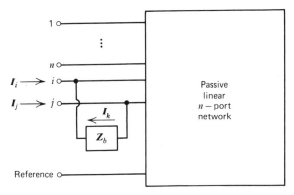

Figure 9-6 Type 4 modification.

Similar equations may be written at all ports. Finally, we write:

$$V_j = Z_b I_k + V_i \tag{9-9a}$$

$$z_{j1}I_1 + z_{j2}I_2 + \ldots + z_{ji}(I_i + I_k) + z_{jj}(I_j - I_k) + \ldots$$
$$= Z_b I_k + z_{i1}I_1 + \ldots + z_{ii}(I_i + I_k) + z_{ij}(I_j - I_k) + \ldots \tag{9-9b}$$

Rearranging:

$$0 = (z_{i1} - z_{j1})I_1 + \ldots + (z_{ii} - z_{ji})I_i$$
$$+ (z_{ji} - z_{jj})I_j + \ldots + (Z_b + z_{ii} + z_{jj} - z_{ij} - z_{ji})I_k \tag{9-9c}$$

Collecting these equations:

$$
\begin{bmatrix} V_1 \\ V_2 \\ \vdots \\ V_n \\ \hline 0 \end{bmatrix} =
\left[\begin{array}{c c}
 & (z_{1i} - z_{1j}) \\
 & (z_{2i} - z_{2j}) \\
[Z]_{\text{old}} & \vdots \\
 & \\
\hline
(z_{i1} - z_{j1}) \ldots & Z_b + z_{ii} + z_{jj} - 2z_{ij}
\end{array} \right]
\begin{bmatrix} I_1 \\ I_2 \\ \vdots \\ I_n \\ \hline I_k \end{bmatrix} \tag{9-10}
$$

Eliminating I_k:

$$[Z]_{\text{new}} = [Z]_{\text{old}} - \frac{1}{Z_b + z_{ii} + z_{jj} - 2z_{ij}} \begin{bmatrix} z_{1i} - z_{1j} \\ \vdots \\ z_{ni} - z_{nj} \end{bmatrix} [(z_{i1} - z_{j1}) \ldots (z_{in} - z_{jn})] \tag{9-11}$$

Again, see problem 9-5 for clarification.

303

These modifications may be used to build $[Z]$ from scratch. An example is helpful.

Example 9-2

Build $[Z]$ for the circuit of example 9-1 by the methods just described.

Solution

If we are to avoid a fifth type of modification (new bus to new bus) we must start with a Type 1 modification, since there are no old busses initially. There is a need to renumber the busses to match the order in which they are created. Therefore:

Old	New
3	(1)
2	(2)
1	(3)

1. We start with the 3 Ω element:

A Type 1 modification produces:

$$[Z]_{new} = [3]$$

2. Next add the 1 Ω element:

304

A Type 2 modification produces:

$$[Z]_{new} = \begin{bmatrix} 3 & 3 \\ 3 & 4 \end{bmatrix}$$

3. Next add the 2 Ω element:

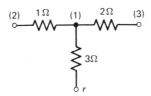

A Type 2 modification produces:

$$[Z]_{new} = \begin{bmatrix} 3 & 3 & 3 \\ 3 & 4 & 3 \\ 3 & 3 & 5 \end{bmatrix}$$

4. Now return to the old number system, requiring that we interchange rows and columns (i.e., row, column (1) becomes row, column 3, etc.)

$$\therefore \quad [Z]_{new} = \begin{bmatrix} 5 & 3 & 3 \\ 3 & 4 & 3 \\ 3 & 3 & 3 \end{bmatrix}$$

5. In example 9-1, bus 3 was not brought out for consideration. Therefore $I_3 \equiv 0$. We write:

$$\begin{bmatrix} V_1 \\ V_2 \\ V_3 \end{bmatrix} = \begin{bmatrix} 5 & 3 & 3 \\ 3 & 4 & 3 \\ 3 & 3 & 3 \end{bmatrix} \begin{bmatrix} I_1 \\ I_2 \\ I_3 \end{bmatrix} \rightarrow 0$$

$$= \begin{bmatrix} 5 & 3 \\ 3 & 4 \\ 3 & 3 \end{bmatrix} \begin{bmatrix} I_1 \\ I_2 \end{bmatrix}$$

If we are not interested in V_3:

$$\begin{bmatrix} V_1 \\ V_2 \end{bmatrix} = \begin{bmatrix} 5 & 3 \\ 3 & 4 \end{bmatrix} \begin{bmatrix} I_1 \\ I_2 \end{bmatrix}$$

and finally:

$$[Z] = \begin{bmatrix} 5 & 3 \\ 3 & 4 \end{bmatrix}$$

(9-12)

which agrees with the result computed in example 9-1. Check with equation (9-3).

9-3 The Sequence Fault Impedance Matrices

Since the sequence networks for a power system are linear and passive, the methods discussed in section 9-2 are directly applicable. The ports we choose to create are those busses at which we wish to apply faults; in general we will assume that all busses in an n bus system are eligible. The ith bus is the one at which the fault is to be applied; "i" can of course range from 1 to n. The reference terminal is designated as "r." We define the following notation for sequence quantities.

$\tilde{V}_0$ = Zero sequence bus voltage vector $(n \times 1)$

$\tilde{V}_1$ = Positive sequence bus voltage vector $(n \times 1)$

$\tilde{V}_2$ = Negative sequence bus voltage vector $(n \times 1)$

$\tilde{I}_0$ = Zero sequence fault current vector $(n \times 1)$

$\tilde{I}_1$ = Positive sequence fault current vector $(n \times 1)$

$\tilde{I}_2$ = Negative sequence fault current vector $(n \times 1)$

The ith entry to each of the above, respectively, would be V_i^0, V_i^1, V_i^2, I_i^0, I_i^1, and I_i^2 where the subscripts refer to the bus number and the superscript indicates the sequence. Consult Figure 9-7 for assigned positive conventions; note particularly the current directions that are opposite to those used in section 9-1, but which are more natural for fault analysis and agree with those used in Chapter 8. The required impedance matrices are:

$[Z_0]$ = Zero sequence fault impedance matrix, $(n \times n)$ general entry z_{ij}^0.

$[Z_1]$ = Positive sequence fault impedance matrix, $(n \times n)$ general entry z_{ij}^1.

$[Z_2]$ = Negative sequence fault impedance matrix, $(n \times n)$ general entry z_{ij}^2.

The zero and negative impedance matrices can be formulated by direct application of the method of section 9-2. However, the positive sequence

306

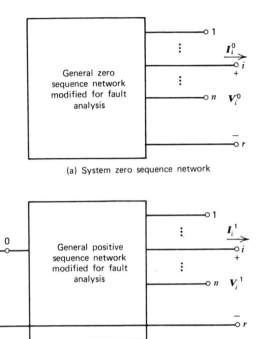

(a) System zero sequence network

(b) System positive sequence network

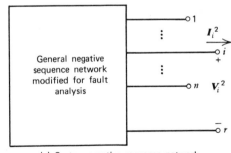

(c) System negative sequence network

Figure 9-7 System sequence networks modified for fault analysis.

network is somewhat more complicated because it contains sources, one for each rotating machine in the network. If we agree to neglect prefault current, this agreement has two implications: first, there are no paths to ground, other than machines, and, second, all line currents are zero. Considering this second fact, we realize that all of the internal voltage sources of generators are equal, both magnitude and phase, and therefore may be combined

307

(in parallel, so to speak) into one equivalent source. The simplified network is shown in Figure 9-7b, with the point "0" locating the "top" of the combined sources. To formulate $[Z_1]$ the source E is set to zero, connecting the point 0 to r. The resulting positive sequence network is then passive and we proceed in the same manner as for the zero and negative cases. The defining equations interrelating these sequence quantities are based on equation (9-1). They are:

$$\tilde{V}_0 = -[Z_0]\tilde{I}_0 \tag{9-13}$$

$$\tilde{V}_1 = \tilde{E} - [Z_1]\tilde{I}_1 \tag{9-14}$$

$$\tilde{V}_2 = -[Z_2]\tilde{I}_2 \tag{9-15}$$

The vector $\tilde{E}$ contains $1\underline{/0°}$ in all entries. The minus signs are necessary because of the reversed positively defined directions of the current. Be sure that you understand precisely what these defined currents are. The external terminations are made at the ith bus to model any one of the four basic fault types discussed in Chapter 8.

The power of this approach is evident when we realize that *all currents are zero until we externally terminate the networks.* Since we terminate only one port (the ith one), then only I_i^0, I_i^1, and I_i^2 are nonzero.

9-4 General Fault Analysis Equations

We are now prepared to terminate the general sequence networks shown in Figure 9-7 at bus i (the faulted bus) as dictated by the fault type. Bear in mind that $I_j = 0 (j = 1, 2, \ldots, n; j \neq i)$. Equations 9-13, 9-14, and 9-15 are our basis for the following development. The appropriate figure in Chapter 8 is useful in visualizing each situation; interpret the single port as the ith port.

BALANCED 3ϕ FAULT (Refer to Figure 8-7)

We observe:

$$\tilde{V}_0 = \tilde{V}_2 = 0 \tag{9-16a}$$

$$\tilde{I}_0 = \tilde{I}_2 = 0 \tag{9-16b}$$

$$V_i^1 = I_i^1 Z_f \tag{9-17a}$$

$$= E - [z_{i1}^1 I_1^1 + \ldots + z_{ii}^1 I_i^1 + \ldots + z_{in}^1 I_n^1] \tag{9-17b}$$

But

$$I_1^1 = I_2^1 = \ldots = I_n^1 = 0 \tag{9-18}$$

except for I_i^1.

Then:

$$V_i^1 = E - (0 + \ldots + z_{ii}^1 I_i^1 + \ldots + 0) \tag{9-17c}$$

$$\therefore \quad I_i^1 = \frac{E}{Z_f + z_{ii}^1} \tag{9-17d}$$

Solving for V_j^1:

$$V_j^1 = E - z_{ji}^1 I_i^1 \tag{9-19a}$$

$$\therefore \quad V_j^1 = \left[1 - \frac{z_{ji}^1}{Z_f + z_{ii}^1} \right] E \tag{9-19b}$$

$$V_j^1 = \left[\frac{Z_f + z_{ii}^1 - z_{ji}^1}{Z_f + z_{ii}^1} \right] E \tag{9-19c}$$

SINGLE LINE TO GROUND FAULT (Refer to Figure 8-10)

We observe

$$I_j^0 = I_j^1 = I_j^2 = 0; \qquad j = 1, \ldots, n \tag{9-19a}$$
$$j \neq i$$

and

$$I_i^0 = I_i^1 = I_i^2 \tag{9-19b}$$

By KVL:

$$V_i^0 + V_i^1 + V_i^2 = 3Z_f I_i^1 \tag{9-20a}$$

$$E - z_{ii}^0 I_i^0 - z_{ii}^1 I_i^1 - z_{ii}^2 I_i^2 = 3Z_f I_i^1 \tag{9-20b}$$

$$\therefore \quad I_i^1 = \frac{E}{z_{ii}^0 + z_{ii}^1 + z_{ii}^2 + 3Z_f} \tag{9-20c}$$

The j bus voltages are:

$$V_j^0 = -z_{ji}^0 I_i^0 \tag{9-21a}$$

$$= -z_{ji}^0 I_i^1 \tag{9-21b}$$

$$= \frac{-z_{ji}^0 E}{z_{ii}^0 + z_{ii}^1 + z_{ii}^2 + 3Z_f} \tag{9-21c}$$

$$V_j^1 = E - z_{ji}^1 I_i^1 \tag{9-22a}$$

$$= \left[1 - \frac{z_{ji}^1}{z_{ii}^0 + z_{ii}^1 + z_{ii}^2 + 3Z_f} \right] E \tag{9-22b}$$

$$= \left[\frac{z_{ii}^0 + z_{ii}^1 + z_{ii}^2 + 3Z_f - z_{ji}^1}{z_{ii}^0 + z_{ii}^1 + z_{ii}^2 + 3Z_f} \right] E \tag{9-22c}$$

$$V_j^2 = -z_{ji}^2 I_i^2 \tag{9-23a}$$

$$= -z_{ji}^2 I_i^1 \tag{9-23b}$$

$$= \frac{-z_{ji}^2 E}{z_{ii}^0 + z_{ii}^1 + z_{ii}^2 + 3Z_f} \tag{9-23c}$$

LINE TO LINE FAULT (Refer to Figure 8-13)

We observe:

$$\tilde{V}_0 = \tilde{I}_0 = 0 \tag{9-24}$$

Equation (9-19a) still holds. From equation 8-15c:

$$E - z_{ii}^1 I_i^1 = Z_f I_i^1 - z_{ii}^2 I_i^2 \tag{9-25a}$$

Realizing $I_i^2 = -I_i^1$ and manipulating:

$$I_i^1 = \frac{E}{z_{ii}^1 + z_{ii}^2 + Z_f} \tag{9-25b}$$

$$V_j^1 = E - z_{ji}^1 I_i^1 \tag{9-26a}$$

$$= \left[\frac{z_{ii}^1 + z_{ii}^2 + Z_f - z_{ji}^1}{z_{ii}^1 + z_{ii}^2 + Z_f} \right] E \tag{9-26b}$$

Furthermore

$$V_j^2 = -z_{ji}^2 I_i^2 \tag{9-27a}$$

$$= +z_{ji}^2 I_i^1 \tag{9-27b}$$

$$= \frac{z_{ji}^2 E}{z_{ii}^1 + z_{ii}^2 + Z_f} \tag{9-27c}$$

DOUBLE LINE TO GROUND FAULT (Refer to Figure 8-16)

We observe:

$$V_i^1 = V_i^2 \tag{9-28}$$

Equation (9-19a) still holds. Also:

$$I_i^0 + I_i^1 + I_i^2 = 0 \tag{9-29}$$

$$I_i^0 = \frac{-V_i^1}{3Z_f + z_{ii}^0} \tag{9-30a}$$

$$I_i^1 = \frac{E - V_i^1}{z_{ii}^1} \tag{9-30b}$$

$$I_i^2 = \frac{-V_i^1}{z_{ii}^2} \tag{9-30c}$$

Substituting into (9-29):

$$\frac{V_i^1}{3Z_f + z_{ii}^0} + \frac{V_i^1 - E}{z_{ii}^1} + \frac{V_i^1}{z_{ii}^2} = 0 \tag{9-31a}$$

Solving for V_i^1:

$$V_i^1 = \frac{z_{ii}^2(3Z_f + z_{ii}^0)E}{z_{ii}^1 z_{ii}^2 + z_{ii}^1(3Z_f + z_{ii}^0) + z_{ii}^2(3Z_f + z_{ii}^0)} \tag{9-31b}$$

We define:

$$\Delta = z_{ii}^1 z_{ii}^2 + z_{ii}^1(3Z_f + z_{ii}^0) + z_{ii}^2(3Z_f + z_{ii}^0) \tag{9-32}$$

Then from (9-30):

$$I_i^0 = \frac{-z_{ii}^2 E}{\Delta} \tag{9-33a}$$

$$I_i^1 = \frac{(z_{ii}^2 + 3Z_f + z_{ii}^0)E}{\Delta} \tag{9-33b}$$

$$I_i^2 = \frac{-(3Z_f + z_{ii}^0)E}{\Delta} \tag{9-33c}$$

The voltages are:

$$V_j^0 = -z_{ji}^0 I_i^0 \tag{9-34a}$$

$$= \frac{z_{ji}^0 z_{ii}^2 E}{\Delta} \tag{9-34b}$$

$$V_j^1 = E - z_{ji}^1 I_i^1 \tag{9-35a}$$

$$= \frac{[\Delta - z_{ji}^1 (z_{ii}^2 + 3Z_f + z_{ii}^0)]E}{\Delta} \tag{9-35b}$$

$$V_j^2 = -z_{ji}^2 I_i^2 \tag{9-36a}$$

$$= \frac{z_{ji}^2 (3Z_f + z_{ii}^0)E}{\Delta} \tag{9-36b}$$

The foregoing development has been algebraically and notationally complicated. To summarize our results, note that we are primarily interested in V_j values. A flow chart that summarizes application of these equations in a general fault analysis computer program is given in Figure 9-8.

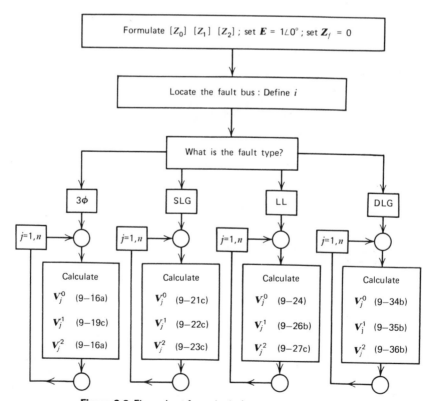

Figure 9-8 Flow chart for calculating sequence voltages.

9-5 Line Current Calculations

The key to solving for all network quantities is to know the voltage at each bus under faulted conditions. We solved this problem in the previous section. However, we are also interested in the currents flowing in all lines throughout the system. Now that we may assume the voltages are known, the calculation is straightforward. Consider the situation shown in Figure 9-9.

$$I_k^m = \frac{V_i^m - V_j^m}{Z_k^m} \tag{9-37a}$$

$$I_{-k}^m = \frac{V_j^m - V_i^m}{Z_k^m} \tag{9-37b}$$

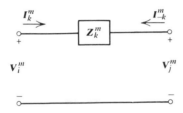

Figure 9-9 Current in the kth branch of the mth sequence network.

where

m = sequence; $m = 0, 1, 2$.

I_k^m = mth sequence current flowing in branch k from bus i to bus j.

I_{-k}^m = mth sequence current flowing in branch k from bus j to bus i.

Z_k^m = mth sequence impedance for branch k; branches are lines and transformers.

V_i^m = mth sequence voltage at ith bus.

V_j^m = mth sequence voltage at jth bus.

It would appear that the notation I_{-k}^m is superfluous since equation (9-37) indicates that I_{-k}^m is the negative of I_k^m. However, as we shall see, this is not always true; wye-delta transformer connections complicate the problem.

A method for locating these busses is diagrammed in the flow chart of Figure 9-10. The notation used requires definition.

313

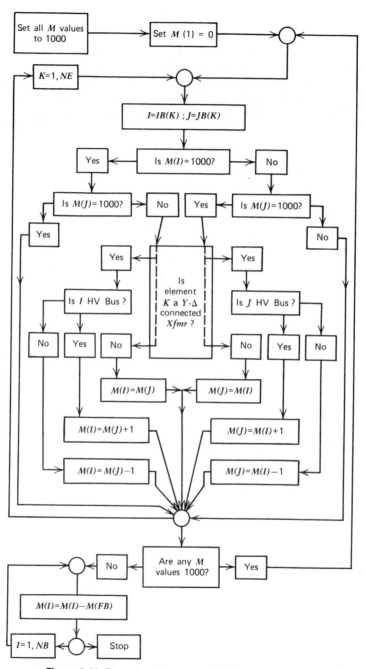

Figure 9-10 Flow chart to account for $Y\text{-}\Delta$ connections.

314

$M(I)$ = The phase adjusting parameter, one for each bus.

$IB(K) = I$ = Terminating bus for element K.

$JB(K) = J$ = Terminating bus for element K.

NE = No. of elements; lines plus transformers.

K = Element number; connected between busses I and J.

FB = Faulted bus.

NB = No. of busses.

The method proceeds as follows. All M values are initially set to 1000, to identify them as unknown. One value, $M(1)$, is set to zero arbitrarily. We next cycle over all the elements, locating the other terminus (J) of those connected to bus one. If lines, Y-Y, or Δ-Δ connected transformers are encountered, $M(J)$ is set to $M(1)$; if Δ-Y connections are encountered, $M(J)$ is set to $M(1) \pm 1$, depending on which is the HV bus. We then check to see if we have determined the proper M value ($M \neq 1000$) for all busses. If not, the cycle is repeated. Note that elements that have both or neither terminal M values determined are skipped. Why?

After all M values are assigned, they are referenced to the faulted bus (i.e., $M(FB) = 0$). We are now prepared to correct the sequence phase values. If:

$$V_j^1 = V_j^1 \underline{/\phi_j^1} \tag{9-38b}$$
$$V_j^2 = V_j^2 \underline{/\phi_j^2} \tag{9-38b}$$

Then the phase values are reset to:

$$\phi_{j_{new}}^1 = \phi_{j_{old}}^1 + M(J)(\pi/6) \tag{9-39a}$$
$$\phi_{j_{new}}^2 = \phi_{j_{old}}^2 - M(J)(\pi/6) \tag{9-39b}$$

ϕ in radians.

The current quantities are:

$$I_k^1 = I_k^1 \underline{/\theta_k^1} \tag{9-40a}$$
$$I_k^2 = I_k^2 \underline{/\theta_k^2} \tag{9-40b}$$

Then:

$$\theta_{k_{new}}^1 = \theta_{k_{old}}^1 + M(I)(\pi/6) \tag{9-41a}$$
$$\theta_{k_{new}}^2 = \theta_{k_{old}}^2 - M(I)(\pi/6) \tag{9-41b}$$

Similarly:

$$\theta^1_{-k_{\text{new}}} = \theta^1_{-k_{\text{old}}} + M(J)(\pi/6) \tag{9-41c}$$

$$\theta^2_{-k_{\text{new}}} = \theta^2_{-k_{\text{old}}} - M(J)(\pi/6) \tag{9-41d}$$

The zero sequence quantities do not require adjustment. We are now prepared to transform to phase values. Equations (2-53) and (2-55) apply:

$$\tilde{V}_{abc} = [T]\,\tilde{V}_{012} \tag{2-53}$$

$$\tilde{I}_{abc} = [T]\tilde{I}_{012} \tag{2-55}$$

where

$\tilde{V}_{012}$ = sequence phase voltage vector (3×1) at any point of interest.

$\tilde{I}_{012}$ = sequence line current vector (3×1) in any line of interest.

The above methods are properly implemented by computer. The material presented in Chapter 8 and the first four sections of this chapter should provide readers with sufficient background to enable them to write their own fault analysis computer program. We summarize the major steps with appropriate references in Table 9-1. The next section presents an example solved by a computer program that has implemented the methods previously discussed.

Table 9–1 Major steps in a fault analysis program.

STEP	REFERENCE
Collect, read in, and store machine, transformer, and line data in per unit on common bases.	Chapters 3, 4, 5, 6, and 8.
Formulate the sequence impedance matrices	Sections 9-1 and 2.
Define the faulted bus and Z_f. Specify type of fault to be analyzed.	
Compute the sequence voltages.	Figure 9-9.
Compute the sequence currents.	Equation (9-37).
Correct for wye-delta connections.	Figure 9-10.
Transform to phase values.	Equations (2-53) and (2-55).
Print out results.	

9-6 An Example Application

The example system presented in Chapter 8 was designed to illustrate calculations of specific network quantities. Even in this relatively simple system a comprehensive hand calculation of all quantities for all four fault types is quite lengthy and involved. We spent considerable time in Chapter 8 (examples 8-1 through 8-6) calculating partial results. Computer calculated results for a comprehensive study of this system are presented in the following example.

Example 9-3

Calculate voltages at all busses and currents in all elements for the system of example 8-1 subjected to each of the four basic fault types located at bus 3.

Solution

The problem was solved by computer using a fault analysis program entitled FALTCALC. Results are presentsed in Figures 9-11 through 9-22. A few comments concerning interpretation of the output should prove helpful.

Calculated quantities are displayed in a self-explanatory labeled form. Complex impedances are printed out in rectangular form (read "R X"). All values are in per unit unless otherwise labeled. The output appears in 6 major sections identified as:

A. Input data
B. Fault impedances
C. Sequence voltages
D. Sequence currents
E. Phase voltages
F. Phase currents

Specific comments follow.

A. *Input data*
Per unit (or percent) values must be on common MVA and voltage bases. The input data are divided into 5 sections, which are:

1. General data
2. Rotating machine data
3. Line data
4. Transformer data
5. Fault specification data

317

Specific comments follow:

1. *General data.* Refer to Figure 9-11.
   NB—Number of busses
   NG—Number of rotating machines
   NL—Number of transmission lines
   NT—Number of transformers

The variable NBKR is the number of circuit breakers in the system. The reference bus is the point at which the base line voltage is defined. Base quantities at all other busses are determined by the methods explained in Chapters 3 and 5.

2. *Rotating Machine Data.* Refer to Figure 9-11

   BUS—Bus number
   Z0—Complex zero sequence impedance in rectangular form, resistance; reactance. Values are in percent.
   Z1—Complex positive sequence impedance in rectangular form, resistance; reactance. Values are in percent. (Note: Negative sequence impedances are assumed equal to positive sequence values.)
   ZN—Complex neutral impedance in rectangular form, resistance; reactance. Values are in percent.

3. *Line Data.* Refer to Figure 9-11.

   BUS—Terminating bus number
   BUS—Terminating bus number
   Z0—Complex zero sequence impedance in rectangular form, resistance; reactance. Values are in percent.
   Z1—Complex positive sequence impedance in rectangular form, resistance; reactance. Values are in percent. (Note: Negative sequence impedances are assumed equal to positive sequence values.)

4. *Transformer Data.* Refer to Figure 9-12.

   HV BUS—HV bus number
   CONN—HV connection code: $0 = $ wye; $1 = $ delta
   LV BUS—LV bus number
   CONN—LV connection code: $0 = $ wye; $1 = $ delta
   Z0—Complex zero sequence impedance in rectangular form, resistance; reactance. Values are in percent.
   Z1—Complex positive sequence impedance in rectangular form, resistance; reactance. Values are in percent. (Note: Negative

sequence impedance values are assumed equal to positive sequence values.)

ZN-HV—Complex HV neutral impedance to ground in rectangular form, resistance; reactance. Values are in percent. (Wye only; enter zero for delta.)

ZN-LV—Complex LV neutral impedance to ground in rectangular form, resistance; reactance. Values are in percent. (Wye only; enter zero for delta.)

5. *Fault Specification Data*

ZF—Complex external fault impedance, resistance; reactance. Normally enter zero.

FB—Bus at which fault is to be applied.

N5—Code which defines fault type to be considered (0 = No; 1 = Yes) where:

N5(1)—Balanced 3ϕ

N5(2)—Single Line to Ground

N5(3)—Line to Line

N5(4)—Double Line to Ground

B. *Fault Impedances* (Refer to Figure 9-14)

The standard fault impedance matrices as defined in equations 9-13, 9-14,

***** INPUT DATA *****

IMPEDANCES IN PERCENT

NB	NG	NL	NT	NBKR	REF. BUS	LINE KV BASE	3 PH MVA BASE
5	2	3	2	8	1	230.0	100.0

** ROTATING MACHINES **

BUS	Z0		Z1		ZN	
4	0.0	5.000	0.0	20.000	0.0	3.000
5	0.0	5.000	0.0	20.000	0.0	3.000

** LINES **

BUS	BUS	Z0		Z1	
3	1	0.0	30.000	0.0	10.000
2	1	0.0	30.000	0.0	10.000
2	3	0.0	30.000	0.0	10.000

Figure 9-11 Input data to FALTCALC for Example 9-3.

319

and 9-15 are printed out in standard matrix form. Values are in rectangular (R X) form and are in per unit.

```
                                                    ** TRANSFORMERS **

HV BUS        CONN      LV BUS       CONN                  Z0                         Z1
          (0-Y,1-D)              (0-Y,1-D)

    1          0          4          0         0.0      5.000         0.0      5.000
    2          0          5          1         0.0      5.000         0.0      5.000

                                                        ZN-HV                      ZN-LV

                                              0.0        0.0          0.0        0.0
                                              0.0        0.0          0.0        0.0
```

Figure 9-12 Continuation of FALTCALC input data for Example 9-3.

```
                 THE FAULT OCCURS AT BUS NUMBER      3


                 CASE(S) RUN ARE:


                     BALANCED THREE PHASE

                     SINGLE LINE TO GROUND

                     LINE TO LINE

                     DOUBLE LINE TO GROUND



         THE EXTERNAL FAULT IMPEDANCE IS      0.0       0.0
```

Figure 9-13 Fault specifications data for Example 9-3.

C. *Sequence Voltages* (Refer to Figures 9-15, 9-17, 9-19, and 9-21)
Zero, positive, and negative sequence voltages at all busses are printed out in polar form (magnitude in per unit, angle in degrees).

D. *Sequence Currents* (Refer to Figures 9-15, 9-17, 9-19, and 9-21)
Zero, positive, and negative sequence currents through all elements are printed out in polar form (magnitude in per unit, angle in degrees). "From Bus" busses are listed sequentially in numerical order; all elements at a given bus are represented in the "To Bus" listings.

320

E. *Phase Voltages* (Refer to Figures 9-16, 9-18, 9-20, and 9-22)
Line to neutral phase voltages; otherwise the same as C.

F. *Phase Currents* (Refer to Figures 9-16, 9-18, 9-20, and 9-22)
Phase (line) currents; otherwise the same as D.

ZERO SEQUENCE IMPEDANCE MATRIX

```
0.0    0.1079 0.0    0.0216 0.0    0.0648 0.0    0.0795 0.0    0.0000
0.0    0.0216 0.0    0.0443 0.0    0.0329 0.0    0.0159 0.0    0.0001
0.0    0.0648 0.0    0.0329 0.0    0.1989 0.0    0.0477 0.0    0.0000
0.0    0.0795 0.0    0.0159 0.0    0.0477 0.0    0.0955 0.0    0.0000
0.0    0.0000 0.0    0.0001 0.0    0.0000 0.0    0.0000 0.0    0.1398
```

POSITIVE SEQUENCE IMPEDANCE MATRIX

```
0.0    0.1396 0.0    0.1101 0.0    0.1249 0.0    0.1117 0.0    0.0881
0.0    0.1101 0.0    0.1395 0.0    0.1248 0.0    0.0881 0.0    0.1116
0.0    0.1249 0.0    0.1248 0.0    0.1748 0.0    0.0999 0.0    0.0999
0.0    0.1117 0.0    0.0881 0.0    0.0999 0.0    0.1293 0.0    0.0705
0.0    0.0881 0.0    0.1116 0.0    0.0999 0.0    0.0705 0.0    0.1293
```

NEGATIVE SEQUENCE IMPEDANCE MATRIX

```
0.0    0.1396 0.0    0.1101 0.0    0.1249 0.0    0.1117 0.0    0.0881
0.0    0.1101 0.0    0.1395 0.0    0.1248 0.0    0.0881 0.0    0.1116
0.0    0.1249 0.0    0.1248 0.0    0.1748 0.0    0.0999 0.0    0.0999
0.0    0.1117 0.0    0.0881 0.0    0.0999 0.0    0.1293 0.0    0.0705
0.0    0.0881 0.0    0.1116 0.0    0.0999 0.0    0.0705 0.0    0.1293
```

Figure 9-14 Sequence fault impedance matrices.

*** BALANCED THREE PHASE FAULT ***

SEQUENCE QUANTITIES

VOLTAGES

BUS	V0	ANG0	V1	ANG1	V2	ANG2
1	0.0	0.0	0.286	0.0	0.0	0.0
2	0.0	0.0	0.286	0.0	0.0	0.0
3	0.0	0.0	0.0	0.0	0.0	0.0
4	0.0	0.0	0.429	0.0	0.0	0.0
5	0.0	0.0	0.429	-30.000	0.0	0.0

CURRENTS

BUS	TO BUS	I0	ANG0	I1	ANG1	I2	ANG2
1	4	0.0	0.0	2.857	-90.000	0.0	0.0
1	3	0.0	0.0	2.859	-90.000	0.0	0.0
1	2	0.0	0.0	0.002	90.000	0.0	0.0
2	3	0.0	0.0	2.861	-90.000	0.0	0.0
2	1	0.0	0.0	0.002	-90.000	0.0	0.0
2	5	0.0	0.0	2.856	90.000	0.0	0.0
3	1	0.0	0.0	2.859	90.000	0.0	0.0
3	2	0.0	0.0	2.861	90.000	0.0	0.0
4	0	0.0	0.0	2.857	-90.000	0.0	0.0
4	1	0.0	0.0	2.857	-90.000	0.0	0.0
5	0	0.0	0.0	2.856	-60.000	0.0	0.0
5	2	0.0	0.0	2.856	-120.000	0.0	0.0

Figure 9-15 Sequence quantities for a balanced 3ϕ fault at bus 3. Data for Example 9-3.

PHASE QUANTITIES

VOLTAGES

BUS	VA	ANGA	VB	ANGB	VC	ANGC
1	0.286	0.0	0.286	-120.000	0.286	120.000
2	0.286	0.0	0.286	-120.000	0.286	120.000
3	0.0	0.0	0.0	0.0	0.0	0.0
4	0.429	0.0	0.429	-120.000	0.429	120.000
5	0.429	-30.000	0.429	-150.000	0.429	90.000

CURRENTS

BUS	TO BUS	IA	ANGA	IB	ANGB	IC	ANGC
1	4	2.857	90.000	2.857	-30.000	2.857	-150.000
1	3	2.859	-90.000	2.859	150.000	2.859	30.000
1	2	0.002	90.000	0.002	-30.000	0.002	-150.000
2	3	2.861	-90.000	2.861	150.000	2.861	30.000
2	1	0.002	-90.000	0.002	150.000	0.002	30.000
5	5	2.856	90.000	2.856	-30.000	2.856	-150.000
3	1	2.859	90.000	2.859	-30.000	2.859	-150.000
3	2	2.861	90.000	2.861	-30.000	2.861	-150.000
4	0	2.857	90.000	2.857	-30.000	2.857	-150.000
4	1	2.857	-90.000	2.857	150.000	2.857	30.000
5	0	2.856	-60.000	2.856	-60.000	2.856	-180.000
5	2	2.856	-120.000	2.856	120.000	2.856	0.000

Figure 9-16 Phase quantities for a balanced 3ϕ fault at bus 3. Data for example 9-3.

*** SINGLE LINE TO GROUND FAULT ***

SEQUENCE QUANTITIES

VOLTAGES

BUS	V0	ANG0	V1	ANG1	V2	ANG2
1	0.118	180.000	0.772	0.0	0.228	180.000
2	0.060	180.000	0.772	0.0	0.228	180.000
3	0.363	180.000	0.681	0.0	0.319	180.000
4	0.087	180.000	0.818	0.0	0.182	180.000
5	0.0	0.0	0.818	-30.000	0.182	-150.000

CURRENTS

BUS	TO BUS	I0	ANG0	I1	ANG1	I2	ANG2
1	4	0.621	90.000	0.911	-90.000	0.911	-90.000
1	3	0.815	-90.000	0.911	-90.000	0.911	-90.000
1	2	0.193	90.000	0.0	0.0	0.0	0.0
2	3	1.008	-90.000	0.912	-90.000	0.912	-90.000
2	1	0.193	-90.000	0.0	0.0	0.0	0.0
2	5	1.201	90.000	0.910	90.000	0.910	90.000
3	1	0.815	90.000	0.911	90.000	0.911	90.000
3	2	1.008	90.000	0.912	90.000	0.912	90.000
4	0	0.621	90.000	0.911	-90.000	0.911	-90.000
4	1	0.621	-90.000	0.911	-90.000	0.911	-90.000
5	0	0.0	0.0	0.910	60.000	0.910	120.000
5	2	0.0	0.0	0.910	-120.000	0.910	-60.000

Figure 9-17 Sequence quantities for a single line to ground fault at bus 3. Data for Example 9-3.

PHASE QUANTITIES

VOLTAGES

BUS	VA	ANGA	VB	ANGB	VC	ANGC
1	0.427	0.0	0.950	-114.268	0.950	114.268
2	0.485	0.0	0.928	-111.003	0.928	111.003
3	0.0	0.0	1.023	-122.124	1.023	122.125
4	0.549	0.0	0.956	-115.058	0.956	115.058
5	0.744	-42.241	0.744	-137.768	1.000	90.005

CURRENTS

BUS	TO BUS	IA	ANGA	IB	ANGB	IC	ANGC
1	4	2.442	90.000	0.289	-90.000	0.289	-90.000
1	3	2.637	-90.000	0.096	90.000	0.096	90.000
1	2	0.195	-90.000	0.193	90.000	0.193	90.000
2	3	2.832	-90.000	0.096	-90.000	0.096	-90.000
2	1	0.195	-90.000	0.193	-90.000	0.193	-90.000
2	5	3.021	90.000	0.291	90.000	0.291	90.000
3	1	2.637	90.000	0.096	-90.000	0.096	-90.000
3	2	2.832	90.000	0.096	90.000	0.096	90.000
4	0	2.442	-90.000	0.289	-90.000	0.289	-90.000
4	1	2.442	-90.000	0.289	-90.000	0.289	90.000
5	0	1.577	-90.000	1.576	-90.000	0.0	0.0
5	2	1.577	-90.000	1.577	90.000	0.0	0.0

Figure 9-18 Phase quantities for single line to ground fault at bus 3. Data for Example 9-3.

SEQUENCE QUANTITIES

VOLTAGES

BUS	V0	ANG0	V1	ANG1	V2	ANG2
1	0.0	0.0	0.643	0.0	0.357	0.0
2	0.0	0.0	0.643	0.0	0.357	0.0
3	0.0	0.0	0.500	0.0	0.500	0.0
4	0.0	0.0	0.714	0.0	0.286	0.0
5	0.0	0.0	0.714	-30.000	0.286	30.000

CURRENTS

BUS	TO BUS	I0	ANG0	I1	ANG1	I2	ANG2
1	4	0.0	0.0	1.428	90.000	1.428	-90.000
1	3	0.0	0.0	1.429	-90.000	1.429	90.000
1	2	0.0	0.0	0.001	90.000	0.001	-90.000
2	3	0.0	0.0	1.430	-90.000	1.430	90.000
2	1	0.0	0.0	0.001	-90.000	0.001	90.000
2	5	0.0	0.0	1.428	90.000	1.428	-90.000
3	1	0.0	0.0	1.429	90.000	1.429	-90.000
3	2	0.0	0.0	1.430	90.000	1.430	-90.000
4	0	0.0	0.0	1.428	-90.000	1.428	-90.000
4	1	0.0	0.0	1.428	-90.000	1.428	90.000
5	0	0.0	0.0	1.428	-60.000	1.428	-60.000
5	2	0.0	0.0	1.428	-120.000	1.428	120.000

Figure 9-19 Sequence quantities for a line to line fault at bus 3. Data for Example 9-3.

PHASE QUANTITIES

VOLTAGES

BUS	VA	ANGA	VB	ANGB	VC	ANGC
1	1.000	0.0	0.558	-153.658	0.558	153.658
2	1.000	0.0	0.558	-153.641	0.558	153.641
3	1.000	0.0	0.500	-180.000	0.500	180.000
4	1.000	0.0	0.623	-143.405	0.623	143.405
5	0.892	-13.907	0.892	-166.093	0.429	90.000

CURRENTS

BUS	TO BUS	IA	ANGA	IB	ANGB	IC	ANGC
1	4	0.0	0.0	2.474	0.000	2.474	-180.000
1	3	0.0	0.0	2.476	-180.000	2.476	0.000
1	2	0.0	0.0	0.002	0.000	0.002	-180.000
2	3	0.0	0.0	2.478	-180.000	2.478	0.000
2	1	0.0	0.0	0.002	-180.000	0.002	0.000
2	5	0.0	0.0	2.473	0.000	2.473	-180.000
3	1	0.0	0.0	2.476	0.000	2.476	-180.000
3	2	0.0	0.0	2.478	0.000	2.478	-180.000
4	0	0.0	0.0	2.474	0.000	2.474	-180.000
4	1	0.0	0.0	2.474	-180.000	2.474	0.000
5	2	1.428	0.000	1.428	0.000	2.856	-180.000
5	2	1.428	180.000	1.428	-180.000	2.856	0.000

Figure 9-20 Phase quantities for a line to line fault at bus 3. Data for Example 9-3.

SEQUENCE QUANTITIES

VOLTAGES

BUS	V0	ANG0	V1	ANG1	V2	ANG2
1	0.113	0.0	0.534	0.0	0.248	0.0
2	0.058	0.0	0.534	0.0	0.248	0.0
3	0.347	0.0	0.347	0.0	0.347	0.0
4	0.083	0.0	0.627	0.0	0.198	0.0
5	0.0	0.0	0.627	-30.000	0.198	30.000

CURRENTS

BUS	TO BUS	I0	ANG0	I1	ANG1	I2	ANG2
1	4	0.595	-90.000	1.864	90.000	0.992	-90.000
1	3	0.781	-90.000	1.866	-90.000	0.993	90.000
1	2	0.185	-90.000	0.001	90.000	0.0	0.0
2	3	0.966	90.000	1.867	-90.000	0.994	90.000
2	1	0.185	90.000	0.001	-90.000	0.0	0.0
2	5	1.151	-90.000	1.864	90.000	0.992	-90.000
3	1	0.781	-90.000	1.866	90.000	0.993	-90.000
3	2	0.966	-90.000	1.867	90.000	0.994	-90.000
4	0	0.595	-90.000	1.864	90.000	0.992	-90.000
4	1	0.595	90.000	1.864	-90.000	0.992	90.000
5	0	0.0	0.0	1.864	60.000	0.992	-60.000
5	2	0.0	0.0	1.864	-120.000	0.992	120.000

Figure 9-21 Sequence quantities for a double line to ground fault at bus 3. Data for Example 9-3.

*** DOUBLE LINE TO GROUND FAULT ***

PHASE QUANTITIES

VOLTAGES

BUS	VA	ANGA	VB	ANGB	VC	ANGC
1	0.895	0.0	0.372	-138.298	0.372	138.298
2	0.840	0.0	0.415	-143.389	0.415	143.389
3	1.042	0.0	0.0	0.0	0.0	0.0
4	0.909	0.0	0.496	-131.583	0.496	131.583
5	0.747	-16.693	0.746	-163.304	0.429	89.989

CURRENTS

BUS	TO BUS	IA	ANGA	IB	ANGB	IC	ANGC
1	4	0.277	90.000	2.680	-22.635	2.680	-157.365
1	3	0.092	-90.000	2.759	153.820	2.759	26.180
1	2	0.185	-90.000	0.186	-89.433	0.186	-90.567
2	3	0.092	90.000	2.847	150.482	2.847	29.518
2	1	0.185	90.000	0.186	90.567	0.186	89.433
2	5	0.279	-90.000	2.938	-32.683	2.938	-147.317
3	1	0.092	90.000	2.759	-26.180	2.759	-153.820
3	2	0.092	-90.000	2.847	-29.518	2.847	-150.482
4	0	0.277	90.000	2.680	-22.635	2.680	-157.365
4	1	0.277	-90.000	2.680	157.365	2.680	22.635
5	0	1.615	27.857	1.616	-27.894	2.856	-179.988
5	2	1.615	-152.125	1.615	152.125	2.856	0.0

Figure 9-22 Phase quantities for a double line to ground fault at bus 3. Data for Example 9-3.

9-7 Summary

Computation of fault currents in power systems is best done by computer. Sufficient information has been presented to readers to enable them to create their own computer program, a project which is recommended to the serious student. The major steps are summarized in Table 9-1.

Computer formulation of the impedance matrices can be accomplished by programming the four modifications presented in section 9-1. Care must be taken to avoid the fifth type of modification (new bus to new bus), which requires infinite entries into $[Z]$. This situation may be avoided two ways: number the busses so as to avoid the problem, or first insert a fictitious large impedance from one of the new busses to reference, changing its status to old bus for the next step. Complete the modification using the old bus to new bus method. The procedure is followed "in triplicate," constructing the zero, positive, and negative sequence impedance matrices. If we are willing to use the same values for positive and negative sequence machine impedances:

$$[Z_1] = [Z_2] \tag{9-42}$$

325

Thus, it is unnecessary to store negative sequence values in separate arrays, simplifying the program and reducing the computer storage requirements significantly. The error introduced by this approximation is usually not significant.

The methods previously discussed neglect the prefault, or load, component of current; that is, the usual assumption is that currents throughout the system were zero prior to the fault. This is almost never strictly true; however, the error produced is small since the fault currents are generally much larger than the load currents. Also the load currents and fault currents are near 90° displaced in phase from each other, making their sum nearer equal to the larger component than would have been the case if the currents were in phase. In addition, selection of precise values for all prefault currents is somewhat speculative, since there is no way of predicting what the loaded state of the system is when a fault occurs.

Occasionally it is important to consider load currents. When this is desirable, a power flow study is made to calculate currents throughout the system. These values are superimposed on (i.e., added to) results from the fault study. This refinement is recommended when the fault currents are small (roughly no more than a factor of five times the load current).

A term that has wide industrial use and acceptance is the "fault level" or "fault MVA" at a bus. It relates to the amount of current that can be expected to flow out of a bus into a three phase fault. As such, it is an alternate way of providing positive sequence impedance information. We define:

$$\text{Fault level in MVA at bus } i = V_{i_{\text{pu nominal}}} I_{i_{\text{pu fault}}} S_{3\phi\text{base}} \qquad (9\text{-}43a)$$

$$= (1) \frac{1}{z_{ii}^1} S_{3\phi\text{base}} \qquad (9\text{-}43b)$$

$$= \frac{S_{3\phi\text{base}}}{z_{ii}^1} \qquad (9\text{-}43c)$$

Fault study results may be further refined by approximating the effect of dc offset using the methods discussed in section 8-8.

The basic reason for making fault studies is to provide data that can be used to size and set protective devices adequately. The role of such protective devices is to detect and remove faults to prevent or minimize damage to the power system. Our next concern is with systems, devices, and methods used for the protection of the power system. This important area is the subject of Chapters 10 and 11, and we proceed with some measure of confidence that we can calculate fault electrical quantities as needed.

Bibliography

[1] Anderson, Paul M., *Analysis of Faulted Power Systems*, Iowa State Press, Ames, Iowa, 1973.

[2] Brown, Homer E., *Solution of Large Networks by Matrix Methods*, John Wiley and Sons, Inc., 1975.

[3] IEEE., *Recommended Practice for Protection and Coordination of Industrial and Commercial Power Systems*, Wiley-Interscience, New York, 1975.

[4] Manufacturer's Publication, *Short-Circuit Current Calculations for Industrial and Commercial Power Systems*, General Electric, Publication GET-3550.

[5] Manufacturer's Publication, *Short-Circuit Currents in Low and Medium Voltage A-C Power Systems*, General Electric, Publication GET-1470D.

[6] Neuenswander, John R., *Modern Power Systems*, International Textbook Co., Scranton, Pa., 1971.

[7] Stagg, Glenn W., and El-Abiad, Ahmed H., *Computer Methods in Power System Analysis*, McGraw-Hill, Inc., New York, 1968.

[8] Stevenson, Jr., William D., *Elements of Power System Analysis*, 3rd edition, McGraw-Hill, Inc., New York, 1975.

[9] Weedy, B. M., *Electric Power Systems*, 2nd edition. John Wiley and Sons Ltd., London, 1972.

Problems

9-1. Consider the sequence networks shown in Figure 8-4 for the system of example 8-1. Evaluate z_{33}^1 using equation (9-2) for this system. Check your answer with the value presented in Figure 9-14.

9-2. Repeat problem 9-1 to find z_{15}^1.

9-3. Repeat problem 9-1 to find z_{33}^0.

9-4. Repeat problem 9-1 to find z_{15}^0.

9-5. Consider the matrix equations

$$\begin{bmatrix} \tilde{V}_A \\ 0 \end{bmatrix} = \begin{bmatrix} [Z_{AA}][Z_{AB}] \\ [Z_{BA}][Z_{BB}] \end{bmatrix} \begin{bmatrix} \tilde{I}_A \\ \tilde{I}_B \end{bmatrix}$$

327

which are:

$$\tilde{V}_A = [Z_{AA}]\tilde{I}_A + [Z_{AB}]\tilde{I}_B$$

$$0 = [Z_{BA}]\tilde{I}_A + [Z_{BB}]\tilde{I}_B$$

Show that:

$$\tilde{V}_A = [Z]\tilde{I}_A$$

where

$$[Z] = [Z_{AA}] - [Z_{AB}][Z_{BB}]^{-1}[Z_{BA}]$$

$\tilde{V}_A$ is $n \times 1$

$\tilde{I}_A$ is $n \times 1$

$\tilde{I}_B$ is $m \times 1$

$[Z_{AA}]$ is $n \times n$

$[Z_{AB}]$ is $n \times m$

$[Z_{BA}]$ is $m \times n$

$[Z_{BB}]$ is $m \times m$

9-6. Formulate $[Z_1]$ by the method of section 9-2 for the system of example 8-1. Refer to Figure 8-4. Check results with values presented in Figure 9-14.

9-7. Repeat problem 9-6, formulating $[Z_0]$.

9-8. In the system of example 8-1 consider a fault at bus 3. Impedance data is presented in Figure 9-14. For a balanced three phase fault, use equation (9-19c) to compute V_3^1. $Z_f = 0$. Check your answers with Figure 9-15.

9-9. Repeat Problem 9-8 for a single line to ground fault. Use equations (9-21c), (9-22c), and (9-23c) to compute V_3^0, V_3^1, and V_3^2, respectively. Check your answers with Figure 9-17.

9-10. Repeat problem 9-8 for a line to line fault. Use equations (9-26b) and (9-27c) to calculate V_3^1 and V_3^2. Check your answers with Figure 9-19.

9-11. Repeat problem 9-8 for a double line to ground fault. Use equations (9-34b), (9-35b), and (9-36b) to compute V_3^0, V_3^1, and V_3^2, respectively. Check your answers with Figure 9-21.

9-12. Calculate the sequence line currents from bus 1 to bus 4 for a balanced three phase fault at bus 3 in the system of example 8-1.

328

Consider the sequence voltages as given (consult Figure 9-15 for values). Check results with Figure 9-15.

9-13. Repeat problem 9-12 for a single line to ground fault. Check results with Figure 9-17.

9-14 Repeat problem 9-12 for a line to line fault. Check results with Figure 9-19.

9-15. Repeat problem 9-12 for a double line to ground fault. Check results with figure 9-21.

9-16. Calculate the phase voltages from the sequence voltages at bus 1 in the system of example 8-1 subjected to a single line to ground fault at bus 3. Obtain sequence values from Figure 9-17 and check results in Figure 9-18.

9-17. Explain the phase angle values presented for zero, positive, and negative sequence voltages at bus 5 in Figure 9-17.

9–18. Calculate V_a, V_b, and V_c in kV at bus 1 for the situation represented in Figure 9-17.

9-19. Calculate I_a, I_b, and I_c in amperes from bus 1 to bus 3 in the transmission line TL13 for the situation represented in Figure 9-17.

POWER SYSTEM PROTECTION 1: DEVICES

"Therefore be at peace with God, whatever you conceive Him
to be. And whatever your labors and aspirations, in the
noisy confusion of life, keep peace in your soul."

Max Ehrmann, DESIDERATA

Calculation of currents and voltages throughout a faulted power system was discussed in some detail in Chapters 8 and 9. Recall that the purpose of these calculations was to provide values to size and set protection equipment. In this chapter we discuss some protective devices, including instrument transformers, relays, circuit breakers, and fuses. We shall present some electrical protection techniques for transformers and rotating machines.

Faults can damage or disrupt power systems in several ways:

- Faults typically allow abnormally large currents to flow, resulting in overheating of power system components.

- The fault is usually a short circuit and exists as an electrical arc in a fluid (such as air). The extremely high temperatures in arcs will vaporize any known substance, causing equipment destruction and fire.

- Faults can lower, or raise, system voltages outside of their acceptable ranges.

- Faults can cause the three phase system to become unbalanced, causing three phase equipment to operate improperly.

- Faults block the flow of power.

- Faults can cause the system to become unstable, and "break-up" (i.e., lose synchronism).

It is important to state the objectives of power system protection schemes. Ideally, a protection scheme should:

- Detect and isolate faults instantaneously at any point in the system.

- Accomplish the above, keeping as much of the system interconnected as possible.

- Since many faults are self-clearing, restore the system to its original configuration as soon as possible.

- Clearly discriminate between normal and abnormal system conditions so that protective devices never operate unnecessarily.

It is virtually impossible to completely satisfy all these objectives all the time. However, some imaginative engineering has produced techniques that function admirably.

We begin our study of system protection by examining some basic protective components.

10-1 Detection of System Variables: Instrument Transformers

Detecting faults and other abnormal power system operating conditions requires that we monitor power system variables—that is, current, voltage, power, and impedance. If we measure current and voltage, we can process these signals to obtain information regarding power and impedance. Since power system voltages and currents are in the kilovolt and kiloamp range, it is necessary to use signals that are proportional to system values for safety, economy, and convenience of measurement.

Transformers of special design called "instrument transformers" are commonly used for this purpose. They have several advantages:

- They are simple, economical, and reliable.
- They provide electrical isolation from the power system.
- They are accurate and will tolerate loading to some extent.

Instrument transformers are of two basic types: voltage, or potential, transformers (PT), and current transformers (CT). A three phase installation of PT's is shown in Figure 10-1. Refer to Figure 10-2a for a typical symbol used for the PT. For system protection applications extremely high accuracy in PT's is usually not required. Therefore it is usually reasonable to model the PT as an ideal transformer.

$$V_2 = \frac{N_2}{N_1} V_1 \tag{10-1}$$

Equation 10-1 reveals that:

1. V_2 is a scaled version of V_1 (Scaled down)
2. V_2 is in phase with V_1.

A standard secondary voltage rating is 120 volts. Commonly available voltage ratios are 1, 2, 2.5, 4, 5, 20, 40, 60, 100, 200, 300, 400, 600, 800, 1000, 2000, 3000, and 4500:1.

Optimally, a PT would be terminated in an infinite impedance, drawing no current. Practically, it will be terminated in a potential sensitive device, such as a voltmeter, whose impedance is high, but not infinite. Loading effects on instrument transformers are generally referred to as "burden."

Figure 10-1 Three Westinghouse 69 kV potential transformers installed in grounded wye. (Photograph courtesy of Georgia Power Co.)

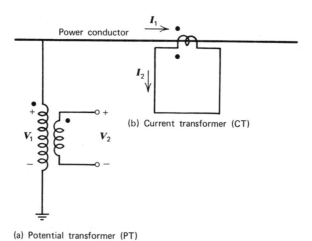

(b) Current transformer (CT)

(a) Potential transformer (PT)

Figure 10-2 Instrument transformer representation.

The term "burden," like "load," is not precisely defined and can mean current, power, or impedance, depending on context. Secondaries are generally rated in the vicinity of 50 VA.

A key consideration is to properly account for the polarity (dot) markings. In some applications the dots may be ignored, but in others an error involving dots can have catastrophic consequences. Consider the following example.

Example 10-1

Examine the open delta PT connection illustrated in Figure 10-3. Assume:

$$V_{AB} = 230000\underline{/0°}$$

$$V_{BC} = 230000\underline{/-120°}$$

$$V_{CA} = 230000\underline{/+120°}$$

and both PT's have a voltage rating of 240 kV : 120 V. Calculate V_{ab}, V_{bc}, and V_{ca} if: (a) the dots are as shown in Figure 10-3; (b) in Figure 10-3 the dot near "b" is moved to "c."

(a) $\dfrac{N_1}{N_2} = \dfrac{240,000}{120} = \dfrac{2000}{1}$

$$\therefore \quad V_{ab} = \frac{1}{2000}(230000\underline{/0°}) = 115\underline{/0°}$$

$$V_{bc} = \frac{1}{2000}(230000\underline{/-120°}) = 115\underline{/-120°}$$

By KVL:

$$V_{ca} = -(V_{ab} + V_{bc})$$

$$= -(115\underline{/-60°}) = 115\underline{/+120°}$$

(b) Again

$$V_{ab} = 115\underline{/0°}$$

But now

$$V_{bc} = -115\underline{/-120°} = 115\underline{/60°}$$

Therefore:

$$V_{ca} = -(V_{ab} + V_{bc}) = 199\underline{/-150°}$$

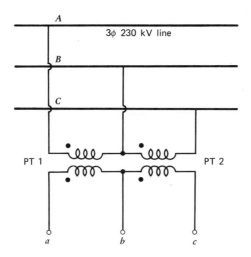

Figure 10-3 Open delta PT connection.

Figure 10-4 Three Westinghouse 200/5 ampere current transformers installed on a 12 kV feeder circuit. (Photograph courtesy of Georgia Power Co.)

Example 10-1

The output of the PT bank is *not* balanced 3φ. The importance of tracking the dot locations should be clear.

A typical installation of current transformers is shown in Figure 10-4.

Frequently the current transformer (CT) may also be modeled as ideal. Refer to Figure 10-2b for the symbol to be used. Its optimum termination would be a perfect short circuit. The CT is terminated in current sensing devices whose terminating impedances, while low, are not zero. A more precise equivalent circuit for the CT is shown in Figure 10-5a. We define

X_2 = Transformer leakage reactance referred to secondary.

X_e = Transformer excitation reactance referred to secondary (saturable reactance)

X_L = Secondary terminating reactance (leads, relay, ammeter, etc.)

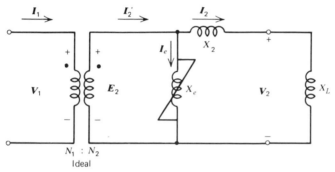

(a) CT equivalent circuit

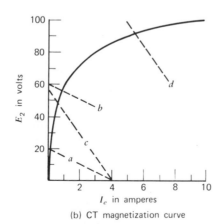

(b) CT magnetization curve

Figure 10-5 Accounting for saturation in CT's.

337

We write:

$$I_2' = \frac{N_1}{N_2} I_1 \tag{10-2}$$

$$E_2 = \frac{N_2}{N_1} V_1 \tag{10-3}$$

$$I_2 = I_2' - I_e \tag{10-4a}$$

$$E_2 = jI_2(X_2 + X_L) \tag{10-5}$$

Substituting (10-2) and (10-4) into (10-5):

$$E_2 = j(X_2 + X_L)\left[\frac{N_1}{N_2} I_1 - I_e\right] \tag{10-6a}$$

Since all impedance elements are inductive:

$$E_2 = (X_2 + X_L)\left[\frac{N_1 I_1}{N_2} - I_e\right] \tag{10-6b}$$

$$I_2 = I_2' - I_e \tag{10-4b}$$

A second constraint that must be satisfied is the magnetization curve shown in Figure 10-5b. To predict I_2 for a given I_1 with known termination (X_L), equation (10-6b) has two unknowns (E_2 and I_e). The second relation between E_2 and I_e is the magnetization curve (E_2 versus I_e: Figure 10-5b. If we plot equation (10-6b) on the same coordinates as the magnetization curve, the intersection of the two plots will represent a solution for E_2 and I_e. Knowing I_e we can now solve for I_2. The "CT error" is the deviation of I_2 from I_2' as expressed as a percentage of I_2'.

$$\text{CT error} = \frac{I_2' - I_2}{I_2'} \tag{10-7a}$$

$$= \frac{I_e}{I_2'} \tag{10-7b}$$

An example should illustrate the use of the above equations.

Example 10-2

A CT has a rated current ratio of 500 : 5 amperes, $X_2 = 0.5$ ohm, and has the magnetization curve given in Figure 10-5b. Compute I_2 and the CT error for the following cases:

Example 10-2

(a) $X_L = 4.5 \ \Omega; \ I_1 = 400 \ \text{A (Load Current)}$
(b) $X_L = 4.5 \ \Omega; \ I_1 = 1200 \ \text{A (Fault Current)}$
(c) $X_L = 13.5 \ \Omega; \ I_1 = 400 \ \text{A (Load Current)}$
(d) $X_L = 13.5 \ \Omega; \ I_1 = 1200 \ \text{A (Fault Current)}$

Solution

(a) Equation (10-6b) becomes:

$$E_2 = 5(4 - I_e)$$

Its plot is the line a. We read $I_e \cong 0.1$

$\therefore$ $I_2 = 3.9$ amperes

$$\text{CT error} = \frac{0.1}{4.0} = 2.5\%$$

(b) Equation (10-6b) becomes:

$$E_2 = 5(12 - I_e)$$

Its plot is the line b. We read $I_e \cong 0.8$.

$\therefore$ $I_2 = 11.2$ amperes

$$\text{CT error} = \frac{0.8}{12} = 6.7\%$$

(c) Equation (10-6b) becomes:

$$E_2 = 14(4 - I_e)$$

Its plot is the line c. We read $I_e \cong 0.6$.

$\therefore$ $I_2 = 3.4$ amperes

$$\text{CT error} = \frac{0.6}{4.0} = 15\%$$

(d) Equation (10-6b) becomes:

$$E_2 = 14(12 - I_e)$$

Its plot is line d. We read $I_e \cong 5.4$.

$\therefore$ $I_2 = 6.6$ amperes

$$\text{CT error} = \frac{5.4}{12} = 45\%$$

We conclude from these results that CT error increases with increasing CT current and is further increased by high terminating impedance. Even at short circuit the CT error is not zero, due to the internal leakage reactance of the CT. The problem is further developed in the following example.

Example 10-3

The CT of example 10-2 is used to drive a current sensitive device that will operate at current levels at or above 8 amperes. Will the device detect the 1200 ampere fault current if its input impedance is:
(a) $X_L = 4.5\,\Omega$?
(b) $X_L = 13.5\,\Omega$?

Solution

If we assume the CT is ideal, we claim I_2 is 12 amperes, and predict the device would detect the 1200 ampere primary current (in fact, any current down to 800 amps), independent of X_L. Using the more precise analysis presented in Example 10-2:
(a) $X_L = 4.5\,\Omega$. Refer to example 10-2b. We computed $I_2 = 11.2$ amperes. Therefore the fault is detected.
(b) $X_L = 13.5\,\Omega$. Refer to example 10-2d. We computed $I_2 = 6.6$ amperes. The fault is *not* detected. The assumption that the CT was ideal in this case would have resulted in a failure to detect a faulted system.

An additional consideration is that the terminating CT impedance is complex, not pure reactance. Usually this effect is ignored for lack of precise data and the fact that the calculation is somewhat conservative, predicting higher CT error than is actually the case. If higher accuracy is required, equation (10-6b) cannot be used.

CT's are rated in terms of a standard 5 ampere secondary; commonly available ratios are shown in Table 10-1. Two basic types of CT's are Class H (high leakage reactance, wound and through type transformers) and Class L (low leakage reactance, busing transformers). There are two standard accuracy classes: $2\frac{1}{2}\%$ and 10%. These are defined as CT errors at some specified voltage and at some specific multiple of (frequently 20 times) normal current. Burden is interpreted as the terminating secondary impedance in ohms (sometimes given in volt-amperes at a specified voltage or current).

Table 10-1 Some standard CT ratios.

CURRENT RATIO	TURNS RATIO
50 : 5	1 : 10
100 : 5	1 : 20
200 : 5	1 : 40
400 : 5	1 : 80
600 : 5	1 : 120
800 : 5	1 : 160
1200 : 5	1 : 240

10-2 Relays

A key component in any power system protection scheme is the relay. A relay is a device that, based on information received from the power system, performs one or more switching actions. The "information" referred to consists of signals proportional to magnitudes and phase angles of power sytem voltages and currents, typically the output of instrument transformers. The relay "decides" to close (or open) one or more sets of normally open (or closed) contacts. The switching action typically energizes the trip coil of a circuit breaker, which then opens the power circuit. It, in effect, "relays" the decision to trip to the circuit breaker, which actually performs the act—thus its name.

Relays must have the following characteristics:

- Reliability—The nature of the problem is that the relay may be idle for periods extending into years and then be required to operate with fast response, as intended, the first time. The penalty for failure to operate properly may run into millions of dollars.

- Selectivity—The relay must not respond to abnormal, but harmless, system conditions such as switching transients or sudden changes in load.

- Sensitivity—The relay must not fail to operate, even in borderline situations, when operation was planned.

- Speed—The relay should make the decision to act as close to instantaneously as possible. If intentional time delay is available, it should be predictable and precisely adjustable.

- Instantaneous—The term means no intentional time delay.

There are several possible ways to classify relays: by function, by construction, by application. Relays are one of two basic types of construction: electromagnetic or solid state. The electromagnetic type relies on the development of electromagnetic forces on movable members, which provide switching action by physically opening or closing sets of contacts. The solid state variety provides switching action with no physical motion by changing the state of a serially connected solid state component from nonconducting to conducting (or vice versa). Electromagnetic relays are older and more widely used; solid state relays are more versatile, potentially more reliable, and faster.

A typical electromagnetic plunger type relay is shown in Figure 10-6a. The plunger is movable. Note that the spring pulls the plunger down when the coil (c-c') is de-energized. The de-energized position is referred to as

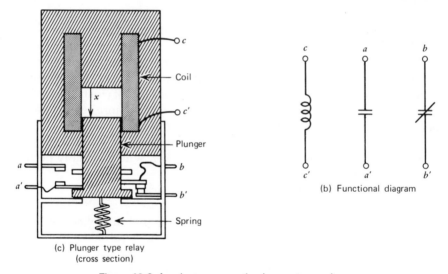

(b) Functional diagram

(c) Plunger type relay
(cross section)

Figure 10-6 An electromagnetic plunger type relay.

normal, so that contacts a-a' are normally open and contacts b-b' normally closed. It is conventional to represent the relay as shown in Figure 10-6b. When the coil is energized, an upward magnetic force will be exerted on the plunger. The magnetic force varies directly with the square of the coil current. When the magnetic force overpowers the restraining spring, the plunger moves upwards, closing the air gap x, which greatly increases the magnetic force. The threshold value of current that starts the plunger moving is referred to as the pickup current. The minimum current that can hold the plunger closed against the spring force is called the dropout current. The relay can be designed to operate on dc or ac currents, the latter type responding to rms current. Such a device can be designed to respond to voltage as well as current if the coil impedance is high.

A solid state circuit that is functionally equivalent to the plunger relay is shown in Figure 10-7. A dc input signal V_i is produced, which is proportional to system current or voltage. When $V_i < V$ the base-emitter $(b-e)$

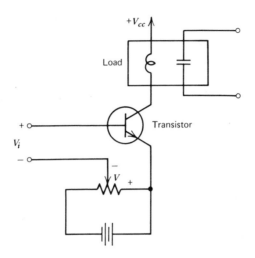

Figure 10-7 Solid state instantaneous overcurrent relay.

junction of the transistor is reverse biased forcing the transistor into its high resistance state. Current flow to the load—in this case shown as an electromagnetic relay coil—is therefore blocked. When $V_i > V$ the base emitter junction is forward biased, producing low resistance from collector to emitter, and allowing current flow through the load. The potentiometer allows for adjustable V, controlling the pickup value. If the input signal is ac, it is rectified to dc.

Both devices are essentially level detectors; they perform a switching action when the input voltage (or current) meets or exceeds a specific and adjustable value. Because they can respond to ac, they can be driven directly from instrument transformers and since such relays can respond to excessive current (or voltage) and have no intentional time delay, they are frequently referred to as instantaneous overcurrent (or overvoltage) relays. Such devices can also be used to detect low levels, in which case they are designated as undercurrent or undervoltage relays. Current devices are designed to have low input impedances (sometimes assumed to be zero) and

343

voltage sensitive devices have high input impedances (sometimes assumed to be infinite), for compatibility with instrument transformers.

A second type of relay is designed to operate only after an adjustable and known time delay. An electromechanical design is shown in Figure 10-8. An ac signal applied to the input coil produces a magnetic field that is

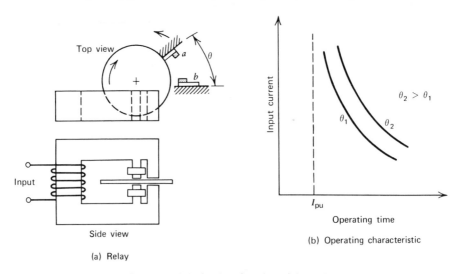

Top view

$\theta_2 > \theta_1$

Input

I_{pu}

Operating time

Side view

(b) Operating characteristic

(a) Relay

Figure 10-8 Induction disc time delay relay.

perpendicular to a conducting aluminum disc. Currents are induced in the disc and interact with the magnetic field so as to produce a torque on the disc, which can rotate. Rotation of the disc is restrained by a spiral spring, pulling the movable contact (a) back against a stop. When the input coil current exceeds the pick up value, contact (a) must travel over the angle θ to meet the fixed contact (b). The greater the coil current, the greater the torque, and therefore the more rapid the disc rotation. Consequently, the operating characteristic is as shown in Figure 10-8b. Adjustment of time delay is achieved by mechanical adjustment of the angle θ, with the larger θ's corresponding to longer time delays.

The time delay may be introduced electrically by several methods; the RC circuit shown in Figure 10-9a can be used to drive the solid state relay in Figure 10-7. If:

$$v_i = V_0 u(t) \tag{10-8a}$$

Then

$$v_0 = V_0[1 - e^{-t/\tau}]u(t) \tag{10-8b}$$

344

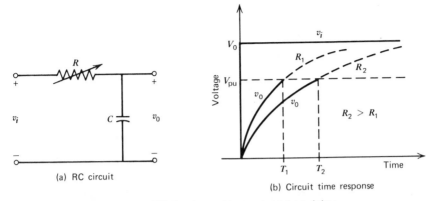

(a) RC circuit

(b) Circuit time response

Figure 10-9 RC circuit used to produce time delay.

where

$$\tau = RC \qquad\qquad (10\text{-}8c)$$

The general response v_0 is indicated in Figure 10-9b.

Example 10-4

Suppose $C = 10\,\mu f$ and $100\,k\Omega \leq R \leq 1\,M\Omega$. In the circuit of Figure 10-9, if $v_i = 2u(t)$ and $V_{pu} = 1\,V$ find T_{delay} for
(a) $R = 100\,k\Omega$
(b) $R = 1\,M\Omega$

Solution

(a) $\tau = RC$

$$= 1 \text{ second}$$

$$v_0 = 2[1 - e^{-t/1}]]_{t=T_{delay}} = 1$$

$$\therefore \quad 1 - e^{-T_{delay}} = 0.5$$

$$e^{-T_{delay}} = 0.5$$

$$e^{T_{delay}} = 2$$

$$\therefore \quad T_{delay} = \ln 2$$

$$= 0.693 \text{ seconds}$$

345

(b) $\tau = RC$

$\quad = 10$ seconds

$\quad v_0 = 2[1 - e^{-t/10}]]_{t=T_{\text{delay}}} = 1$

$\quad e^{T_{\text{delay}}/10} = 2$

$\quad T_{\text{delay}}/10 = \ln 2$

$\quad \therefore \quad T_{\text{delay}} = 6.93$ seconds

The time characteristics of this solid state relay are similar to those of the electromagnetic induction disc relay, shown in Figure 10-8b. A typical electromagnetic relay is the Westinghouse CO-7, shown in Figure 10-10.

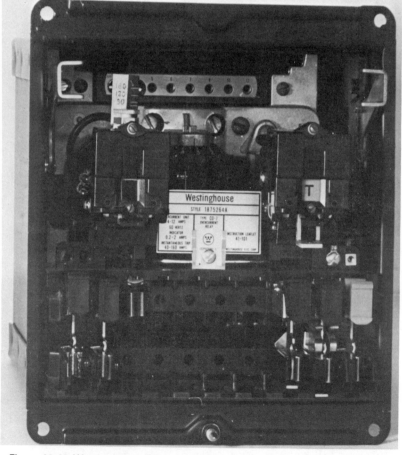

Figure 10-10 Westinghouse CO-7 time delay overcurrent relay. (Photograph by Auburn Engineering Learning Resources Center.)

Relays of this type have two basic adjustable settings:

Current Tap Setting (CTS): the pickup current in amperes.

Time Dial Setting (TDS): the setting which adjusts the amount of intentional time delay.

Corresponding relay characteristics are shown in Figure 10-11. This general type is referred to as a time delay overcurrent relay. Overvoltage, undercurrent, and undervoltage time delay relays are also variations of this type.

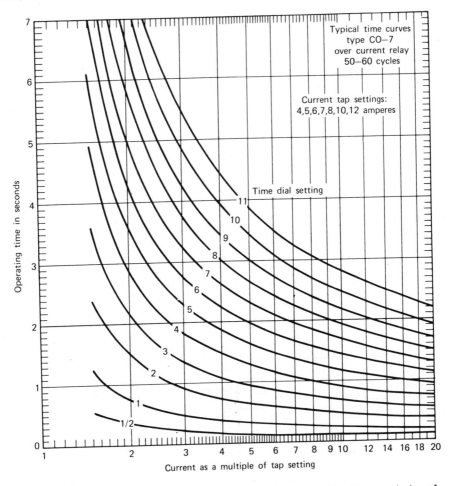

Figure 10-11 CO-7 time delay overcurrent relay characteristics. Used by permission of Westinghouse Electric Corporation.

Example 10-5

Refer to Example 10-2b. Suppose the load reactance ($X_L = 4.5\ \Omega$) is the input impedance to the relay of Figure 10-8. If the tap setting is 5 amperes and time dial setting is 2, determine the operating time from Figure 10-11.

Solution

From Example 10-2b:

$$I_2 = I_{relay} = 11.2 \text{ amperes}$$

$$\frac{I_{relay}}{I_{pickup}} = \frac{11.2}{5} = 2.24$$

From Curve 2, Figure 10-11, we read:

$$T_{operating} = 1.25 \text{ seconds}$$

A third kind of electromechanical relay is the balanced beam type, illustrated in Figure 10-12. The contacts are kept normally open by a soft restraining spring, or gravity. Current flowing in the operating coil will pull the left end of the beam down, and close the contacts. The relay is extremely sensitive to operating current, so much so that this can be undesirable. A restraint coil is typically provided as shown in Figure 10-12a. Observe that:

$$I_c = I_a - I_b \tag{10-9a}$$

In a typical application, all three currents are in phase so that:

$$I_c = I_a - I_b \tag{10-9b}$$

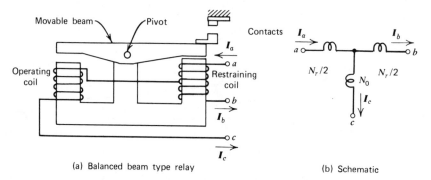

(a) Balanced beam type relay

(b) Schematic

Figure 10-12 Balanced beam relay with restraint.

If $I_a = I_b$, $I_c = 0$, and a force is developed to pull down the right side of the beam and open the contacts. Only when $I_a \neq I_b$ will $I_c \neq 0$, and the relay have a tendency to operate. The electromagnetic forces are proportional to the square of the mmf's. The condition for relay operation is then:

$$(N_0 I_c)^2 > \left(\frac{N_r}{2} I_a + \frac{N_r}{2} I_b\right)^2 \tag{10-10a}$$

where

$N_0 =$ No. of turns on the operating coil

$N_r =$ No. of turns on the restraint coil

Let

$$N_r = k N_0 \tag{10-11}$$

where $\quad 0 < k < 1$

Then from (10-10a):

$$(N_0 I_c)^2 > \left(\frac{k N_0}{2} (I_a + I_b)\right)^2 \tag{10-10b}$$

Taking the square root:

$$I_c > \frac{k}{2}(I_a + I_b) \tag{10-12}$$

Using equation (10-9b)

$$I_a - I_b > \frac{k}{2}(I_a + I_b) \tag{10-13}$$

or

$$I_a > \frac{2+k}{2-k} I_b \tag{10-14a}$$

Since the roles of I_a and I_b are interchangeable:

$$I_b > \frac{2+k}{2-k} I_a \tag{10-14b}$$

See Figure 10-13 for a graphic illustration of relay operating zones. Because operation depends on a *difference* current, such relays are typically applied in schemes referred to as *differential* protection. Differential relaying is

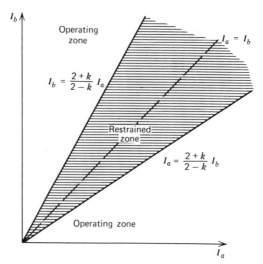

Figure 10-13 Operating zones for balanced beam relay.

commonly used for transformer protection and is discussed in more detail in section 10-5. An electronic differential relay is also available.

Up to this point the relays discussed have been of the single input variety and therefore can respond to current or voltage levels. It is possible to design a device that responds to two inputs. Consider the device in Figure 10-14. Two windings, located on a stator and 90° separated in space, are supplied with ac currents I_1 and I_2. A cup shaped aluminum rotor is subjected to an electromagnetic torque whose average value is:

$$T = KI_1I_2 \sin(\alpha_2 - \alpha_1) \tag{10-15}$$

where

$$I_1 = I_1/\underline{\alpha_1}$$

$$I_2 = I_2/\underline{\alpha_2}$$

The motion of the rotor can be used to open (or close) sets of contacts. Note that if I_2 *leads* I_1 (up to 180°), the torque is in the direction shown whereas if I_1 leads I_2 the torque reverses. The backward motion can be restrained by mechanical stops. The device is effectively a two phase motor.

As with previous relays the input signals can be either voltages or currents and these signals may be passed through phase shifting networks. Therefore a more general formulation of equation (10-15) is

$$T = K_c AB \cos(\theta - \tau) \tag{10-16}$$

350

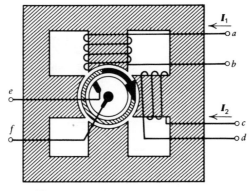

(a) Induction cup relay construction

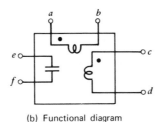

(b) Functional diagram

Figure 10-14 Induction cup relay.

where

A = Rms value of input #1 (current or voltage)

B = Rms value of input #2 (current or voltage)

τ = Adjustable phase constant; $-180° \leq \tau \leq +180°$

θ = Phase shift between input signals

K_c = Constant of proportionality

The signal T may be formulated electronically as indicated in the block diagram of Figure 10-15. The relay parameters τ and K_c are independently adjustable, and controlled by setting the phase adjust network and low pass filter gain. Such relays have wide application in transmission line protection as we see in Chapter 11.

Operation of most relays can be predicted by use of the following general relay equation:

$$T = K_a A^2 + K_b B^2 + K_c AB \cos(\theta - \tau) - K_s \qquad (10\text{-}17a)$$

351

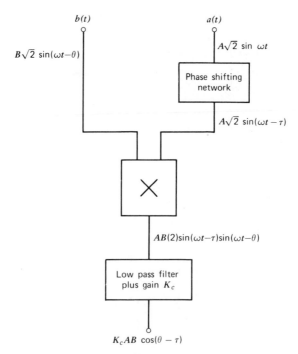

Figure 10-15 Block diagram for electronic two input relay.

where

K_a, K_b, and K_c are adjustable gain constants

K_s = Adjustable spring constant

τ = Adjustable phase constant

A = Rms value of input signal proportional to a system voltage or current.

B = A second input, similar to A

θ = Phase shift between A and B.

Operation is predicted by:

$T > 0$; relay operates   (10-17b)

$T < 0$; relay does not operate   (10-17c)

$T = 0$; border between operating and nonoperating regions   (10-17d)

Such relays may also be of the time delay type, requiring additional data to predict operating time.

The output of a relay is a switching action. The switching action triggers a much larger device, the circuit breaker, which is the topic of our next section.

10-3 Circuit Interrupters: Fuses and Circuit Breakers

The problem of breaking continuity in an electrical circuit might seem trivial to those of us accustomed to flipping a wall switch to turn off a light as we leave a room, or to us as we examine the switching mechanism of an ordinary flashlight. Indeed, at low voltage and current levels, the problem *is* trivial since the surrounding air has excellent insulating properties. However, as we consider voltages and currents in the kilovolt and kilo-ampere range the problem becomes complicated. When we add the constraint that the circuit is to be broken *quickly* (say within a few milliseconds), the problem requires a major engineering effort for its solution.

When two current contacts first part, a voltage instantly appears between the contacts. Because the contacts start infinitesimally close together, large voltage gradients appear in the medium between the contacts even at moderate voltages. If this gradient exists any appreciable time, ionization will occur in the medium, making it a gaseous conductor, and current flow will continue. This current flow will immediately heat the conduction path to extremely high temperatures radiating intense light and heat. This visible conduction path, consisting of ionized hot gas, is called an electrical arc. The heat produced will sustain the ionization, complicating the interruption of current flow. The arc heat can also damage the contacts and other parts of the interrupting device. These problems must be solved in the design of a practical circuit breaker.

To appreciate the problem consider the simple dc circuit of Figure 10-16a. Assume the switch is opened at $t = 0$. What is the voltage, v_{ab}, at $t = 0^+$? Obviously $v_{ab}(0^+)$ is 100 volts. When the switch contacts are close together, say, 0.01 mm, the voltage gradient is 100 kV/cm, more than enough to spark over. In this case, however, the arc formed extinguishes itself as the contacts separate to a greater distance; the supply circuit is capable of supplying at most one ampere to the arc, which at 100 volts means that at most 100 watts will be available for arc heating.

It is wrong to assume that voltages no greater than the supply value will be encountered in switching electrical circuits. For example, consider the

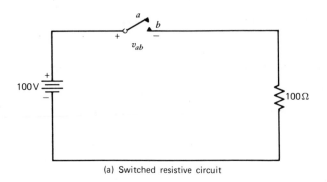

(a) Switched resistive circuit

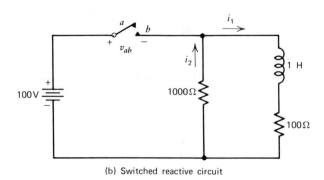

(b) Switched reactive circuit

Figure 10-16 Switching in electrical circuits.

circuit of Figure 10-16b. Assume that the switch has been closed for an appreciable time, and is opened at $t = 0$. Observe that:

$$i_1(0^-) = \frac{100}{100}$$

$$= 1 \text{ A}$$

At $t = 0^+$, current continuity through the inductor requires that

$$i_1(0^+) = i_1(0^-) = 1 \text{ A}$$

and

$$i_2(0^+) = i_1(0^+) = 1 \text{ A}$$

$$\therefore \quad v_{ab}(0^+) = 100 + 1(1000)$$

$$= 1100 \text{ volts}$$

354

Electric and magnetic field effects are always present to some extent in every circuit, and can cause transient switching voltages to appear that are far in excess of normal operating voltages. In large power systems voltages approaching the megavolt range and currents in hundreds of kiloamperes may be encountered, providing enormous amounts of power for the creation of electrical arcs. Power interrupting devices must cope with these arcs, allowing for their formation and subsequent extinction.

One of the simplest and oldest protective devices is the fuse. It is designed so that i^2R heating in the link at or above a predetermined current is sufficient to melt the link and interrupt the current. Since the current must persist long enough for the link to absorb the heat of fusion, a fuse inherently has an associated time delay. Typical time-current curves are shown in Figure 10-17. It is desirable to avoid the crosshatched region, since the link may be partially melted and operation could subsequently be unpredictable. Four factors are usually considered for a particular application.

1. Voltage Rating—The value at or above the nominal system rms line voltage. The idea is that a blown fuse should be able to withstand this voltage. Normal transient overvoltages are allowed for.

2. Continuous Current Rating—The fuse should pass this current indefinitely without blowing.

3. Interrupting Current Rating—This value should equal or exceed the largest rms current that the supply circuit is capable of providing. This rating is the largest current that the fuse is capable of interrupting.

4. Time Response—This time is predicted from the fuse's time-current curves.

Fuses are relatively economical, do not require relays or instrument transformers, and are reliable. They are available in a large range of sizes, and can be designed as "one shot" devices, or reusable devices with replaceable links. They have the obvious disadvantages that they are not suitable for remote control and multiple switching operation. Some common fuse types are shown in Figure 10-18.

Explanation of some terminology with regard to power switchgear follows:

Disconnect Switch—a switch designed to disconnect power apparatus drawing essentially no current.

355

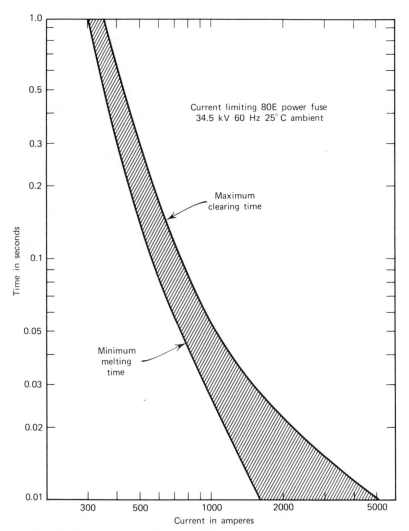

Figure 10-17 Time-current characteristics for a power fuse. Data courtesy of General Electric Co.

Load-break Switch—a switch designed to interrupt load level current, but not the much larger fault currents.

Circuit Breaker—a switch designed to interrupt fault currents.

The circuit breaker is a mechanical device capable of breaking and reclosing a circuit under all conditions, including when the system is

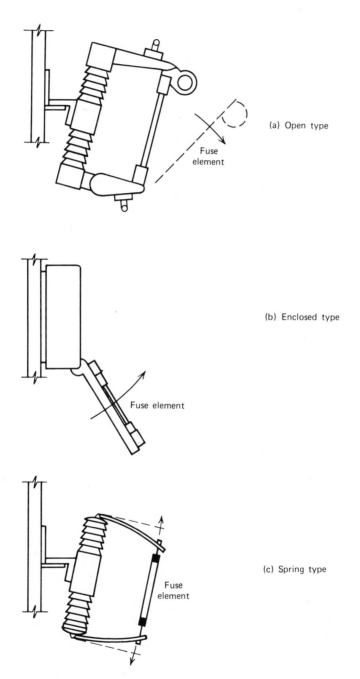

(a) Open type

Fuse element

(b) Enclosed type

Fuse element

(c) Spring type

Fuse element

Figure 10-18 Some types of power fuse mountings.

faulted and currents are at their greatest values. When contacts part an arc forms. The role of the circuit breaker is to extinguish this arc as quickly as possible. The initial arc path is essentially directly between the contacts. It is desired that the arc be forced into a sinuous path, where it is elongated, cooled, and finally extinguished. Simplifying the process is the fact that the arc current is ac, and will go through a natural zero twice during its 60 Hz cycle.

The arc is forced into the elongated path by interaction with a magnetic field (magnetic blow-out breakers), or a stream of air (air blast breakers). Refer to Figure 10-19. The arc is sometimes interrupted in air, in oil (oil

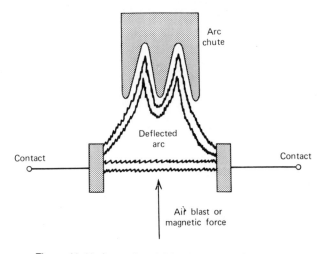

Figure 10-19 Arc extinguishing in a circuit breaker.

circuit breakers; OCB's), in SF_6 (Sulphur hexaflouride), or in a vacuum. A 500 kV breaker is shown in Figure 10-20.

The forces necessary to pull apart heavy contacts rapidly are substantial and the required energy for this action is usually stored in springs or in compressed air tanks. Compressed air is also necessary for air blast breakers. Remember that the electrical system is an unreliable power source at the time that breaker action is required. To achieve higher withstand voltage capability, sets of individual breaking contacts are arranged in series within a breaker. Higher current capacity is achieved by shunting the main break contacts with a power resistor, which is removed after the primary arc is quenched.

Some of the more important power circuit breaker ratings are as follows.

Figure 10-20 Westinghouse three-phase sulfur hexafluoride, 550-kV, 3000 amp power circuit breaker. (Photograph courtesy of Georgia Power Co.)

Voltage

- Nominal Rated—The breaker should be used in systems having this rms line voltage rating or less.

- Low Frequency rms Voltage—The 60 Hz maximum insulation withstand value.

- Impulse Crest Voltage—A measure of the insulation strength to withstand a voltage pulse of standard parameters.

Current

- Continuous—Maximum continuous rms 60 Hz current that the breaker can carry without exceeding the allowable temperature rise.

- Momentary—Maximum rms asymmetrical current that the breaker can stand without damage. Usually 1.6 times the interrupting rating.

359

- Interrupting Rating—Maximum rms symmetrical current that the breaker can safely interrupt.

- Interrupting Time—The time (usually given in cycles on a 60 Hz base; 60 cycles equals one second) from the instant the trip coil is energized until the fault current is cleared.

Sometimes apparent power (MVA) ratings are supplied; these are related to the corresponding current values. The trend is away from this practice since specification of current ratings is more direct.

For distribution circuits (in the 2.4 to 46 kV range) circuit breakers with built-in instrument transformers and relays are available. Such devices are known as reclosers. Since most faults are self-clearing, reclosers operate on the principle that if the circuit is de-energized a short time there is a good chance that whatever caused the fault was vaporized and the ionized arc path has dissipated.

Refer to Figure 10-21. A recloser carries normal load current when a fault occurs. The recloser senses overcurrent and breaks the circuit, instantaneously or after an intentional time delay. After a short preset interval

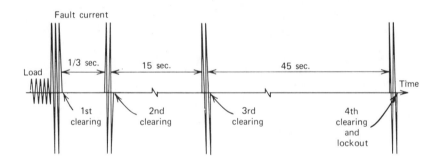

Figure 10-21 Recloser current versus time.

[usually 1/3 second (20 cycles)] the recloser "recloses." If the fault persists, the recloser again trips, and waits a second preset interval (usually 15 seconds). The recloser, being a persistent little devil, closes again, only to trip if the fault is still there. After a third open interval (usually 45 seconds), the recloser closes one last time. If the fault persists, the recloser opens and "locks out," requiring manual resetting. If any reclosure is successful, the entire cycle is reset automatically, requiring from 30 to 240 seconds.

For low voltage applications molded case circuit breakers are available. These devices have a dual trip operation: one, a thermal trip with time delay similar to fuses and two, an electromagnet instantaneous trip, sometimes with an adjustable instantaneous trip threshold current.

10-4 Sequence Filters

An electrical power system normally operates in a balanced 3ϕ sinusoidal steady state mode. In terms of symmetrical components this implies that zero and negative sequence quantities are zero, or at least small. Presence of substantial amounts of these quantities indicate system unbalance and consequently some sort of system malfunction. It is useful to design circuits that respond to these quantities; the outputs of such circuits can then be used to drive relays that are used to initiate actions to protect the system. Such circuits are referred to as "sequence filters." There are four basic types that are of most importance and we discuss examples of each. For convenience of notation we use capital subscripts for primary (system) quantities and lower case subscripts for secondary (relaying) quantities.

ZERO SEQUENCE CURRENT FILTER

Refer to Figure 10-22a. We recall from equation (4-14) that:

$$I_0 = 1/3(I_a + I_b + I_c) \tag{10-18}$$

The CT's behave essentially as current sources so that the filter output produces:

$$I_a + I_b + I_c = 3I_0 \tag{10-19}$$

ZERO SEQUENCE VOLTAGE FILTER

Refer to Figure 10-22b. We recall that from (4-13):

$$V_0 = 1/3(V_a + V_b + V_c) \tag{10-20}$$

The PT's behave as voltage sources so that the filter output produces:

$$V_a + V_b + V_c = 3V_0 \tag{10-21}$$

Note that polarity markings on all transformers are critical. If a minus sign error is made, we do *not* produce V_0 or I_0, either in magnitude or phase.

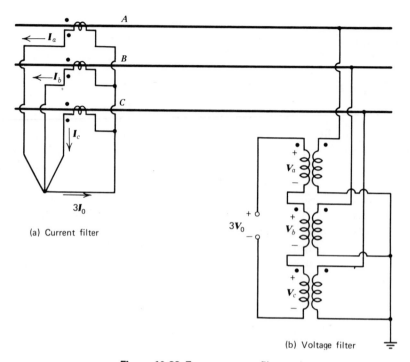

(a) Current filter

(b) Voltage filter

Figure 10-22 Zero sequence filters.

Example 10-6

A 12 kV 10 MVA load cannot tolerate more than 10% zero sequence current and should be interrupted when this limit is exceeded. Design a zero sequence current filter to detect this condition.

Solution

$$I_{\text{rated}} = \frac{10}{0.012\sqrt{3}} = 481 \text{ amps}$$

$$\therefore \quad I_{0_{\text{max}}} = 48.1 \text{ amps}$$

Choose $500:5$ amp CT's. Use the circuit of Figure 10-22a. At $I_{0_{\text{max}}}$ the filter output is $3(0.481) = 1.443$ amps. We choose an overcurrent relay with its pickup current set at 1.4 amperes.

NEGATIVE SEQUENCE VOLTAGE FILTER

Consider Figure 10-23a. Let us apply "pure" zero sequence:

$$V_{an} = V_{bn} = V_{cn} = V_0 \qquad\qquad (10\text{-}22a)$$

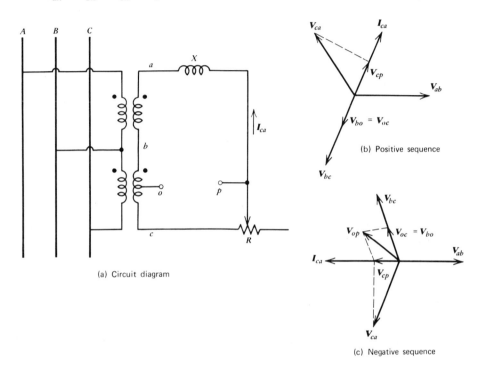

(a) Circuit diagram

(b) Positive sequence

(c) Negative sequence

Figure 10-23 Negative sequence voltage filter.

Then:

$$V_{ab} = V_{an} - V_{bn} = 0 \qquad\qquad (10\text{-}22b)$$

$$V_{bc} = V_{bn} - V_{cn} = 0 \qquad\qquad (10\text{-}22c)$$

$$V_{ca} = V_{cn} - V_{an} = 0 \qquad\qquad (10\text{-}22d)$$

We conclude that the PT's do not respond to zero sequence values. Now let us apply "pure" positive sequence. Refer to the phasor diagram in Figure 10-23b:

$$V_{ab} = V_{bc} = V_{ca} = V_1\sqrt{3} \qquad\qquad (10\text{-}23)$$

Adjust R until $V_{op} = 0$. This will occur when I_{ca} (and V_{cp}) lags V_{ca} by 60°. The point "o" is a center tap on the coil bc. The appropriate value of R calculates to $X/\sqrt{3}$. The filter is now properly adjusted and "blind" to positive sequence.

Now consider the application of "pure" negative sequence voltage. Refer to Figure 10-23c.

$$V_{ab} = V_{bc} = V_{ca} = V_2\sqrt{3} \tag{10-24}$$

By KVL:

$$V_{op} = V_{oc} + V_{cp}$$

$$= \frac{V_{ab}}{2}\sqrt{3}\underline{/150°}$$

$$= (3/2)V_2\underline{/150°} \tag{10-25}$$

By superposition, in general

$$V_{op} = 0 + (3/2)V_2 + 0 \tag{10-26}$$

We now connect a voltage sensitive relay between terminals o-p to detect the negative sequence voltage.

NEGATIVE SEQUENCE CURRENT FILTER

Study the circuit of Figure 10-24a. The three coupled inductors have mutual reactance $X_{ac} = X_{bc} = X$ and self reactance $X_s = kX$. Replace the relay coil with an ammeter for adjustment purposes. Now drive the filter with "pure" positive sequence and adjust R until the ammeter reads zero. The balance condition is:

$$I_c R = jI_a X - jI_b X \tag{10-27}$$

It follows that:

$$R = \sqrt{3}X \tag{10-28}$$

Examine the phasor diagram in Figure 10-24b.

The filter is now properly adjusted and "blind" to positive sequence. Now apply a general 3ϕ input. By KVL:

$$(I_c - I_0 - I)R = jI_a X - jI_b X + jIkX \tag{10-29a}$$

$$(I_0 + aI_1 + a^2I_2 - I_0 - I)\sqrt{3}X = jX(I_0 + I_1 + I_2 - I_0 - a^2I_1 - aI_2) + jIkX \tag{10-29b}$$

$$\sqrt{3}(aI_1 + a^2I_2 - I) = j(1 - a^2)I_1 + j(1 - a)I_2 + jkI \tag{10-29c}$$

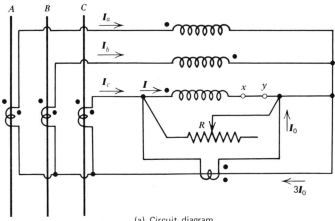

(a) Circuit diagram

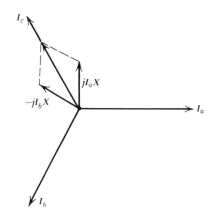

(b) Positive sequence current input

Figure 10-24 Negative sequence current filter.

But

$$\sqrt{3}a = j(1 - a^2) \tag{10-30}$$

$$\therefore \quad \sqrt{3}a^2 I_2 - \sqrt{3}I = j(1 - a)I_2 + jkI \tag{10-31a}$$

But

$$\sqrt{3}a^2 - j(1 - a) = 2\sqrt{3}a^2 \tag{10-32}$$

$$\therefore \quad (\sqrt{3} - jk)I = 2\sqrt{3}a^2 I_2 \tag{10-31b}$$

$$I = \frac{2\sqrt{3}a^2}{\sqrt{3} - jk} I_2 \tag{10-31c}$$

365

We connect a current sensitive relay between terminals x-y and therefore can detect negative sequence current.

10-5 Transformer Protection

Each major component of the power system represents a substantial capital investment, and, as such, merits consideration of protecting the component against damaging currents and voltages. A basic concern is with internal short circuits. Power transformers are frequently protected against such eventualities by a technique referred to as "differential protection." The method is illustrated in Figure 10-25.

$$I_1' = I_1/a_1 \tag{10-32a}$$

$$I_2' = I_2/a_2 \tag{10-32b}$$

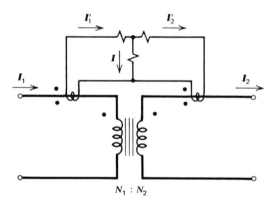

Figure 10-25 Differential protection for transformers.

where

$a_1 = $ primary CT turns ratio

$a_2 = $ secondary CT turns ratio

By KCL:

$$I = I_1' - I_2' \tag{10-33a}$$

$$= I_1/a_1 - I_2/a_2 \tag{10-33b}$$

Suppose we choose:

$$\frac{a_2}{a_1} = \frac{N_1}{N_2} \tag{10-34}$$

Then equation (10-33b) becomes:

$$I = \frac{1}{a_1}\left(I_1 - \frac{I_2}{N_1/N_2}\right) \tag{10-33c}$$

Now if the transformer were ideal:

$$I_2 = \frac{N_1 I_1}{N_2} \tag{10-35}$$

and

$$I = 0$$

If a short circuit occurs anywhere inside the CT's, equation (10-35) is not satisfied and from (10-33c):

$$I \neq 0 \tag{10-36}$$

The relay in Figure 10-25 will detect current unbalance and can be used to activate a circuit breaker that can de-energize the transformer.

The scheme is quite sensitive; in fact this can be a drawback, since the relay can trip the breaker in situations that are not harmful to the transformer. The relay is usually of the balanced beam type and is supplied with restraining coils to desensitize it. Refer back to section 10-2. If we assume a center tapped restraining coil as shown in Figure 10-12, the operating condition is predictable from equation (10-14). Also refer to Figure 10-13. The ratio N_0/N_r (operating turns to restraining turns) may be selected to control relay sensitivity.

Example 10-7

A 100 kVA 2400/240 volt step down transformer is to be differentially protected. Choose appropriate CT Ratios. Also determine the ratio N_r/N_0 if the relay is to tolerate a mismatch in current of up to 20% of I_1.

Solution

$$I_{1_{\text{rated}}} = \frac{100}{2.4} = 41.7 \text{ amps}$$

$\therefore \quad I_{2_{\text{rated}}} = 417 \text{ amps}$

$\therefore \quad$ Select primary CT 50:5

and secondary CT 500:5

Assume $I_2 = 0.8I_1$. At this condition the relay should be on the verge of operation and (10-14a) becomes an equation:

$$\frac{2-k}{2+k}(I_1) = 0.8I_1$$

$$(2-k) = (2+k)0.8$$

$$1.8k = 0.4$$

$$k = 0.222$$

Actual power transformer performance departs from ideal in several respects; for example, the primary current includes the exciting current, whereas the secondary current does not, meaning that equation (10-35) is not precisely satisfied. One scheme used to counteract this unbalance is harmonic restraint. The magnetizing current is nonsinusoidal, with a particularly large third harmonic component. See Figure 10-26. The filter $F1$ is designed to pass fundamental and block third harmonic components of the differential current I to the operating coil R_0. The filter $F2$ passes third harmonic and blocks fundamental components to the restraint coil R_r. Therefore, when the differential current results from magnetizing current, there is sufficient third harmonic to restrain operation. However, the relay can still detect strong 60 Hz differential current and operate.

Another problem encountered involves "inrush" current—that is, transient current waveforms that appear when a transformer is first energized. To understand the causes of initial high current consider Figure 10-27a. Suppose we switch the transformer on at a voltage maximum.

$$\lambda = \int_0^t e\, dt \tag{10-37}$$

λ and i are as shown. Note the current excursion Δi. Now suppose we switch at a voltage zero as shown in Figure 10-27b. Now observe Δi. Note that extremely high currents are initially possible. A differential relay will see this inrush current as differential current and will trip out the transformer

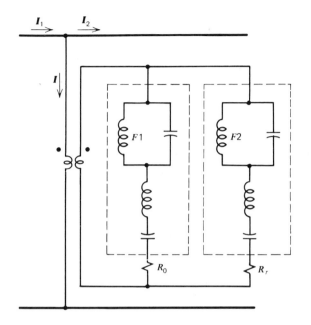

Figure 10-26 Magnetizing current compensation by Method of Harmonic restraint.

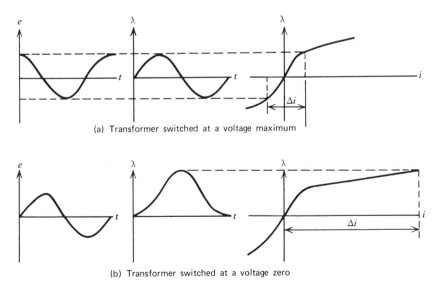

(a) Transformer switched at a voltage maximum

(b) Transformer switched at a voltage zero

Figure 10-27 Initial e, λ, and i in a transformer primary.

unless the protection scheme is modified to account for this effect. Several solutions are possible:

- Since this current is also nonsinusoidal, harmonic restraint may be used.

- The strong dc component that is present may be used to desensitize the relay.

- The relay may be temporarily desensitized with a time delay relay.

- The relay may be temporarily desensitized, but only if at full voltage. This method has the advantage of distinguishing inrush current from fault current.

This last mentioned method is shown in Figure 10-28. When a transformer is normally energized, the voltage is at or near rated voltage. A high speed voltage relay (HSVR) will pick up, closing its contacts. A time delay voltage relay (TDVR) will also pick up after a preset time delay. While both relay contacts are closed, inrush current will be shunted through the small resistor R, thereby desensitizing the relay. Finally the TDVR contacts will

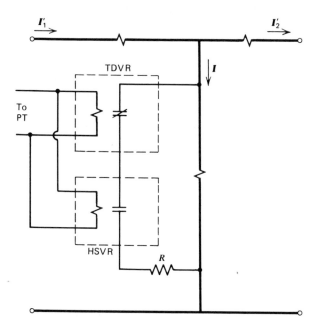

Figure 10-28 Desensitizing the differential relay to account for inrush current.

open, restoring maximum sensitivity to the relay. If we were to close in on a fault, the relay HSVR would see low voltage and fail to pick up, thereby opening the shunt path.

These techniques are applicable to transformers in 3ϕ banks, and are popular methods of protection. A basic problem is insuring that phases of all currents are correct. Study the 3ϕ connections as shown in Figure 10-29. Note that the Y-Δ transformer connection produces $30°$ phase shifts in the line currents. This may be compensated for by Δ-Y (opposite) connections in the CT secondaries. To simplify the language let us understand "primary CT current" to mean secondary CT current available from the CT's connected to the transformer's primary and "secondary CT current" to mean secondary CT current available from the transformer secondary CT's. Also let the terms "primary and secondary current" refer to *transformer* primary and secondary current.

1. Determine the primary and secondary currents (ABC, abc). A phasor diagram is useful for cataloging information. Consider all transformers to be ideal.

2. The rule is to connect the CT secondaries opposite to the way the transformer windings are connected. Start at the primary. In our example, the transformer primary is Δ so connect the primary CT's in Y (arbitrarily, but symmetrically).

3. Determine the primary CT currents ($A'B'C'$).

4. Connect the secondary CT's opposite the transformer secondary (in our example Δ, since the transformer secondary is Y) in such a way that the CT line currents ($a'c'$, $b'a'$, $c'b'$) are in phase with A', B', C', respectively. This is the key (and most difficult) step.

5. Connect to the differential relays as shown in Figure 10-29a.

Some thought and study will show that this arrangement can detect all types of internal faults, but is immune to external faults. For example, consider an external single line to ground fault on the primary. The bank will block zero sequence because of the Δ, and therefore primary and secondary CT's see no zero sequence current. If we consider the same fault on the secondary, the secondary CT's will not pass zero sequence currents because of their Δ connection and primary CT's because of no transformer primary zero sequence current. The scheme may be modified to deal with inrush and magnetizing current as previously discussed.

Other types of relays are used for transformer protection, such as gas pressure detectors and gas analyzers. Also, modification of the previously

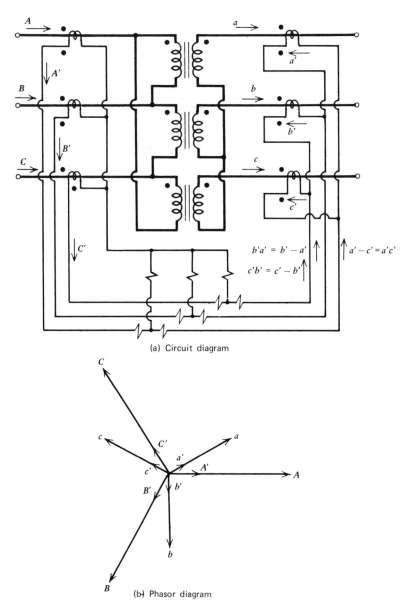

(a) Circuit diagram

(b) Phasor diagram

Figure 10-29 Differential relays applied to 3ϕ transformer banks.

mentioned techniques are necessary for special purpose transformers, such as tap changing, phase shifting, power rectifier, and multi-winding transformers.

10-6 Motor and Generator Protection

Large 3ϕ rotating machines are major components of a power system and must be protected from a variety of hazards. The basic danger to a machine is excessive heating, which in turn may cause insulation and structural damage. Mechanical forces and electrical voltages of destructive intensities are also possible. Because motors and generators have similar protection needs, we have grouped them together. A listing of some of the possible hazards appears in Table 10-2. Machines are rugged devices and frequently the detection of a potentially damaging condition is used only to trigger an alarm.

Table 10-2 Hazards to rotating machines.

HAZARD	SYNCHRONOUS GENERATORS	SYNCHRONOUS MOTORS	INDUCTION MOTORS
Stator overload	×	×	×
Stator winding shorts	×	×	×
Stator voltage unbalance	×	×	×
Bearing overheating	×	×	×
Loss of excitation	×	×	
Field ground fault	×	×	
Loss of synchronism	×	×	
Undervoltage		×	×
External faults	×	×	
Motoring	×		
Overvoltage	×		
Overspeed	×		

A basic hazard is stator overheating due to prolonged excessive stator currents, or failure of the cooling system. Generally, thermocouples are imbedded in the stator slots and detection of excessive temperature is used to annunciate an alarm. The machine is typically not removed from service, but action is taken to remove the source of overload, either automatically or by station operator. For some motors, long-time inverse-time overcurrent relays or fuses are sometimes used. Fuses should be of the "slow blow" variety to tolerate the large currents experienced at starting.

Various types of internal faults, such as turn-to-turn, phase-to-phase, and phase to ground faults, may occur in a machine. Differential relaying techniques are typically used because of their selectivity and speed of response.

373

When unbalanced voltages exist at the stator terminals one can decompose them into their corresponding symmetrical components. The positive sequence voltages produce a revolving field that revolves in the same direction and with the same speed ω as the rotor. Since no relative motion is present, there are no induced voltages or currents in the rotor. However, the negative sequence quantities produce a magnetic field revolving in a direction opposite from the rotor with relative speed 2ω. This means that large double frequency voltages and currents are induced into the rotor windings and iron. The attendant i^2R losses serve to overheat the rotor. Negative sequence quantities may be detected by sequence filters and, if in excess of predetermined limits, actuate relays that typically annunciate alarms. For small unbalance this is the only action taken, relying on the operator to take corrective action. For larger unbalances the unit's main breaker may be tripped.

Bearing and lubricating oil temperatures are continuously monitored with thermocouples and relays are used to actuate an alarm when they are excessive. Unattended units may be taken off line.

Loss of excitation in synchronous machines generally causes undesirable, and usually damaging, operation. In motors, the rotor will drop below synchronous speed and run as an induction motor on its amortisseur windings. If these windings were designed for starting duty only, continuous operation will cause them to overheat. Synchronous generators are normally overexcited in order to deliver reactive power (Q) to the system. When excitation is lost, this is impossible; in fact Q flow reverses and the machine *absorbs* considerable Q. This can have serious consequences on the system, even to the extent of causing instability. The following example will demonstrate.

Example 10-8

A synchronous generator operates as shown in Figure 10-30.

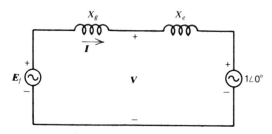

Figure 10-30 Generator operation.

Example 10-8

$$X_g = 0.8$$

$$X_e = 0.2$$

$$E_f = 1.414\underline{/45°}$$

$$I = \frac{1+j-1}{j1}$$

$$= 1\underline{/0°}$$

$$V = 1 + 1(j0.2)$$

$$= 1 + j0.2$$

$$= 1.02\underline{/11.3°}$$

$$S = VI^*$$

$$= 1.02\underline{/11.3°}$$

$$= 1 + j0.2$$

or

$$Q = +0.2$$

$$Z = \frac{V}{I} = 1 + j0.2$$

Now assume loss of excitation $(E_f = 0)$

$$I = -\frac{1\underline{/0}}{j}$$

$$= +j$$

$$V = \frac{0.8}{1.0}(1)$$

$$= 0.8\underline{/0°}$$

$$S = VI^*$$

$$= -j0.8$$

$$Q = -0.8$$

$$Z = -j0.8$$

To summarize:

	Before	After
V	1.02	0.80
Q	+0.20	−0.80
Z	$1 + j0.2$	$−j0.80$
I	1.0	1.0

In the previous example we observe that:

- V dropped.

- The flow of Q reversed.

- The impedance went from basically resistive to capacitive.

- The current magnitude did not change, indicating that overload protection would not operate.

These are general results for loss of excitation. A relay scheme can be designed that is sensitive to Z. When it detects strong capacitive Z, along with a significant voltage drop, loss of excitation will be assumed and the unit tripped off line. More is said about relay schemes that respond to Z in Chapter 11.

Synchronous machine fields are typically ungrounded; therefore, if any point in the field circuit is grounded, no ground currents will flow. However, once a ground fault occurs, the rest of the winding is electrically stressed, and the probability of a second ground fault is increased. If the second ground fault occurs, the consequences are:

- The arc will burn the rotor at the fault point.

- Parts of the rotor winding may carry abnormally high currents.

- The magnetic field will become unbalanced, creating forces, torques, and vibrations of strengths capable of damaging the rotor and stator.

Usually detection of the first ground fault is used to annunciate an alarm, leaving it to the operator's discretion as to whether to remove the machine from service. The second fault requires immediate action, usually opening the field breaker and taking the unit off line.

Generators are usually protected against loss of synchronism indirectly by overspeed protection associated with the prime mover. Overfrequency relays are also employed. Motors are usually protected indirectly through

Example 10-8

the combined action of the undervoltage relays and loss of excitation protection. For applications where this is a foreseeable problem, a relay that counts power reversals over a present time interval may be employed.

Operating motors on sustained low voltage may cause them to draw excessive current, run at slow speed, and in the case of synchronous motors, prevent synchronization. Although the overload relays will eventually operate, such protection is usually too slow, and undervoltage time delay relays are advisable. The voltage and time delay settings depend on the details of the particular application.

Generators, and also motors, when subjected to external faults will supply currents many times in excess of their ratings. If the fault is out in the power system (on a transmission line, for example), it is important to let the line breaker clear the fault and keep the generator in service. However, in case of line breaker failure, or a fault between the line breaker and generator breaker it is necessary that the generator protection scheme detect and interrupt the fault.

Usually it is undesirable to allow real power flow in generators to reverse (motoring). The hazard involved in motoring is to the generator prime mover. The problems involved are:

- Turbine blade overheating, with steam turbines

- Cavitation, with hydraulic turbines

- Backfire, and fire hazard, with diesel engines

The problem can be corrected with a directional unit that detects real power reversal. A small amount of reverse power is usually tolerated to account for the machine's rotational losses.

Sometimes generators are protected against excessive stator voltage. Generally the exciter-generator-regulator loop has a response fast enough to obviate the need for additional voltage control. As backup protection, an overvoltage relay may be set to trip the field breaker. Generator overspeed protection is provided by the prime mover governor. Backup may be provided by an overfrequency relay set to respond to about 5% overspeed. The action taken is basically to trip the generator off line and shut down the prime mover.

Rotating machines are protected from all of the foregoing hazards, and more. It should occur to the reader that many protective relays functionally overlap and perform backup protection for each other. The two cardinal principles to keep in mind are that any scheme chosen must reliably protect the machine from all foreseeable hazards, and at the same time, must never unnecessarily remove the unit from service.

10-7 Summary

We have discussed the basic components used for power system protection. Instrument transformers and relays detect and process power system currents and voltages, and "decide" whether to break a circuit or not. Circuit breakers and fuses perform the actual circuit interruption.

Methods used for protecting machines and transformers were presented. Transformers must be protected from overloads, but protection schemes must permit inrush current surges. Internal faults must be detected and cleared. Rotating machines must also be protected from overloads, but in addition require protection from a variety of other hazards.

The transmission lines that interconnect generators, transformers, and loads deserve special consideration for two reasons. First, since lines span large geographical distances and are exposed to the elements, they experience most of the faults. Second, line protection schemes are confronted with most of the coordination problems, insuring that a fault causes minimum disruption in system operation. Line protection is the topic of the next chapter.

Bibliography

[1] Anderson, Paul M., *Analysis of Faulted Power Systems*, Iowa State Press, Ames, Iowa, 1973.

[2] Brown, H. E., and Person, C. E., "Digital Calculations of Single-Phase to Ground Faults." *Trans. AIEE*, vol. 79 (pt. 3): pp. 657–60, 1960.

[3] Brown, H. E., Person, C. E., Kirchmayer, L. K., and Stagg, G. W., Digital Calculation of Three-Phase Short Circuits by Matrix Method. Trans. AIEE 79 (pt. 3): 1277–82, 1960.

[4] Brown, Homer E., *Solution of Large Networks by Matrix Methods*, John Wiley and Sons, Inc., 1975.

[5] Elgerd, Olle I., *Electric Energy Systems Theory: An Introduction*, McGraw-Hill Inc., New York, 1971.

[6] El-Abiad, A. H., "Digital Calculation of Line-to-Ground Short Circuits by Matrix Methods." *Trans. AIEE*, vol. 79 (pt. 3): pp. 323–32, 1960.

[7] IEEE., *IEEE Standard Dictionary of Electrical and Electronics Terms*, Wiley-Interscience, New York, 1972.

[8] IEEE., *Recommended Practice for Protection and Coordination of Industrial and Commercial Power Systems*, Wiley-Interscience, New York, 1975.

[9] Manufacturer's Publication., *Distribution-System Protection Manual*, McGraw-Edison Power Systems Division, Publication 71022.

[10] Manufacturer's Publication., *Short-Circuit Current Calculations for Industrial and Commercial Power Systems*, General Electric, Publication GET-3550.

[11] Manufacturer's Publication., *Short-Circuit Currents in Low and Medium Voltage A-C Power Systems*, General Electric, Publication GET-1470D.

[12] Mason, C. Russell, *The Art and Science of Protective Relaying*, John Wiley and Sons, Inc., 1956.

[13] Neuenswander, John R., *Modern Power Systems*, International Textbook Co., Scranton, 1971.

[14] Stagg, Glenn W., and El-Abiad, Ahmed H., *Computer Methods in Power System Analysis*, McGraw-Hill, Inc., New York, 1968.

[15] Stevenson, Jr., William D., *Elements of Power Systems Analysis*, 3rd edition. McGraw-Hill, Inc., New York, 1975.

[16] Weedy, B. M., *Electric Power Systems*, 2nd edition. John Wiley and Sons Ltd., London, 1972.

[17] Westinghouse Electric Corporation, *Electrical Transmission and Distribution Reference Book*, 4th edition. East Pittsburgh, Pa., 1950.

Problems

10-1. For the PT and CT shown in Figure 10-2 the ratios are $N_1/N_2 = 2000$ and $I_1 : I_2 = 4000 : 5$ A. Suppose the power conductor is phase a of a 3ϕ transmission line operating at 230 kV; 300 MVA, pf 0.8 lagging, power flowing from left to right. ($V_1 = V_{an} = 133,000\underline{/0°}$). Compute V_2 and I_2.

10-2. For the CT of Figure 10-5, $X_2 = 0.5\ \Omega$ and its ratio is $500 : 5$ A. If $I_1 = 500$ A calculate I_2 and the CT error if:
(a) $X_L = 0$
(b) $X_L = 4.5$ ohms
(c) $X_L = 9.5$ ohms

10-3. Repeat Problem 10-2 for $I_1 = 1500$ A.

10-4. Suppose the C0-7 relay is supplied with 20 amperes. Determine the operating time for the following settings.
(a) CTS = 4A; TDS = 10
(b) CTS = 5A; TDS = 8
(c) CTS = 6A; TDS = 1/2
(d) CTS = 8A; TDS = 2

10-5. For the differential relay it is desired that a 30% mismatch in currents be tolerated (i.e., $I_a/0.7 > I_b > 0.7 I_a$). Determine the turns ratio N_r/N_0.

10-6. Consider the general relay equation (10-17). $K_b = K_c = 0$; $A = I$ will stimulate an overcurrent device. If $K_a = 0.1$, the current pickup value can be controlled by adjusting K_s. Determine K_s for the following pickup values:

(a) 1A
(b) 2A
(c) 5A
(d) 10A

10-7. Consider equation (10-17). $K_b = K_c = 0$; $A = V$ will simulate an undervoltage relay if $K_a < 0$ and $K_s < 0$. Find an appropriate value for K_a if the device is to pick up for $V \le 100$ volts and $K_s = -0.3$.

10-8. A normal wall switch would be described as a:
(a) disconnect?
(b) load-break switch?
(c) circuit breaker?

10-9. A 3ϕ 2400 volt 800 kVA rotating machine cannot tolerate over about 10% negative sequence current (i.e., 10% of full load current). A 0.5 A overcurrent relay is to be used in the negative sequence current filter of Figure 10-24. Pick the closest acceptable current ratio from those available in Table 10-1.

10-10. In Figure 10-29 each transformer is 1ϕ rated at 4800/480 volts and 150 kVA. The high voltage side is wye connected and is the secondary. The primary CT's are rated at 600:5 A. Determine the correct secondary CT ratings.

11

POWER SYSTEM PROTECTION 2: LINES

"With all its sham, drudgery and broken
dreams, it is still a beautiful world."
Max Ehrmann, DESIDERATA

Because a line is spread over a significant geographic area, it is exposed to a variety of hazards. Causes of line faults include lightning, wind, ice, snow, salt spray, birds, airplanes, and automobiles. It is not surprising that lines experience many more faults than other components, and that their protective relaying schemes are required to operate much more frequently. Recall that one of our objectives for system protection was to keep as much of the system interconnected as possible when clearing faults. A basic consideration with regard to line protection is therefore proper coordination of circuit breakers and relays. We begin with overcurrent protection.

11-1 Overcurrent Protection for Short Lines

Many medium voltage systems are protected using overcurrent methods. This approach is feasible in situations where fault currents are much greater than load currents. Time delay overcurrent relays are typically used; controlled time delay is important so that relays may be coordinated properly. By coordination we mean that relays operate in a sequence so as to interrupt service as little as possible when clearing a fault. To study this point consider the simple radial system illustrated in Figure 11-1. A fault at

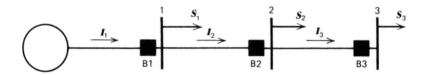

Figure 11-1 An example 13.8 kV system.

bus 3 will cause breakers B1, B2, and B3 to tend to operate. If B1 opens before B3, busses 2 and 3 will unnecessarily be de-energized. Therefore, we wish to set the relays controlling B1 and B2 so that B2 operates before B1 and B3 before B2. We shall investigate the particulars in Example 11-1.

Example 11-1

For the 13.8 kV system of Figure 11-1 determine relay settings to protect the system from faults. Pertinent system data is provided in Tables 11-1, 11-2, and 11-3. Assume 3 relays per breaker, one for each phase as shown in Figure 11-2.

Table 11-1 Maximum system loads for the system of Figure 11-1.

	MAXIMUM LOADS	
BUS	S IN MVA	LAGGING PF
1	8.5	0.90
2	3.0	0.90
3	5.0	0.90

Table 11-2 Maximum fault current in amperes for system of Figure 11-1.

FAULT TYPE	BUS 1	BUS 2	BUS 3
3Ø	3120	2808	2496
SLG	2880	2592	2304,
LL	2960	2664	2368
DLG	3040	2736	2432

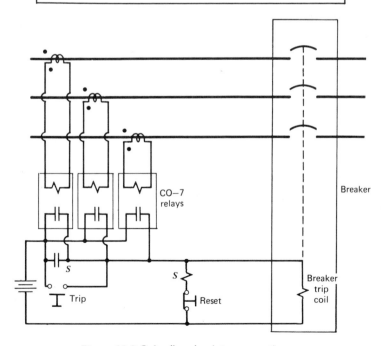

Figure 11-2 Relay/breaker interconnections.

383

Table 11-3 Breaker/relay data for system of Figure 11-1.

BREAKER	RELAY	C.T.R.	BREAKER OPERATING TIME
B1	CO-7	800:5	6 Cycles
B2	CO-7	400:5	6 Cycles
B3	CO-7	400:5	6 Cycles

Solution

Observe that relays are all CO-7's, whose characteristics are provided in Figure 10-11. Two settings are to be determined for each triad of relays: the current tap setting (CTS) and the time dial setting (TDS). We first consider the current tap settings (CTS). It is necessary that the relay not actuate for load currents. We start with B3:

$$I_3 = \frac{5}{0.0138\sqrt{3}}$$

$$= 209 \text{ A}$$

$$I_{3_{relay}} = \frac{209}{400/5}$$

$$= 2.61 \text{ A}$$

The lowest tap setting for the CO-7 is 4 A. Therefore select $CTS = 4$ A for B3. At B2 the relay sees the combined loads S_2 and S_3: Note that $S_2 + S_3 = S_2 + S_3$ because of identical power factors.

$$\therefore \quad I_2 = \frac{8}{0.0138\sqrt{3}}$$

$$= 335 \text{ A}$$

$$I_{2_{relay}} = \frac{335}{400/5}$$

$$= 4.18 \text{ A}$$

We select CTS = 5 A for B2. At B1 the relay sees the combined loads S_1, S_2, and S_3.

$$\therefore \quad I_1 = \frac{16.5}{0.0138\sqrt{3}}$$

$$= 690 \text{ A}$$

$$I_{1_{relay}} = \frac{690}{800/5}$$

$$= 4.31 \text{ A}$$

We select a CTS = 5 A for B1. We are now ready to determine time dial settings (TDS) for our relays. We shall coordinate using the largest fault current, which for this example was produced by the 3∅ fault. Since we wish to clear faults as rapidly as possible select TDS = 1/2 for B3. The B3 relay operating time for a balanced 3∅ fault at bus 3 is determined from the CO-7 characteristics (see Figure 10-11).

$$\frac{I_{3_{relay}}}{CTS_3} = \frac{2496/(400/5)}{4}$$

$$= 7.8$$

From the curves:

$$T_3 = 0.15 \text{ s}$$

To set B2, allow for the breaker operating time (6 cycles = 0.1 second), plus 0.3 second to allow for relay overtravel, error·introduced by calculation approximations, and departure of actual relay operation from published characteristics. Also:

$$\frac{I_{2_{relay}}}{CTS_2} = \frac{4}{5}(7.8)$$

$$= 6.24$$

$$T_2 = 0.15 + 0.1 + 0.3$$

$$= 0.55 \text{ s}$$

From the curves:

$$TDS_2 \cong 2$$

We now turn our interest to setting the B1 relay. A fault at bus 2 produces a $3\emptyset$ fault current of 2808 A. Relay B2 operates as follows:

$$\frac{I_{2_{\text{relay}}}}{\text{CTS}_2} = \frac{2808/(400/5)}{5} = 7.02$$

$$T_2 = 0.53\,\text{s}$$

Therefore

$$T_1 = 0.53 + 0.1 + 0.3$$

$$= 0.93\,\text{s}$$

$$\frac{I_{1_{\text{relay}}}}{\text{CTS}_1} = \frac{2808/(800/5)}{5} = 3.51$$

From the curves, select:

$$\text{TDS}_1 = 2.5$$

The TDS is continuously adjustable.
The relays are now properly set.

Example 11-1 illustrates the basic approach used in overcurrent relay coordination studies. We have already discussed several other considerations that tend to complicate the situation. These factors are:

- Motor starting currents

- Transformer inrush current

- CT error

- The dc offset, both magnitude and rate of decay.

Precise consideration of these factors is complicated and usually prevented by the lack of accurate data. Therefore, rules of thumb based on field experience exist that account for most typical situations (for example, the 0.3 s margin). This is an interesting and important area and the reader interested in pursuing the topic in depth should consult publications that develop the topic in depth (for example, see [8] in the end-of-chapter Bibliography).

11-2 Directional Relays

When systems become complex, coordination of overcurrent relays becomes quite complicated and in some cases impossible. To understand the problem consider the system shown in Figure 11-3. Suppose there is a fault at

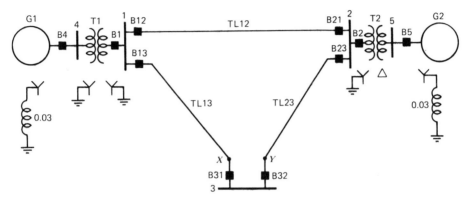

Figure 11-3 Example system from Chapter 8 with the circuit breaker/relay locations shown.

point X. Breakers B31 and B32 will sense the same currents (so will B23; however, we deal with this problem in the next section). We want B31 and B13 to operate to clear the fault. (If B32 operates we lose service at bus 3.) Using time delay overcurrent relays, we can set B31 faster than B32. However, now consider a fault at point Y. Breaker B31 will trip before B32, isolating bus 3 (after B23 trips). It is therefore impossible to coordinate time delay overcurrent relays so that line faults do not result in the loss of a bus in this system.

This problem can be overcome by the use of the directional relay. Recall the general relay equation:

$$T = K_a A^2 + K_b B^2 + K_c AB \cos(\theta - \tau) - K_s \tag{10-17a}$$

For this general two input device to function as a directional relay set:

$$K_a = K_b = K_s = 0 \tag{11-1a}$$

$$A = A\underline{/0°} = V \tag{11-1b}$$

$$B = B\underline{/-\theta} = I \tag{11-1c}$$

$$\tau = 90° \tag{11-1d}$$

where

$V = $ PT secondary voltage proportional to system line to neutral voltage

$I = $ CT secondary current proportional to system line current

$\theta = $ phase angle by which I *lags* V

The general relay equation becomes:

$$T = K_c VI \cos(\theta - 90°) \tag{11-2a}$$

$$= K_c VI \sin \theta \tag{11-2b}$$

Suppose we define a function Z as:

$$Z = \frac{V}{I} \tag{11-3a}$$

$$= \frac{V/\underline{0°}}{I/\underline{-\theta}} \tag{11-3b}$$

$$= Z/\underline{\theta} \tag{11-3c}$$

This function Z is somewhat cryptic; certainly it is not the impedance of a single element. It becomes even more mysterious when we realize that faults, except for $3\emptyset$, will create unbalances between phase currents and voltages. For the moment think of Z simply as a mathematical function defined by equation (11-3). Furthermore:

$$Z = Z/\underline{\theta} \tag{11-4a}$$

$$= Z \cos \theta + jZ \sin \theta \tag{11-4b}$$

$$= R + jX \tag{11-4c}$$

so that

$$R = Z \cos \theta \tag{11-4d}$$

$$X = Z \sin \theta \tag{11-4e}$$

We return to equation (11-2b). Realize that $T \geq 0$ is required for component operation. Therefore

$$K_c VI \sin \theta \geq 0 \tag{11-5}$$

Divide by $K_c I^2$:

$$\frac{V}{I} \sin \theta \geq 0 \tag{11-6a}$$

388

or

$$X \geq 0 \qquad (11\text{-}6b)$$

The understanding is that for any positive X as defined in equation (11-4e) the relay will operate. Refer to Figure 11-4a. The simple system has a

(a) Example system: Relay located at M.

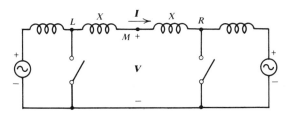

(b) Positive sequence network for system in (a). Switches model faults.

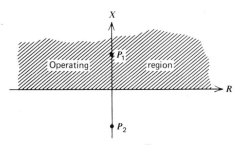

(c) Operating region located in $\overline{Z}$ plane: P_1 locates fault at R. P_2 locates fault at L.

Figure 11-4 The directional relay.

directional relay located a point midway down a transmission line (M). Suppose a $3\emptyset$ fault occurs to the right (at R). Refer to the positive sequence network in Figure 11-4b. Clearly:

$$\frac{V}{I} = +jX \qquad (11\text{-}7a)$$

On the other hand a fault to the left (at L) produces:

$$\frac{V}{I} = -jX \tag{11-7b}$$

In the Z plane we are operating at P_1 in the former case and at P_2 in the latter (refer to Figure 11-4c). Therefore, the relay responds to faults to the right but not to the left. In this sense the relay is "directional" and can be very useful for coordination purposes.

Examine Figure 11-5. Observe that both the overcurrent *and* the directional relay must "see" the fault before the breaker is allowed to trip. Observe that the operating direction is controlled by the device polarity and indicated by dots; consequently, polarity markings (dots) on all devices are critical and their meaning must be clearly understood.

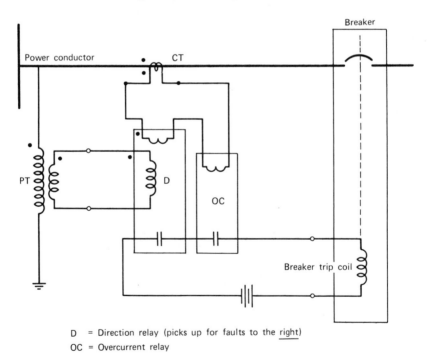

D = Direction relay (picks up for faults to the right)
OC = Overcurrent relay

Figure 11-5 Basic arrangement used for application of directional relays. One phase shown.

If we return to the example system of Figure 11-3, the six breakers B12, B21, B13, B31, B23, and B32 should be controlled by directional relays arranged to respond to faults on the line side of the breaker. A fault at X will now be cleared by B31 but not by B32 since the associated directional relay is "blind" to faults in that direction and will block operation.

The directional relay is also sometimes called a reactance relay for obvious reasons.

11-3 Impedance (Distance) Relays

When considering the example system shown in Figure 11-3, we observe a further difficulty. A fault at X will be sensed by relays controlling B31 and B23 that will detect substantially the same current. The directional relays at both locations will permit operation. How then can we insure that B31, and not B23, will clear the fault at X?

The problem can be solved by adjusting our general two input relay to respond to impedance, as opposed to voltage or current. The "impedance" involved is that defined in equation (11-3). Consider the situation illustrated in Figure 11-6a. The corresponding positive sequence network is shown in Figure 11-6b.

Suppose instrument transformers are located at bus R and produce V and I signals as usual. We define, as in section 11-2,

$$Z = \frac{V}{I} \qquad\qquad (11\text{-}3a)$$

$$= R + jX \qquad\qquad (11\text{-}4c)$$

In the normal loaded condition:

$$Z = 1.0 + j0.3 \qquad\qquad (11\text{-}8a)$$

Graphically, in the Z plane, we are located at point a. See Figure 11-6c. Imagine a fault to be modeled as a switch shorting the line. For a fault at the far bus (F) we close S_F. Now:

$$Z = j0.2 \qquad\qquad (11\text{-}8b)$$

and we move to point b. We recall from Chapter 4 that the line series impedance is directly proportional to the distance from the sending end. Therefore, a fault halfway down the line would halve the line impedance and be modeled by the switch location S_H. Closing S_H:

$$Z = j0.1 \qquad\qquad (11\text{-}8c)$$

391

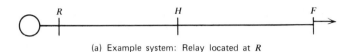

(a) Example system: Relay located at R

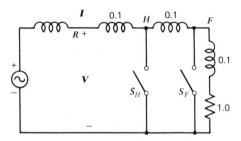

(b) Positive sequence network for example system

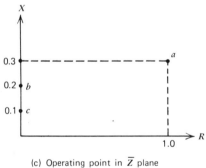

(c) Operating point in $\bar{Z}$ plane

Figure 11-6 Distance relaying concepts.

located at point c in Figure 11-6c. We observe several points of general significance here:

- Under normal conditions we are relatively far from the R-X origin with a large R component.

- Faults move us in close to the origin

- Z, under faulted conditions, is predominantly X (100% in our example).

- Faults produce *positive* X; this is a consequence of our assigned positive directions for **V** and **I**.

If we can obtain a relay that responds to **Z**, we can exploit these facts to design a protection scheme for lines.

Consider the universal relay equation from section 10-2.

$$T = K_a A^2 + K_b B^2 + K_c AB \cos(\theta - \tau) - K_s \qquad (10\text{-}17a)$$

Again recall that the performance of any relay may be predicted from this equation and that positive torque (T) operates the relay. We set:

$$K_c = K_s = 0 \qquad (11\text{-}9a)$$

$$A = V \qquad (11\text{-}9b)$$

$$B = I \qquad (11\text{-}9c)$$

Inequality (10-17b) becomes:

$$K_a V^2 + K_b I^2 \geq 0 \qquad (11\text{-}10)$$

To determine the border between the operating and nonoperating region set:

$$K_a V^2 + K_b I^2 = 0 \qquad (11\text{-}11a)$$

$$\therefore \quad \frac{V^2}{I^2} = \frac{-K_b}{K_a} \qquad (11\text{-}11b)$$

and since in general

$$Z = \frac{V}{I} \qquad (11\text{-}3)$$

then

$$Z_r = \sqrt{\frac{-K_b}{K_a}} \qquad (11\text{-}12)$$

where Z_r is the critical value of Z at which the relay is on the verge of operation. Either K_a or K_b is to be negative, but not both. Realizing this, equation (11-12) states that Z is a constant (Z_r). The locus of constant Z plotted in the **Z** plane is a circle. If K_a is negative, the relay operating region is inside the circle; if K_b is negative, the operating region is outside the circle. See Figure 11-7 for details. Operation on small impedance $(K_a < 0; Z < Z_r)$ is the arrangement that we assume for the rest of our study.

The "reach" of a distance relay refers to how far down a line the relay will respond to faults. A reach of 100% means that the relay is set to detect

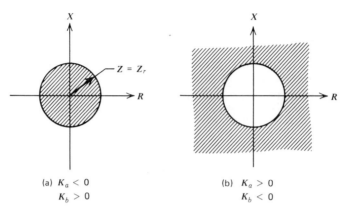

(a) $K_a < 0$
$K_b > 0$

(b) $K_a > 0$
$K_b < 0$

Figure 11-7 Distance relay characteristics in **Z** plane. Operating region is crosshatched.

faults at any point down to the far end. Referring back to the line of Figure 11-6, setting the relay for 100% reach would correspond to $Z_r = 0.2$; 50% reach to $Z_r = 0.1$. Observe that the normal load point is well outside the operating circle, in either case.

It is common practice to use more than one (typically three) distance relays per phase with progressively longer reaches and corresponding longer time delays. The associated Z regions are referred to as "zones of protection." The zone one relay is typically set for about 80% reach and instantaneous operation. The zone two and three relays are typically set for longer reaches and time delays: 120%, 12 to 18 cycles; and 250%, 60 cycles, respectively. The idea is to have each point in the system fall within at least two zones of protection so that there is primary and backup protection everywhere. A typical arrangement is shown in Figure 11-8. The corresponding Z plane characteristics are shown in Figure 11-9. Note the normal operating point and the line impedance locus. Faults along the transmission line will transfer the system operating point to this locus, with the closer faults nearer the origin. Reaches greater than 100% are only logical if the next bus is terminated in a single additional line. If multiple lines are involved, relay operation is unpredictable without specific calculated data.

Example 11-2

Consider the example system shown in Figure 11-3. Recall from Chapter 10 that the three lines are identical with positive sequence impedance $Z_1 = j0.1$. Assume that the six line breakers are controlled by zone distance and

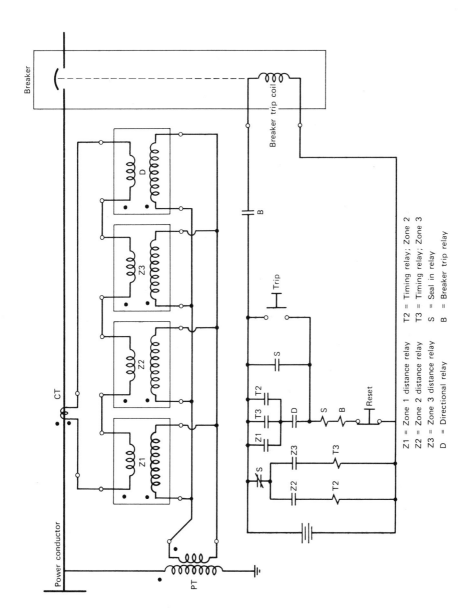

Figure 11-8 Three zone distance relay scheme. One phase only.

Z1 = Zone 1 distance relay   T2 = Timing relay; Zone 2
Z2 = Zone 2 distance relay   T3 = Timing relay; Zone 3
Z3 = Zone 3 distance relay   S = Seal in relay
D = Directional relay   B = Breaker trip relay

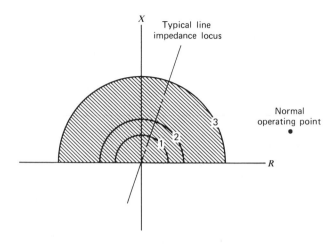

Figure 11-9 Operating region for three zone distance relay with directional restraint corresponding to the arrangement of Figure 11-8.

directional relays as shown in Figure 11-8. The reaches for the three zones are to be:

Zone 1—80%
Zone 2—120%
Zone 3—250%

Consider only 3ϕ faults.

(a) Determine the settings (Z_r) for all distance relays in per unit.
(b) If the PT's are rated $133 \, kV : 115 \, V$ and the CT's $400 \, A : 5 \, A$ convert the settings into ohms.
(c) Discuss relay operations for a fault at point X, assuming X is 10% down the line TL31 from bus 3.

Solution

(a) The reaches are given. Therefore:

Zone 1: $Z_r = 0.1 \times 80\% = 0.08$
Zone 2: $Z_r = 0.1 \times 120\% = 0.12$
Zone 3: $Z_r = 0.1 \times 250\% = 0.25$

Because of the system's symmetry all six sets of relays have identical settings.

(b) Recall that:

$$V_{LN_{base}} = \frac{230}{\sqrt{3}}$$

$$= 133\,\text{kV}$$

$$I_{L_{base}} = \frac{100}{0.23\sqrt{3}}$$

$$= 251\,\text{A}$$

Equivalent instrument transformer secondary quantities are:

$$V_{base} = 133\left(\frac{115}{133}\right)$$

$$= 115\,\text{V}$$

$$I_{base} = 251\left(\frac{5}{400}\right)$$

$$= 3.14\,\text{A}$$

$$\therefore \quad Z_{base} = \frac{115}{3.14}$$

$$= 36.7\,\Omega$$

Therefore, the settings are:

Zone 1: $Z_r = 0.08\,(36.7) = 2.93$ ohms
Zone 2: $Z_r = 0.12\,(36.7) = 4.40$ ohms
Zone 3: $Z_r = 0.25\,(36.7) = 9.16$ ohms

(c) Locate point X on the diagram in Figure 11-3. We comment on all line breaker operations:

B31—Fault is in Zone 1. Instantaneous operation.
B32—Directional unit blocks operation.
B23—Fault is in Zone 2. Delayed operation. B31 should trip first, preventing B23 from tripping.
B21—Fault duty is light. Fault in Zone 3, if detected at all.
B12—Directional unit blocks operation.
B13—Fault is in Zone 2 (just outside of Zone 1). Delayed operation.

The line breakers B13 and B31 clear the fault as desired. In addition, the breakers B1 and B4 must be coordinated with B13 so that the trip sequence is B13, B1, and B4 from fastest to slowest. Likewise B13, B31, and B23 should be faster than B2 and B5.

11-4 Modified Impedance Relay

It is possible to incorporate directional capability into the impedance relay. This is accomplished by the addition of a voltage, IZ_α, to V that is applied to the relay. From Figure 11-10, we write:

$$V_r = V - IZ_\alpha \tag{11-13}$$

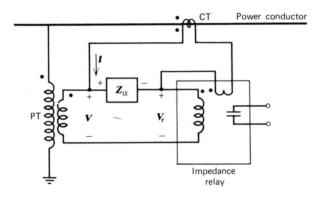

Figure 11-10 Modified impedance relaying arrangement.

If we define:

$$V = V\underline{/0} \tag{11-14a}$$

$$I = I\underline{/-\theta} \tag{11-14b}$$

$$Z_\alpha = Z_\alpha\underline{/\alpha}; \quad -\pi/2 \le \alpha \le \pi/2 \tag{11-14c}$$

$$= R_\alpha + jX_\alpha \tag{11-14d}$$

Then

$$V_r = V - IZ_\alpha\underline{/\alpha - \theta} \tag{11-15a}$$

$$= V - IZ_\alpha \cos(\alpha - \theta) - jIZ_\alpha \sin(\alpha - \theta) \tag{11-15b}$$

Computing the squared magnitude:

$$V_r^2 = [V - IZ_\alpha \cos(\alpha - \theta)]^2 + [IZ_\alpha \sin(\alpha - \theta)]^2 \tag{11-16a}$$

$$= V^2 - 2IZ_\alpha V \cos(\alpha - \theta) + I^2 Z_\alpha^2 \tag{11-16b}$$

We recall equation (11-11b) from section 11-3:

$$\frac{V_r^2}{I^2} = \frac{-K_b}{K_a} \tag{11-11b}$$

$$= Z_r^2$$

Observe that V_r, not V, is the correct voltage since that is what is applied to the relay in this situation. Divide equation (11-16b) by I^2 and substitute (11-12):

$$Z_r^2 = \frac{V^2}{I^2} - 2Z_\alpha \frac{V}{I} \cos(\alpha - \theta) + Z_\alpha^2 \tag{11-17a}$$

Recognizing that $Z = V/I$:

$$Z_r^2 = Z^2 - 2Z_\alpha Z \cos(\alpha - \theta) + Z_\alpha^2 \tag{11-17b}$$

Recall that

$$Z^2 = R^2 + X^2 \tag{11-4d}$$

$$R = Z \cos \theta \tag{11-4e}$$

$$X = Z \sin \theta \tag{11-4f}$$

Then

$$Z_r^2 = R^2 + X^2 - 2Z_\alpha Z \left[\cos \alpha \, \cos \theta + \sin \alpha \, \sin \theta\right] + Z_\alpha^2 \tag{11-18a}$$

$$Z_r^2 = R^2 + X^2 - 2Z_\alpha R \cos \alpha - 2Z_\alpha X \sin \alpha + Z_\alpha^2 \tag{11-18b}$$

Rearranging:

$$(R - Z_\alpha \cos \alpha)^2 + (X - Z_\alpha \sin \alpha)^2 = Z_r^2 \tag{11-19a}$$

or

$$(R - R_\alpha)^2 + (X - X_\alpha)^2 = Z_r^2 \tag{11-19b}$$

Equation (11-19b) is that of a circle in the Z plane with center coordinates R_α, X_α and radius Z_r as shown in Figure 11-11a. If we set $Z_\alpha = Z_r$ and α to match the angle of the line impedance locus, the operating region will be located as shown in Figure 11-11b. Such relays are sometimes referred to as "Mho" elements. Note that the Mho relay is inherently directional. This arrangement is superior to the basic impedance type because it centers the operating region about the line impedance locus and obviates the need for a separate directional element.

As with the basic impedance type, multiple Mho relays may be adjusted to sense faults at different distances down the line and thus provide zone

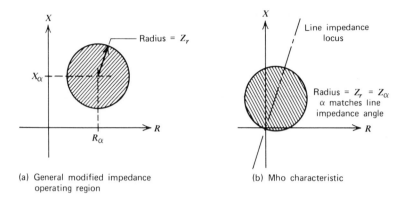

(a) General modified impedance
   operating region

(b) Mho characteristic

Figure 11-11 Modified impedance characteristics.

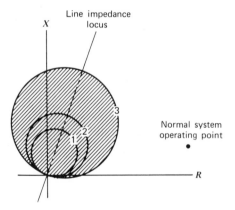

Figure 11-12 Modified impedance zone
protection.

protection. The operating characteristics are shown in Figure 11-12. Second and third zone operations are permitted only after an intentional time delay.

11-5 Distance Relay Response to Unbalanced Faults

Up to this point we have thought only in terms of balanced three phase faults. In this case Z of equation (11-3) is the line positive sequence impedance, which is directly proportional to line length. Although this was shown only for radial lines, it is also true for loop systems (see problem 11-9

400

for proof). Is the same generalization true for unbalanced faults; that is, is the impedance as defined by equation (11-3) directly proportional to distance to the fault location?

To answer this question consider the general situation shown in Figure 11-13. Shown is a line fed from both ends subjected to a SLG fault at point F located k down the line from X where the relays are located. We write:

$$V_0 + V_1 + V_2 = k(Z_0 I_0 + Z_1 I_1 + Z_2 I_2) \tag{11-20a}$$

$$= k(Z_0 I_0 + Z_1 I_1 + Z_2 I_2 + Z_1 I_0 - Z_1 I_0) \tag{11-20b}$$

Converting to phase quantities:

$$V_a = k Z_1 I_a + k(Z_0 - Z_1) I_0 \tag{11-21}$$

The impedance that is detected by the phase relay is:

$$Z_a = \frac{V_a}{I_a} \tag{11-22a}$$

$$= k Z_1 + k(Z_0 - Z_1)\frac{I_0}{I_a} \tag{11-22b}$$

The answer to our question is therefore "No, Z_a is a more complicated function of k." Observe that if $Z_1 = Z_0$, Z_a *is* directly proportional to the distance to the fault. An example would be useful.

Example 11-3

For the situation depicted in Figure 11-13 assume the following:

$$Z_0 = 3Z_1$$

$$z_{11}^0 = z_{22}^0 = 0.3 Z_1$$

$$z_{12}^0 = 0.2 Z_1$$

$$z_{11}^1 = z_{11}^2 = z_{22}^1 = z_{22}^2 = 0.4 Z_1$$

$$z_{12}^1 = z_{12}^2 = 0.3 Z_1$$

Evaluate the impedance Z_a sensed by the phase a relay for a single line to ground fault on phase a.

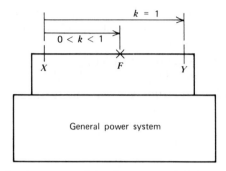

(a) Single line, loop system. Relays located at X; fault located at F.

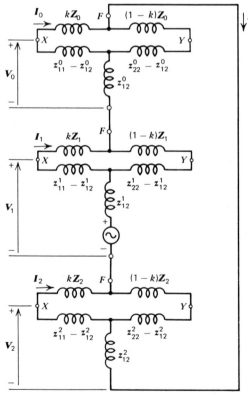

(b) Sequence networks interconnected for a SLG fault at F.

Figure 11-13 Distance relaying response to SLG faults.

402

Solution

By the current divider

$$I_0 = \frac{(1-k)Z_0 + z_{22}^0 - z_{12}^0}{Z_0 + z_{11}^0 - z_{12}^0 + z_{22}^0 - z_{12}^0}I$$

$$= \frac{(1-k)3Z_1 + 0.1Z_1}{3Z_1 + 0.1Z_1 + 0.1Z_1}I = (0.969 - 0.938\,k)I$$

$$I_1 = \frac{(1-k)Z_1 + z_{22}^1 - z_{12}^1}{Z_1 + z_{11}^1 - z_{12}^1 + z_{22}^1 - z_{12}^1}I$$

$$= \frac{(1-k)Z_1 + 0.1Z_1}{Z_1 + 0.1Z_1 + 0.1Z_1}I = (0.917 - 0.833\,k)I$$

$$I_2 = I_1$$

$$I_a = I_0 + I_1 + I_2$$

$$= (2.803 - 2.604\,k)I$$

Recalling equation (11-22b):

$$Z_a = kZ_1 + k(Z_0 - Z_1)\frac{I_0}{I_a} \qquad\qquad \text{(11-22b)}$$

$$Z_a = \left[k + 2k\left[\frac{0.969 - 0.938k}{2.803 - 2.604k}\right]\right]Z_1$$

$\underline{k = 1:}$

$$Z_a = 1.312Z_1$$

$\underline{k = 0.5:}$

$$Z_a = 0.833Z_1$$

Dividing:

$$\frac{0.833Z_1}{1.312Z_1} = 0.635$$

Observe that there is not a simple direct proportion relating distance and impedance in this case. If we set the relay for 100% reach, a fault located halfway down the line ($k = 0.5$) will produce an impedance that is 63.5% (not 50%) of the full length setting.

In general, unbalanced faults on loop systems will not present an impedance to the distance relay that is truly proportional to distance to the fault location. However, the relay can be set precisely for any fault location for which fault study data is available, and in general the impedance will decrease or increase as the fault is placed closer or farther from the relay sensing point. The question is somewhat academic in any case since other approximations in the system models prevent precise calculation of relay operation.

We have to this point assumed that the impedance relays receive instrument transformer secondary versions of *line* currents and *line to neutral* voltages as shown in Figure 11-14a. Such an arrangement is suitable for detection of SLG faults and the associated relays are identified as "ground fault relays." As we shall see, the impedance detected by these relays for LL faults is considerably different. It is therefore impossible to set a single set of relays properly to respond to both line and ground faults. For line (LL) faults an alternate arrangement, shown in Figure 11-14b, is used. The associated relays are referred to as "phase relays," since they detect phase to phase (LL) faults. Only two PT's, connected in open delta, are required. Because of the variables involved, the "ground" arrangement is frequently called the wye connection and the "phase" arrangement the delta connection. Both arrangements will respond to the rarer DLG and 3ϕ faults.

Data for setting such relays are available from fault studies. The wye values are available directly since these are the quantities involved in the symmetrical component transformation. The delta values must be calculated:

$$I_{ab} = \frac{1}{\sqrt{3}}(I_a - I_b) \tag{11-23a}$$

$$I_{bc} = \frac{1}{\sqrt{3}}(I_b - I_c) \tag{11-23b}$$

$$I_{ca} = \frac{1}{\sqrt{3}}(I_c - I_a) \tag{11-23c}$$

$$V_{ab} = \frac{1}{\sqrt{3}}(V_{an} - V_{bn}) \tag{11-23d}$$

$$V_{bc} = \frac{1}{\sqrt{3}}(V_{bn} - V_{cn}) \tag{11-23e}$$

$$V_{ca} = \frac{1}{\sqrt{3}}(V_{cn} - V_{an}) \tag{11-23f}$$

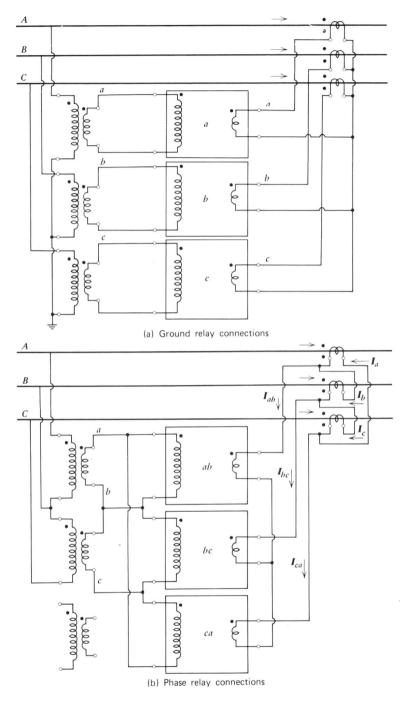

(a) Ground relay connections

(b) Phase relay connections

Figure 11-14 Ground and phase relay connections.

Division by $\sqrt{3}$ is necessary because of a quirk of the pu system ($V_\phi = V_L$ in pu). The ground relays detect:

$$Z_i = \frac{V_i}{I_i} \qquad i = a, b, c \tag{11-24a}$$

and the phase relays detect:

$$Z_{ij} = \frac{V_{ij}}{I_{ij}} \qquad ij = ab, bc, ca \tag{11-24b}$$

An example will be useful.

Example 11-4

Refer back to the example system of Chapters 8 and 9 and also Figure 11-3 in this chapter. Consider ground and phase distance relays to be located at the bus 1 end of transmission line TL13.
(a) Determine the impedances seen by all relays for all four fault types located at bus 3.
(b) Determine the ground and phase impedance relay settings for 100% reach. Discuss relay responses to all fault types.

Solution

(a) Refer to example 9-3. Computer calculated results are presented in Figures 9-15 through 9-22. Specific results that apply to protection of the line TL13 are summarized in Table 11-4. Equations (11-23) were used to calculate the delta values. Ground and phase impedance values were computed from equations (11-24); the results are tabulated in Table 11-5.
(b) The proper Z settings may be read from Table 11-5 since the reach is to be 100%. Assume that all relays are mho elements.

Ground relays:

$$Z_r = Z_\alpha = \frac{0.16}{2} = 0.08$$

$$\alpha = 90°$$

Table 11-4 Currents and voltages for line TL13 at bus 1. Fault located at bus 3.

SEQUENCE	CURRENT			VOLTAGE		
	0	1	2	0	1	2
3ϕ	0	$2.86/-90$	0	0	$0.286/0$	0
SLG	$0.81/-90$	$0.91/-90$	$0.91/-90$	$0.12/180$	$0.77/0$	$0.23/180$
DLG	$0.78/90$	$1.87/-90$	$0.99/90$	$0.11/0$	$0.53/0$	$0.25/0$
LL	0	$1.43/-90$	$1.43/90$	0	$0.64/0$	$0.36/0$
WYE	*a*	*b*	*c*	*an*	*bn*	*cn*
3ϕ	$2.86/-90$	$2.86/150$	$2.86/30$	$0.29/0$	$0.29/-120$	$0.29/+120$
SLG	$2.63/-90$	$0.10/90$	$0.10/90$	$0.426/0$	$0.95/246$	$0.95/114$
DLG	$0.10/-90$	$2.76/154$	$2.76/26$	$0.89/0$	$0.37/220$	$0.37/140$
LL	0	$2.48/180$	$2.48/0$	$1.00/0$	$0.56/206$	$0.56/154$
DELTA	*ab*	*bc*	*ca*	*ab*	*bc*	*ca*
3ϕ	$2.96/-60$	$2.86/180$	$2.86/60$	$0.29/30$	$0.29/-90$	$0.29/150$
SLG	$1.58/-90$	0	$1.58/-90$	$0.69/47$	$1.00/-90$	$0.69/133$
DLG	$1.62/-28$	$2.86/180$	$1.62/28$	$0.69/12$	$0.28/-90$	$0.69/168$
LL	$1.43/0$	$2.86/180$	$1.43/0$	$0.88/9$	$0.28/-90$	$0.88/171$

Table 11-5 Impedances seen by phase and ground relays. Bus 1 end of TL13. Fault at bus 3.

CASE	GROUND IMPEDANCE			PHASE IMPEDANCE		
	a	*b*	*c*	*ab*	*bc*	*ca*
3ϕ	$0.10/90$	$0.10/90$	$0.10/90$	$0.10/90$	$0.10/90$	$0.10/90$
SLG	$0.16/90$	$9.50/156$	$9.50/24$	$0.45/137$	∞	$0.45/43$
DLG	$8.90/90$	$0.13/66$	$0.13/114$	$0.43/40$	$0.10/90$	$0.43/140$
LL	∞	$0.23/26$	$0.23/154$	$0.61/9$	$0.10/90$	$0.61/171$

Phase relays:

$$Z_r = Z_\alpha = \frac{0.10}{2} = 0.05$$

$$\alpha = 90°$$

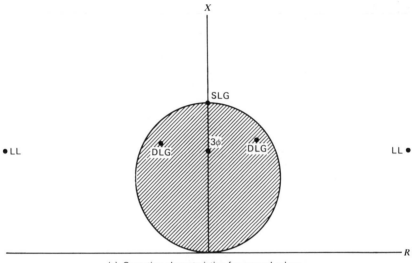

(a) Operating characteristics for ground relays

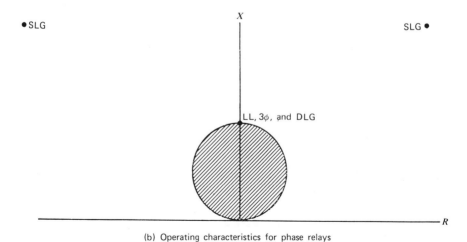

(b) Operating characteristics for phase relays

Figure 11-15 Fault types as sensed by ground and phase relays.

408

The operating characteristics are shown in Figure 11-15. Observe that LL faults are *not* sensed by the ground relays and SLG faults are *not* sensed by the phase relays. Both DLG and 3φ are sensed by both sets of relays.

11-6 Pilot Relaying

The term "pilot relaying" essentially means remote control of circuit breakers; that is, the decision to open or close a circuit breaker is made geographically remote (perhaps many miles) from the breaker location. The basic advantage realized is high speed tripping, which in turn permits:

- Minimum damage to equipment
- Minimum stability problems
- Automatic reclosing

This scheme requires communication channels to carry system voltage and current information to the decision-making location. Three basic channels are used:

- Separate electrical circuits, frequently telephone circuits.
- Power line carrier—The power transmission line itself is used as the communication circuit. Signals are applied to all three phases through a L-C voltage divider network. Frequencies range from 30 to 200 kHz. The signals are confined to the line in question by L-C blocking filters at each end, referred to as "line traps." Refer to Figure 11-16.
- Microwave—Relaying information can be broadcast at microwave frequencies, line-of-sight, between directional dish antennas. This channel suffers from atmospheric disturbance and fade. It has the advantage that considerable information can be broadcast in one beam.

There are several basic schemes used and we discuss one of the simplest: directional comparison.

Examine the situation in Figure 11-17a. We wish a fault at F1 to be cleared by BX (not BL, BR, or BY). If all breakers are controlled by directional relays arranged to look down their respective lines, BL and BY will not operate since the fault appears "behind" them. However, BR can see the fault F1. How can we block the operation of BR?

Figure 11-16 500 kV line trap.

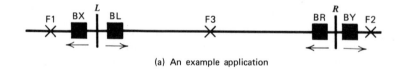

(a) An example application

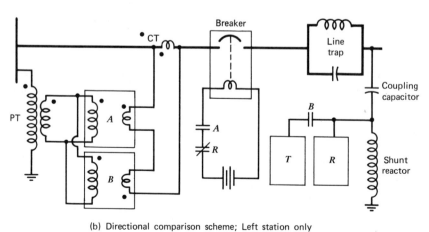

(b) Directional comparison scheme; Left station only

Figure 11-17 Line protection by directional comparison: power line carrier channel.

410

Consider the relay arrangement shown in Figure 11-17b. There are two directional relays, with relay A connected to look down the line towards bus R and relay B to look backwards. A fault at F1 is detected by B but not A. Since contact A remains open, BL will not operate. Contact B closes, allowing the transmitter to broadcast a guard frequency on the line. This signal is detected by a receiver R at both ends of the line, opening the normally closed R contacts, and blocking the operation of BR (and BL also). A fault F2 produces similar action with the operations at left and right stations reversed. A fault within the protected zone (F3) is detected by both A relays and neither of the B relays, blocking the guard frequency, and allowing BL and BR to trip.

Directional comparison greatly simplifies coordination problems and permits high speed tripping. The need for precise relay settings and corresponding calculated system fault currents and voltages is eliminated. Similar schemes use the advantages of pilot relaying to overcome the disadvantages of simple distance relaying. Pilot relaying is amenable to computer control of the protection system and this is an extremely interesting developing area of study. We have introduced only the fundamentals, which should serve as a base for further specialized study.

11-7 Summary

It is important to clear faults from a power system in a manner that interrupts service as little as possible. The design of protection systems to accomplish this is therefore of interest. Time delay overcurrent relays are useful devices in this regard. The directional, or reactance, relay is another component that can be used to block undesired breaker operation.

Impedance, or distance, relays are used on HV and EHV lines and simplify coordination by limiting the range at which a particular relay/breaker set can sense a fault. Zone protection provides redundancy of protection that maintains coordination and allows for equipment failure. Pilot relaying permits high speed breaker operation that in turn results in fast fault clearing and automatic breaker reclosing.

Because the majority of line faults are self clearing, after a line has been de-energized for several cycles, the breakers are typically reclosed one or more times in the hope that the line can be quickly restored to service. If the breakers continue to close in on a fault, they finally will open and lock out, requiring human inspection, repair, and restoration of service.

System protection is an interesting area within power system engineering and we have considered only the basics. There is much more, both in breadth and in depth, to be learned.

411

Bibliography

[1] Anderson, Paul M., *Analysis of Faulted Power Systems*, Iowa State Press, Ames, Iowa, 1973.

[2] Brown, H. E., and Person, C. E., "Digital Calculations of Single-Phase to Ground Faults," *Trans. AIEE*, vol. 79 (pt. 3): pp. 657–60, 1960.

[3] Brown, H. E., Person, C. E., Kirchmayer, L. K., and Stagg, G. W., "Digital Calculation of Three-Phase Short Circuits by Matrix Method," *Trans. AIEE*, vol. 79 (pt. 3): pp. 1277–82, 1960.

[4] Brown, Homer E., *Solution of Large Networks by Matrix methods*, John Wiley and Sons, Inc., 1975.

[5] Elgerd, Olle I., *Electric Energy Systems Theory: An Introduction*, McGraw-Hill Inc., New York, 1971.

[6] El-Abiad, A. H., Digital Calculation of Line-to-Ground Short Circuits by Matrix Method," *Trans. AIEE*, vol. 79 (pt. 3): pp. 323–32, 1960.

[7] IEEE, *IEEE Standard Dictionary of Electrical and Electronics Terms*, Wiley-Interscience, New York, 1972.

[8] IEEE, *Recommended Practice for Protection and Coordination of Industrial and Commercial Power Systems*, Wiley-Interscience, New York, 1975.

[9] Manufacturer's Publication, *Distribution-System Protection Manual*, McGraw-Edison Power Systems Division, Publication 71022.

[10] Manufacturer's Publication, *Short-Circuit Current Calculations for Industrial and Commercial Power Systems*, General Electric, Publication GET-3550.

[11] Manufacturer's Publication, *Short-Circuit Currents in Low and Medium Voltage A-C Power Systems*, General Electric, Publication GET-1470D.

[12] Mason, C. Russell, *The Art and Science of Protective Relaying*, John Wiley and Sons, Inc., 1956.

[13] Neuenswander, John R., *Modern Power Systems*, International Textbook Co., Scranton, Pa., 1971.

[14] Stagg, Glenn W., and El-Abiad, Ahmed H., *Computer Methods in Power System Analysis*, McGraw-Hill, Inc., New York, 1968.

[**15**] Stevenson, Jr., William D., *Elements of Power Systems Analysis*, 3rd edition. McGraw-Hill, Inc., New York, 1975.

[**16**] Weedy, B. M., *Electric Power Systems*, 2nd edition. John Wiley and Sons Ltd., London, 1972.

[**17**] Westinghouse Electric Corporation, *Electrical Transmission and Distribution Reference Book*, 4th edition. East Pittsburgh, Pa., 1950.

Problems

11-1. Rework example 11-1 if the maximum loads are revised to 8, 4, and 5 MVA at busses 1, 2, and 3. Also change the CT ratio at bus 2 to 600:5. Keep all other data the same.

11-2. In example 11-1 why did we select the *largest* fault current?

11-3. Consider the general relay equation (10-17). Suppose $K_a = K_b = K_s = 0$; $A = V$ and $B = I$. Crosshatch the operating region in the Z plane if:

(a) $\tau = 0°$
(b) $\tau = 45°$
(c) $\tau = -45°$
(d) $\tau = -90°$

11-4. Design a relay scheme that will create a sector-shaped operating region in the Z plane bounded by the unit circle, the reactance axis, and a 45° line through the origin. Specify K_a, K_b, K_c, K_s, and τ for each relay used and produce a drawing similar to Figure 11-5 showing the details of relay interconnections.

11-5. Consider example 11-2. Repeat part (c) for a fault at point Y. Assume point Y is 60% down the line TL23 from bus 2 and that relays are set in parts (a) and (b).

11-6. Assume a modified impedance relay with the following settings:

$$K_a = -1$$
$$K_b = 25$$
$$Z_\alpha = 1 + j5$$

Accurately plot the operating characteristics.

11-7. Consider the example 300 km 500 kV line of the Appendix. Assume a CT ratio of 4000:5 A and a PT ratio of 289 kV:115 V sensing line current and phase voltage. Determine Z_r and Z_α values for a mho relay that is to have an 80% reach for 3ϕ faults.

11-8. The line of problem 11-7 has a thermal rating of 2624 MVA. Determine the corresponding Z value in ohms as sensed by the mho relay. Plot this constant Z (a circle) to scale in the Z plane. On the same plot show the mho relay operating region, as computed in problem 11-7.

11-9. Prove that Z of equation (11-3) is directly proportional to distance to a 3ϕ fault on a line connected at both ends into a general power system.

11-10. Refer to example 11-4. Calculate impedances detected by the ground relays at bus 1 for a SLG fault on phase a at bus 3. Check your results against those presented in Tables 11-4 and 11-5.

11-11. Rework problem 11-10 for a LL fault between phases b and c at bus 3. Compute the phase relay impedances.

12

POWER SYSTEM STABILITY

"Be cheerful. Strive to be happy."
Max Ehrmann, DESIDERATA

Synchronous machines operating in parallel are inherently stable. If this were not the case multigenerator systems would be impossible. By stable we mean that the machines can normally recover from small random perturbing forces and still remain synchronized.

We have already discussed steady state stability in some detail for lines and generators. It is possible for a system to be stable in a steady state sense and still experience stability problems when subjected to switching operations. The most severe switching operations include applications of faults, clearing of faults, and inadvertent tripping of lines and generators. This problem is referred to as transient stability and is the main topic of this chapter. There is also a "dynamic" stability problem that is mentioned later.

From a mathematical viewpoint we shall see that the problem involves solving a system of nonlinear differential equations, and therefore requires numerical analysis techniques. It is important to appreciate that the question of stability will be affected by the initial conditions—that is, the loaded condition of the system before switching.

The most severe switching action is the balanced three phase fault, as far as transient stability is concerned. This is fortunate from an analytical viewpoint since it simplifies the complexity of our system model, requiring only the positive sequence network. Another basic simplification is that the time constants of relevance are of such values that ac circuit techniques may still be used; that is, "transient ac" methods as explained in Chapter 6 are applicable. We shall also assume that all machines are lossless. This simplification is defensible for two reasons:

- Practical machines, while not lossless, are highly efficient ($\sim$95%).

- Our results based on this assumption will be conservative (the system will be somewhat more stable than our results predict).

We begin our study of stability by reviewing the steady state stability of synchronous machines as discussed in Chapter 6. The reader is strongly encouraged at this point to reread sections 6-3 and 6-4; the following discussion is a continuation of this material.

12-1 The Basic Synchronous Machine Equation of Motion

Referring to the turbine/generator system of Figure 6-10 we write

$$T_m - T_e = J\frac{d\omega_m}{dt} + B\omega_m \tag{12-1}$$

where

T_m = Mechanical turbine torque in Nm

T_e = Electromagnetic counter-torque in Nm

J = Mass polar moment of inertia of all rotating parts (rotors of generator and turbine) in kg-m^2

B = Damping torque coefficient in Nm-s

ω_m = Rotor shaft velocity in rad/s

The damping torque, $B\omega_m$, is caused by rotor bearing friction, windage, magnetic losses, and any other drag torques that oppose rotation. Compared to other torque terms it is small; for this reason let us neglect this term to simplify our development. Therefore:

$$T_m - T_e = J\frac{d\omega_m}{dt} \tag{12-2a}$$

Multiplying through by ω_m:

$$P_m - P_e = \omega_m J\frac{d\omega_m}{dt} \tag{12-2b}$$

where

$P_m = \omega_m T_m$ = Mechanical turbine power in W. $\tag{12-2c}$

$P_e = \omega_m T_e$ = Electromagnetic power in W. $\tag{12-2d}$

Refer back to Section 6-4 to refresh your understanding of the power angle δ. Recall that it was defined as the phase shift between the machine internal voltage (E) and terminal voltage (V). In this sense δ is measured in *electrical* radians, so let us modify the notation and designate this δ as δ'_e. If we neglect phase leakage reactance and resistance, this same δ is a spatial angle between two spatially revolving magnetic fields inside the machine. The rotor magnetic field is physically produced by dc currents flowing in the rotor windings and therefore revolves at the rotor angular velocity ω_m. The stator magnetic field is physically produced by the combined effect of three phase voltages applied to the three phase stator windings and therefore revolves at angular velocity ω_s where:

$$\omega_s = \frac{2\pi f_e}{N/2} \text{ mechanical rad/s} \tag{12-3}$$

where

> f_e = System electrical frequency in Hz as detected at the generator terminals

> N = No. of generator poles

Normally these two fields are synchronized; that is, revolve at the same angular velocity ($\omega_s = \omega_m$) with a fixed angle δ'_m separating them. Refer to Figure 12-1 for help. Under transient conditions there can be relative motion. In general we write:

$$\omega_m = \omega_s + \frac{d\delta'_m}{dt} \tag{12-4a}$$

where

$$\delta'_m = \frac{\delta'_e}{N/2} \tag{12-4b}$$

We shall find that it is desirable to work with a delta that has a more general definition than that as given in equation (12-3). Again refer to Figure 12-1. We write

$$\omega_m = \omega_{r_m} + \frac{d\delta_m}{dt} \tag{12-5}$$

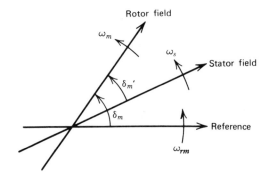

Figure 12-1 Definition for δ'_m and δ_m.

where

> ω_{r_m} = Reference angular velocity in mechanical rad/s of a revolving reference line, which may or may not be revolving at constant velocity.

Substitute (12-5) in (12-2):

$$P_m - P_e = \omega_m J \frac{d}{dt}\left[\omega_{r_m} + \frac{d\delta_m}{dt}\right] \tag{12-6a}$$

$$= \omega_m J \left[\frac{d\omega_{r_m}}{dt} + \frac{d^2\delta_m}{dt^2}\right] \tag{12-6b}$$

It will prove convenient to measure angles in electrical rad/s. Then define:

$$\delta = \frac{N}{2}\delta_m \quad \text{electrical radians} \tag{12-7a}$$

$$\omega_r = \frac{N}{2}\omega_{r_m} \quad \text{electrical rad/s} \tag{12-7b}$$

Substitution into (12-6) produces:

$$P_m - P_e = \frac{\omega_m J}{(N/2)}\left[\frac{d\omega_r}{dt} + \frac{d^2\delta}{dt^2}\right] \tag{12-8}$$

The units of (12-8) are SI. It is useful to convert to per unit. Therefore divide through by the power base $S_{3\phi_{\text{base}}}$, which is typically the generator nameplate $S_{3\phi}$ rating. Then:

$$P_{m_{\text{pu}}} - P_{e_{\text{pu}}} = \frac{\omega_m J}{(N/2)S_{3\phi_{\text{base}}}}\left[\frac{d\omega_r}{dt} + \frac{d^2\delta}{dt^2}\right] \tag{12-9}$$

It is common practice to use a related constant H, instead of J, to account for inertia. The definition follows:

$$H = \frac{\text{Kinetic energy of all rotating parts at synchronous speed}}{S_{3\phi_{\text{Rating}}}} \tag{12-10a}$$

$$= \frac{\frac{1}{2}J\omega_{\text{sync}}^2}{S_{3\phi_{\text{Rating}}}} \tag{12-10b}$$

where

$$\omega_{\text{sync}} = \frac{2\pi f}{N/2} \quad \text{mechanical rad/s} \tag{12-10c}$$

f = nominal system frequency (60 Hz)

Consider the coefficient on the right side of equation (12-9). If $\omega_m = \omega_{sync}$:

$$\frac{\omega_{sync}J}{(N/2)S_{3\phi base}} = \frac{N/2}{\pi f(N/2)}\left[\frac{\frac{1}{2}J\omega_{syne}^2}{S_{3\phi base}}\right] \tag{12-11a}$$

$$= \frac{H}{\pi f} \tag{12-11b}$$

Although under transient conditions $\omega_m \neq \omega_{sync}$, it is close and we write as an approximation:

$$P_m - P_e = \frac{H}{\pi f}\left[\frac{d^2\delta}{dt^2} + \frac{d\omega_r}{dt}\right] \tag{12-12}$$

where revised definitions of all quantities follow:

P_m = turbine mechanical power, in per unit

P_e = generator output power, in per unit

H = inertia constant, in seconds

δ = power angle, measured from rotating reference to generator rotor q axis in electrical radians

ω_r = angular velocity of rotating reference, in electrical radians/s

t = time, in seconds

We drop the "pu" subscript to simplify the notation. Equation (12-12) is referred to as the swing equation and is basic to transient stability studies.

12-2 Simplified Models for the Generator and External System

As we noted in Chapter 6 the synchronous machine is an inherently complicated device. A complete mathematical model is quite involved and requires a lengthy development if it is to be properly presented. Experience shows that certain simplifying assumptions can greatly reduce the model complexity without seriously affecting the accuracy of transient stability calculations. The most important initial assumptions we will make are that

- Magnetic saturation can be ignored.

- Transient ac methods can be used.

- The machine is lossless.

- Only balanced 3ϕ operation is investigated.

The balanced 3ϕ steady state ac electrical performance of a synchronous generator can be predicted from

$$E_f = V + jX_q I_q + jX_d I_d \tag{12-13a}$$

$$I = I_d + I_q \tag{12-13b}$$

where

$E_f = E_f\underline{/\delta} = $ Internal rotor field generated phasor voltage, located on the q axis.

$V = $ Generator terminal phasor voltage.

$I = $ Generator terminal phasor current.

$I_d = $ Direct axis component of I.

$I_q = $ Quadrature axis component of I.

$X_d = $ Direct axis synchronous reactance.

$X_q = $ Quadrature axis synchronous reactance.

All values are in per unit; the currents and voltages are positive sequence values.

It is necessary to separate I into its components I_d and I_q to account for saliency. The corresponding phasor diagram is shown in Figure 12-2.

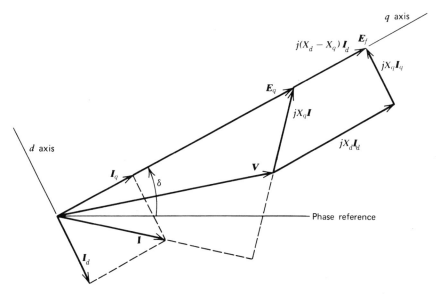

Figure 12-2 Phasor diagram for steady state machine model.

Observe that E_f is located on the q axis, which defines δ. Consider the situation where V and I are known and we wish to evaluate E_f, and therefore δ. We have a problem. Direct use of (12-13) is thwarted because the location of the d and q axes are unknown, which prevents computation of I_d and I_q. Therefore, consider

$$E_f = V + jX_q I_q + jX_d I_d \tag{12-14a}$$

$$= V + jX_q(I - I_d) + jX_d I_d \tag{12-14b}$$

$$= E_a + j(X_d - X_q)I_d \tag{12-14c}$$

where

$$E_q = V + jX_q I \tag{12-15}$$

We observe that E_q *can* be calculated directly since V and I are known. The phasor $j(X_d - X_q)I_d$ falls along the q axis, revealing that $E_q = E_q\underline{/\delta}$, locating the q axis. The component I_d can now be calculated and E_f evaluated.

Under transient conditions, X'_d, the direct axis transient reactance, should be used instead of X_d, since it will account for mutual coupling between the rotor field winding and the stator windings. Since the field is on the d axis only, the corresponding quadrature value X'_q is equal to X_q. Therefore

$$E'_q = V + jX_q I_q + jX'_d I_d \tag{12-16a}$$

$$= E_q + j(X'_d - X_q)I_d \tag{12-16b}$$

where E_q remains as defined in equation (12-15). Typically, $X_q > X'_d$, which produces the situation illustrated in the phasor diagram of Figure 12-3. A

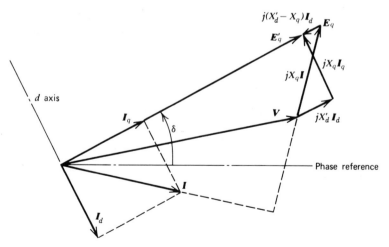

Figure 12-3 Phasor diagram for transient machine model.

particularly simple result occurs if $X_d' = X_q$. Equation (12-16) then becomes

$$\boldsymbol{E}_q' = \boldsymbol{E}_q \tag{12-17a}$$

$$= jX_q\boldsymbol{I} + \boldsymbol{V} \tag{12-17b}$$

$$= jX_d'\boldsymbol{I} + \boldsymbol{V} \tag{12-17c}$$

Our generator may then be modeled by the simple network of Figure 12-4. We use this simplified model for the rest of this chapter.

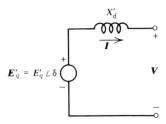

Figure 12-4 Simplified transient generator model.

Looking out from the generator terminals the external system is a vast network of interconnected transformers, lines, and other generators. A disturbance, caused by a fault or switching, can potentially cause transient stability problems for any and all generators in the system. However, in many cases only the generator closest to the fault is seriously affected and it is possible to lump the external system into an equivalent network.

In order to decide just how far into the external system to take such an equivalent it is common practice to talk in terms of an "infinite" bus. Such a bus is that whose voltage and frequency is substantially unaffected by the disturbance under investigation. These points are a matter of engineering judgment and are selected from experience. When the matter is in doubt studies can be made with the bus in question treated as infinite or allowed to vary in voltage and frequency and the differences studied. Assuming that one encounters infinite busses at some point back into the system, prior to encountering other machines, it is reasonable to reduce the sytem to its Thevenin equivalent. The infinite busses are treated as ideal voltage sources. The Thevenin source has the ideal properties of maintaining constant voltage, phase, and frequency. The system equivalent circuit is shown in Figure 12-5 where

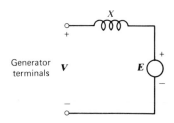

Figure 12-5 System equivalent circuit.

E = Thevenin equivalent system voltage, computed at the terminals of a specified generator, looking into the system. Infinite busses are treated as ideal voltage sources.

X = Thevenin equivalent reactance. Resistance is considered negligible.

For purposes of understanding transient stability we will concentrate on the simplified situation shown in Figure 12-6. We will further assume

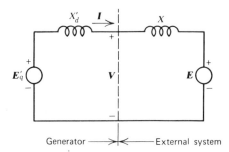

Figure 12-6 Simplified generator/system model for transient stability.

that the transients involved are too fast to allow the generator/turbine speed control loop to respond or to allow the generator voltage control loop to respond and therefore can have no effect on the system frequency. The effect of these assumptions on equation (12-12) is:

$$P_m - P_e = \frac{H}{\pi f} \frac{d^2\delta}{dt^2} \tag{12-18}$$

where

$$P_m = \text{Constant}$$

$$P_e = \frac{E'_q E}{X'_d + X} \sin \delta$$

$\omega_r = \omega_{\text{sync}}$, so that $\dfrac{d\omega_r}{dt} = 0$

424

The functions P_e and P_m are plotted in Figure 12-7. We say that the system is in equilibrium when $P_m = P_e$; that is, $d^2\delta/dt^2 = 0$. Observe that a given value of P_m will result in two equilibrium values for P_e, and therefore $\delta : \delta^s$ and δ^u. Suppose we are operating at δ^s when δ is suddenly increased by a small amount due to some spurious disturbance. Observe that $P_e > P_m$ and that $d^2\delta/dt^2$ in equation (12-18) is negative. Therefore the system will respond by *decreasing* δ and returning to δ^s. Now suppose δ is increased from δ^u. Observe that $P_m > P_e$ and that $d^2\delta/dt^2$ is positive. The system responds by *increasing* δ and moving further from δ^u. For this reason δ^s and δ^u are referred to as stable and unstable equilibria, respectively.

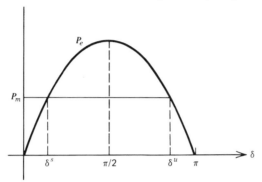

Figure 12-7 Plots of P_m and P_e versus δ.

We are now prepared to discuss transient stability. Refer to Figure 12-8. Suppose the system is operating normally at $P_m = P_{m_1}$ and $\delta = \delta_1$ when P_m is suddenly changed to P_{m_2} at $t = 0$. Because of rotor inertia δ cannot adjust

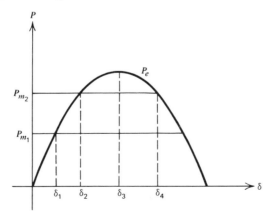

Figure 12-8 System reaction to sudden change in P_m.

425

instantly. Therefore at $t = 0^+$, $\delta = \delta_1$ and $P_m > P_e$. The rotor begins to accelerate, allowing δ to increase. This continues until $\delta = \delta_2$ at which point $P_m = P_e$. However, the rotor cannot stop here due to inertia, and δ must continue to increase. For $\delta > \delta_2$, $P_m < P_e$, producing rotor deceleration. Eventually the rotor stops at δ_3, and swings back toward δ_2. There will be rotor oscillation about δ_2 (forever, as predicted by equation 12-18) that will eventually damp out; the rotor will stabilize at the new equilibrium value δ_2. The greater the step increase in P_m the farther the maximum swing on $\delta(\delta_3)$. Observe that for $\delta > \delta_4$, $P_m > P_e$ and the rotor again is *accelerated*, causing δ to continue to increase. Therefore, stability will be lost if δ swings past δ_4.

Although sudden changes in P_m can theoretically cause transient stability problems, this is not usually of practical concern since P_m typically changes slowly. The same behavior is noted, however, for sudden changes in P_e. This can be caused by intentional or accidental switching in the external system. As an example consider the sytem shown in Figure 12-9.

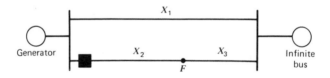

Figure 12-9 Example system.

Example 12-1

Consider the system of Figure 12-9.

$$X_1 = 0.4 \qquad E'_q = 1.2$$

$$X_2 = 0.2 \qquad P_m = 1.5$$

$$X_3 = 0.2 \qquad X'_d = 0.2$$

$$E = 1.0$$

Suppose the system is operating in equilibrium at δ_1 when the breaker inadvertently opens, switching us to P'_e. Evaluate δ_1, δ_2, and δ_4 as shown in Figure 12-10.

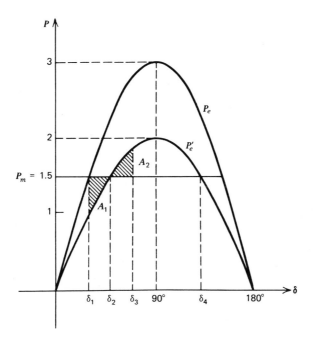

Figure 12-10 System response to change in P_e.

Solution

At $t = 0^-$:

$$X = \frac{(0.2+0.2)(0.4)}{0.2+0.2+0.4}$$

$$= 0.2$$

$$P_e = \frac{EE'_q}{X'_d + X} \sin \delta$$

$$= \frac{(1.2)(1.0)}{0.2+0.2} \sin \delta$$

$$= 3.0 \sin \delta$$

But when $\delta = \delta_1$, $P_{e_1} = P_m$. Therefore

$$3.0 \sin \delta_1 = 1.5$$

$$\delta_1 = 30°$$

427

Now at $t = 0$, the breaker opens and we switch to curve P'_e in Figure 12-10.

$$P'_e = \frac{(1.2)(1.0)}{0.2+0.4} \sin \delta$$

$$= 2.0 \sin \delta$$

To solve for δ_2 and δ_4:

$$P'_e = P_m$$

$$\sin \delta = \frac{1.5}{2.0} = 0.75$$

$$\delta = \sin^{-1}(0.75) = 48.6° \text{ or } 131.4°$$

$$\delta_2 = 48.6°; \qquad \delta_4 = 131.4°$$

Observe that we still have not solved for δ_3 and consequently have not resolved the question of stability. We can, of course, directly solve (12-18). It is of the nonlinear differential type and will require numerical techniques (we investigate this approach later). There are other methods that can provide stability information without actually solving the equation; one is the equal area criterion discussed next.

12-3 Predicting Stability: Equal Area Methods

We need a method for predicting stability. Observe that:

$$\frac{d}{dt}\left[\frac{d\delta}{dt}\right]^2 = 2\frac{d\delta}{dt} \cdot \frac{d^2\delta}{dt^2} \tag{12-19a}$$

Therefore

$$\frac{d^2\delta}{dt^2} = \frac{d(d\delta/dt)^2}{2d\delta} \tag{12-19b}$$

Substituting (12-19b) into (12-18) produces

$$\frac{H}{2\pi f} d\left(\frac{d\delta}{dt}\right)^2 = (P_m - P_e)\, d\delta \tag{12-20}$$

Integrating:

$$\left(\frac{d\delta}{dt}\right)^2 = \frac{2\pi f}{H} \int_{\delta_0}^{\delta} (P_m - P_e)\, d\delta \tag{12-21}$$

Study equation (12-21) closely. It states that the area between the two functions, P_m and P_e, is proportional to the square of $d\delta/dt$, with the angular velocity relative to the reference velocity. Since kinetic energy is also proportional to angular velocity squared we can think of these areas as related to kinetic energy relative to a rotating reference. When $P_m > P_e$, the rotor is accelerating and the rotor acquires an "energy" of A_1 that must be offset by A_2, acquired when the rotor decelerates ($P_m < P_e$).

The integration begins at some initial $\delta = \delta_0$, at which point $d\delta/dt = 0$ and continues out to some arbitrary δ. The usefulness of equation (12-21) is that when its right side is zero, so is $d\delta/dt$, that is, the rotor has "stopped." To understand the application of (12-21) consider the following example.

Example 12-2

Continue Example 12-1 and predict whether the system will be stable or unstable for that situation using equation (12-21). If the system is stable, evaluate δ_3.

Solution

Refer back to example 12-1, Figures 12-9 and 12-10. All angles should be in radians.

$$\delta_1 = 30° = 0.524 \text{ rad}$$

$$\delta_2 = 48.6° = 0.848 \text{ rad}$$

$$\delta_4 = 131.4° = 2.293 \text{ rad}$$

Observe that in the interval δ_1 to δ_2, $P_m > P_e$ and the rotor is accelerating. We compute the corresponding area A_1:

$$A_1 = \int_{\delta_1}^{\delta_2} (P_m - P_e) \, d\delta$$

$$= \int_{0.524}^{0.848} (1.5 - 2 \sin \delta) \, d\delta$$

$$= 1.5\delta + 2 \cos \delta \big]_{0.524}^{0.848} = 0.0773$$

429

The question critical to stability is whether there is enough negative area $(-A_2)$ in the interval $\delta_2 < \delta < \delta_4$ to offset this motion. We compute:

$$-A_{2_{max}} = \int_{\delta_2}^{\delta_4} (P_m - P_e)\, d\delta$$

$$\int_{0.848}^{2.293} (1.5 - 2\sin\delta)\, d\delta$$

$$= 1.5\delta + 2\cos\delta\big]_{0.848}^{2.293} = -0.478$$

Since $A_{2_{max}} > A_1$ the rotor will not swing as far as δ_4 and the system is stable. We will now calculate the greatest rotor swing δ_3. The rotor will be stopped when $A_2 = A_1$. Therefore

$$\int_{\delta_2}^{\delta_3} (P_m - P_e)\, d\delta = -0.0773$$

$$\int_{0.848}^{\delta_3} (1.5 - 2\sin\delta)\, d\delta = -0.0773$$

$$1.5\delta + 2\cos\delta\big]_{0.848}^{\delta_3} = -0.0773$$

Simplifying:

$$1.5\delta_3 + 2\cos\delta_3 = 2.518$$

The equation is nonlinear; we resort to iterative methods for its solution. The result is

$$\delta_3 = 1.218 \text{ rad} \quad (69.8°)$$

Stability problems may also be caused by faults on the system. For example, if a 3ϕ fault occurs at the terminals of a synchronous machine it is completely "decoupled" from the external system and has no hope of maintaining stability unless coupling is quickly restored. Similar problems are caused by faults remote from the generator. To understand this point study the next example.

Example 12-3

Refer to the system of example 12-1 as shown in Figure 12-9. Suppose a balanced 3ϕ fault occurs at point F.
(a) If the fault is not removed, will the generator be stable?
(b) Suppose the fault is cleared by opening the breaker at $\delta = \delta_c = 60°$. Is the system stable? If so, calculate δ_3.

Solution

(a) The fault at F reconfigures the external system positive sequence network to that shown in Figure 12-11. We evaluate its Thevenin equivalent:

$$X_2 = \frac{0.4(0.2)}{0.2+0.4} = 0.133$$

$$E = \frac{0.2}{0.2+0.4}(1.0) = 0.333$$

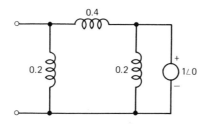

Figure 12-11 External system for Example 12-3 when subjected to a 3ϕ fault at F.

Therefore, under faulted conditions

$$X'_d + X = 0.2 + 0.133$$

$$= 0.333$$

$$P_e = \frac{(1.2)(0.333)}{0.333}\sin\delta$$

$$= 1.2\sin\delta$$

Stability is obviously impossible, since $P_{e_{max}}(1.2)$ is less than $P_m(1.5)$, preventing even steady state stability.

(b) To attack this problem let us define:

$$P_e^0 = 3.0\sin\delta = \text{Prefault } P_e \text{ curve.}$$

$$P'_e = 1.2\sin\delta = \text{Faulted } P_e \text{ curve.}$$

$$P''_e = 2.0\sin\delta = \text{Postfault } P_e \text{ curve.}$$

All three are shown in Figure 12-12. If the fault is cleared at 60° (1.047 rad), the accelerating area is:

$$A_1 = \int_{0.524}^{1.047} (P_m - P'_e) \, d\delta$$

$$= \int_{0.524}^{1.047} (1.5 - 1.2 \sin \delta) \, d\delta$$

$$= 1.5\delta + 1.2 \cos \delta \big]_{0.524}^{1.047}$$

$$= 0.346$$

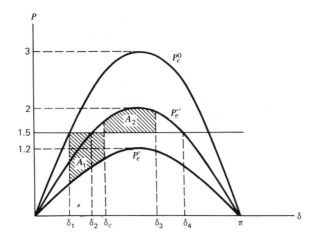

Figure 12-12 P-δ plots for Example 12-3.

When the fault is cleared we switch from P'_e to P''_e. The maximum available decelerating area $(-A_2)$ is

$$-A_{2_{max}} = \int_{\delta_c}^{\delta_4} (P_m - P''_{e_3}) \, d\delta$$

$$= \int_{1.047}^{2.293} (1.5 - 2 \sin \delta) \, d\delta$$

$$= 1.5\delta + 2 \cos \delta \big]_{1.047}^{2.293}$$

$$= -0.453$$

Since $A_{2_{max}} > A_1$ the system is stable. To solve for δ_3:

$$\int_{\delta_c}^{\delta_3} (P_m - P_e'') \, d\delta = -0.346$$

$$\int_{1.047}^{\delta_3} (1.5 - 2 \sin \delta) \, d\delta = -0.346$$

$$1.5\delta + 2 \cos \delta \big]_{1.047}^{\delta_3} = -0.346$$

$$1.5\delta_3 + 2 \cos \delta_3 = 2.225$$

The equation again is nonlinear and requires iterative methods for solution. The result is:

$$\delta_3 = 1.848 \text{ rad.}$$

Observe that the larger δ_c (i.e., the longer it takes to clear the fault) the greater risk of instability. The maximum rotor swing δ_3 will approach δ_4 as δ_c increases. The *critical* clearing angle δ_{cc} is the largest δ_c allowable for stability, at which condition $\delta_3 = \delta_4$. We apply this to an example.

Example 12-4

Calculate the critical clearing angle for the system of example 12-1.

Solution

Again using (12-16):

$$\int_{\delta_1}^{\delta_{cc}} (P_m - P_e') \, d\delta = -\int_{\delta_{cc}}^{\delta_4} (P_m - P_e'') \, d\delta$$

$$\int_{0.524}^{\delta_{cc}} (1.5 - 1.2 \sin \delta) \, d\delta + \int_{\delta_{cc}}^{2.293} (1.5 - 2 \sin \delta) \, d\delta = 0$$

$$1.5\delta + 1.2 \cos \delta \big]_{0.524}^{\delta_{cc}} + 1.5\delta + 2 \cos \delta \big]_{\delta_{cc}}^{2.293} = 0$$

$$\therefore \quad \delta_{cc} = 1.196 \text{ rad} \quad (68.6°)$$

Since these methods involve equating certain areas in P, δ plots, this general approach is referred to as the Equal Area method. It is general for simplified single machine/infinite bus problems and can be stated as follows:

$$A_1 \le A_{2_{max}} \text{ for stability} \tag{12-32}$$

where

A_1 = accelerating area in P, δ coordinates $(P_m > P_e)$.

$A_{2_{max}}$ = maximum decelerating area in P, δ coordinates $(P_m < P_e)$.

12-4 Solution of the Swing Equation

One important drawback to the Equal Area approach is that whereas the critical clearing *angle* may be calculated, the critical clearing *time* remains unknown. There is a need to solve the swing equation for δ as a function of time.

A basic problem is described as follows. The generator is running at synchronous speed, stable, and loaded to some known amount. This state we shall call the prefault condition. At $t = 0$, an arbitrary switching action, called a fault, occurs in the external system. The fault is of the balanced 3ϕ type, requiring only the positive sequence network, and is typically, but not necessarily, a short circuit, as discussed in Chapter 8. The system in this state shall be identified as the faulted system. After some time has elapsed, at $t = T_c$ the fault is "cleared;" that is, a second switching action occurs, intended to remove or isolate the fault. The resulting configuration shall be referred to as the postfault system. This was the situation investigated in examples 12-3 and 12-4.

Since there are three system configurations there are three P_e functions, denoted as P_e^0, P_e', and P_e''; the prefault, faulted, and postfault functions. The prefault function is used to compute the initial $\delta(\delta_1)$. The faulted and postfault functions are to be used in the faulted $(0 < t < T_c)$ and postfault $(T_c < t)$ intervals. We are now prepared to discuss the time solution of this problem.

It is possible to solve the equation on the analog computer. Refer to Figure 12-13. The output δ can be recorded as a function of time. The function generators create P_e' and P_e'', the faulted and postfault P_e functions. Switching from P_e' to P_e'' is controlled by the function relay, which in turn is driven by the comparator. The comparator has a digital output that changes when the sum of inputs 1 and 2 makes the transition from negative to positive. The output of integrator 3 is a positive time ramp. When this signal equals the potentiometer output T_c, switching occurs. Therefore the adjustable T_c value correlates to clearing time. By examining the output δ for

434

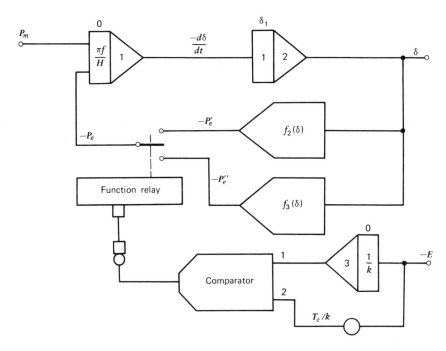

Figure 12-13 Analog solution of swing equation.

various T_c settings, stability can be ascertained. It is easy (and fun) to determine the critical clearing time by trial and error.

Digital methods may also be used to solve the swing equation. One popular method is the Runge–Kutta approach. Consider two first order differential equations in two variables x and y such that:

$$\frac{dx}{dt} = f(x, y) \tag{12-23a}$$

$$\frac{dy}{dt} = g(x, y) \tag{12-23b}$$

We start with known initial conditions x^0 and y^0 and a time step Δt. Compute the following eight constants:

$$k_1^0 = f(x^0, y^0)\, \Delta t \tag{12-24a}$$

$$l_1^0 = g(x^0, y^0)\, \Delta t \tag{12-24b}$$

$$k_2^0 = f(x^0 + \tfrac{1}{2}k_1^0, y^0 + \tfrac{1}{2}l_1^0)\, \Delta t \tag{12-24c}$$

435

$$l_2^0 = g(x^0 + \tfrac{1}{2}k_1^0, \ y^0 + \tfrac{1}{2}l_1^0) \, \Delta t \qquad (12\text{-}24\text{d})$$

$$k_3^0 = f(x^0 + \tfrac{1}{2}k_2^0, \ y^0 + \tfrac{1}{2}l_2^0) \, \Delta t \qquad (12\text{-}24\text{e})$$

$$l_3^0 = g(x^0 + \tfrac{1}{2}k_2^0, \ y^0 + \tfrac{1}{2}l_2^0) \, \Delta t \qquad (12\text{-}24\text{f})$$

$$k_4^0 = f(x^0 + k_3^0, \ y^0 + l_3^0) \, \Delta t \qquad (12\text{-}24\text{g})$$

$$l_4^0 = g(x^0 + k_3^0, \ y^0 + l_3^0) \, \Delta t \qquad (12\text{-}24\text{h})$$

We use these eight constants to estimate the change in x and y as follows:

$$\Delta x^0 = \tfrac{1}{6}(k_1^0 + 2k_2^0 + 2k_3^0 + k_4^0) \qquad (12\text{-}25\text{a})$$

$$\Delta y^0 = \tfrac{1}{6}(l_1^0 + 2l_2^0 + 2l_3^0 + l_4^0) \qquad (12\text{-}25\text{b})$$

The values t, x, and y are updated as follows:

$$t^1 = 0 + \Delta t \qquad (12\text{-}26\text{a})$$

$$x^1 = x^0 + \Delta x^0 \qquad (12\text{-}26\text{b})$$

$$y^1 = y^0 + \Delta y^0 \qquad (12\text{-}26\text{c})$$

We then replace x^0 and y^0 with x^1 and y^1 and recalculate the k's, l's, Δx, and Δy. In general

$$t^{k+1} = (k+1) \, \Delta t \qquad (12\text{-}27\text{a})$$

$$x^{k+1} = x^k + \Delta x^k \qquad (12\text{-}27\text{b})$$

$$y^{k+1} = y^k + \Delta y^k \qquad (12\text{-}27\text{c})$$

The calculations are best done by computer.

Now we will apply this method to the simplified single machine/infinite bus transient stability problem. We assign the following:

$$x = \delta \qquad (12\text{-}28\text{a})$$

$$y = \alpha = \frac{d\delta}{dt} \qquad (12\text{-}28\text{b})$$

It follows that

$$\frac{d\delta}{dt} = f(\delta, \alpha) = \alpha \qquad (12\text{-}29\text{a})$$

$$\frac{d\alpha}{dt} = g(\delta, \alpha) = \frac{\pi f}{H}(P_m - P_e) \qquad (12\text{-}29\text{b})$$

where

$$P_e = P_{max} \sin \delta \qquad (12\text{-}30\text{a})$$

$$= \frac{E_q' E}{X_d' + X} \sin \delta \qquad (12\text{-}30\text{b})$$

436

The initial condition (δ_1) is computed using prefault values:

$$P_m = \frac{E'_q E_1}{X'_d + X_1} \sin \delta_1 \qquad (12\text{-}31a)$$

$$\therefore \quad \delta_1 = \sin^{-1} \frac{P_m(X'_d + X_1)}{E'_q E_1} \qquad (12\text{-}31b)$$

For $0 < t < T_c$:

$$P_e = P'_e = \frac{E'_q E_2}{X'_d + X_2} \sin \delta \qquad (12\text{-}32)$$

For $t > T_c$:

$$P_e = P''_e = \frac{E'_q E_3}{X'_d + X_3} \sin \delta \qquad (12\text{-}33)$$

Consider the following example.

Example 12-5

Consider the system of example 12-1. Solve for and plot δ versus time for several clearing times, and therefore determine the critical clearing time T_{cc} by trial and error. The inertia constant H is 3.0. Use a time step $\Delta t = 0.02$ sec.

Solution

From example 12-3 we recall

$\delta_1 = 0.524$ rad

$P'_e = 1.2 \sin \delta$

$P''_e = 2.0 \sin \delta$

$P_m = 1.5$

The initial conditions are

$\delta^0 = \delta_1 = 0.524$

$\alpha^0 = 0$

$f(\delta, \alpha) = \alpha$

$g(\delta, \alpha) = \dfrac{\pi f}{H}(P_m - P_e)$

$\qquad = 62.83(P_m - P_e)$

437

The problem was programmed and solved by computer. Results appear in Table 12-1 and Figure 12-14. By trial and error we find:

$$0.15 \le T_{cc} \le 0.17$$

$$1.026 \le \delta_{cc} \le 1.307$$

This agrees with $\delta_{cc} = 1.196$ as determined by the equal area method.

Table 12-1 Computer calculated swing curve data for example 14-5.

LAST STABLE RUN. TC = 0.15 SECOND			FIRST UNSTABLE RUN. TC = 0.17 SECOND		
**** OUTPUT DATA ****			**** OUTPUT DATA ****		
TIME	DELTA	ALPHA	TIME	DELTA	ALPHA
0.000	0.524	0.000	0.000	0.524	0.000
0.020	0.535	1.127	0.020	0.535	1.127
0.040	0.569	2.225	0.040	0.569	2.225
0.060	0.624	3.266	0.060	0.624	3.266
0.080	0.699	4.228	0.080	0.699	4.228
0.100	0.792	5.093	0.100	0.792	5.093
0.120	0.902	5.851	0.120	0.902	5.851
0.140	1.026	6.501	0.140	1.026	6.501
FAULT CLEARED			0.160	1.161	7.049
0.160	1.161	7.049	FAULT CLEARED		
0.180	1.298	6.568	0.180	1.307	7.514
0.200	1.423	5.998	0.200	1.452	6.934
0.220	1.537	5.382	0.220	1.584	6.312
0.240	1.639	4.757	0.240	1.704	5.694
0.260	1.728	4.147	0.260	1.812	5.113
0.280	1.805	3.568	0.280	1.909	4.592
0.300	1.871	3.031	0.300	1.996	4.148
0.320	1.926	2.539	0.320	2.076	3.789
0.340	1.973	2.091	0.340	2.149	3.522
0.360	2.010	1.684	0.360	2.217	3.352
0.380	2.040	1.313	0.380	2.283	3.284
0.400	2.063	0.972	0.400	2.349	3.324
0.420	2.079	0.653	0.420	2.417	3.481
0.440	2.089	0.351	0.440	2.489	3.770
0.460	2.093	0.057	0.460	2.569	4.211
0.480	2.092	− 0.235	0.480	2.659	4.830
0.500	2.084	− 0.533	0.500	2.764	5.665
0.520	2.070	− 0.844	0.520	2.887	6.766
0.540	2.050	− 1.176	0.540	3.036	8.197
0.560	2.023	− 1.534	0.560	3.218	10.040
0.580	1.988	− 1.927	0.580	3.441	12.385
0.600	1.946	− 2.358	0.600	3.717	15.322

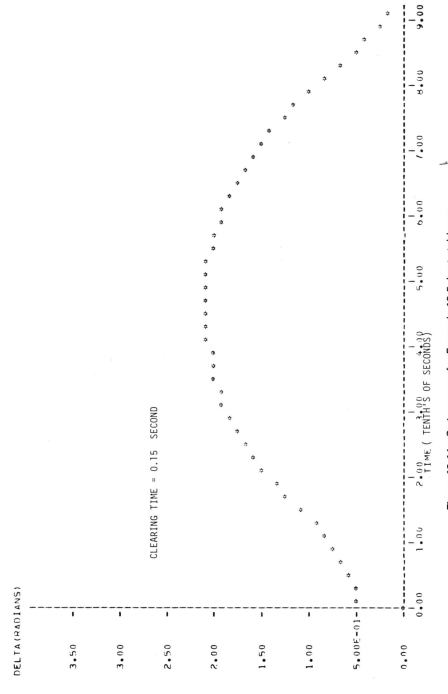

Figure 12-14a Swing curve for Example 12-5. Last stable run.

439

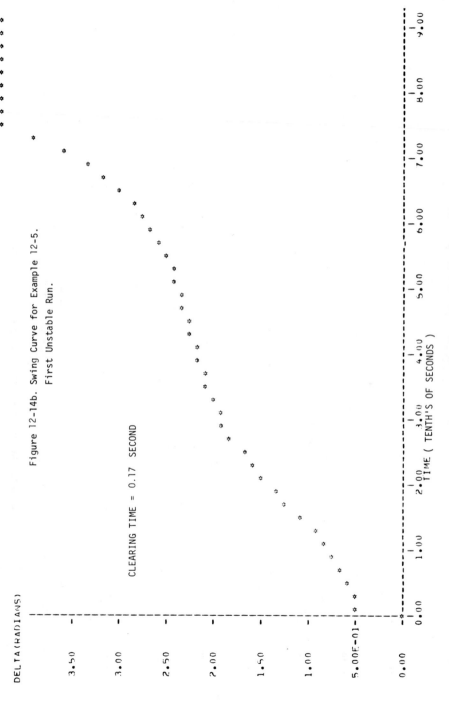

Figure 12-14b. Swing Curve for Example 12-5. First Unstable Run.

CLEARING TIME = 0.17 SECOND

Figure 12-14b Swing curve for Example 12-5. First unstable run.

12-5 Multimachine Systems

The reduction of the external system to a Thevenin equivalent, including an infinite bus, was of course a simplifying approximation. The sources in the external network are also synchronous machines similar to the one under investigation, and are subject to transient stability problems. The more general problem is then to consider a complete system. A system circuit model is shown in Figure 12-15. The n nodes shown are those that represent

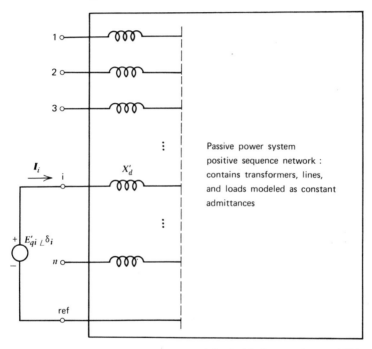

Figure 12-15 Power system circuit model suitable for transient stability analysis.

generators. Modeling the generators with the simple circuit of Figure 12-4 permits a particularly simple approach. We define the n nodes at the tops of the source portion (E'_{qi}) of the equivalent circuit. The generator terminals are internal to the network. These, and all other internal system busses, we shall number consecutively from $n+1$ to $n+m$, where m is the total number of internal nodes. The entire boxed-in network is passive, assuming that we model system loads as constant admittances. We write:

$$\begin{bmatrix} \tilde{I}_s \\ \tilde{0} \end{bmatrix} = [Y'] \begin{bmatrix} \tilde{E} \\ \tilde{V} \end{bmatrix} \tag{12-34a}$$

441

where

$$\tilde{I}_s = \begin{bmatrix} I_1 \\ \vdots \\ I_n \end{bmatrix} = \text{Source current vector } (n \times 1) \qquad (12\text{-}34b)$$

$\tilde{0}$ is $m \times 1$ $\qquad\qquad\qquad (12\text{-}34c)$

$$\tilde{E} = \begin{bmatrix} E_1 \\ \vdots \\ E_n \end{bmatrix} = \text{Generator internal source voltage vector } (n \times 1) \qquad (12\text{-}34d)$$

with general entry

$$E_i = E'_{qi}\underline{/\delta_i}$$

$$\tilde{V} = \begin{bmatrix} V_{n+1} \\ \vdots \\ V_{n+m} \end{bmatrix} = \text{Internal bus voltage vector } (m \times 1) \qquad (12\text{-}34e)$$

$$[Y'] = \begin{bmatrix} [Y_{AA}] & \vdots & [Y_{AB}] \\ \hline [Y_{BA}] & \vdots & [Y_{BB}] \end{bmatrix} \qquad (12\text{-}34f)$$

$$= \text{Partitioned admittance matrix } (n+m) \times (n+m)$$

$[Y_{AA}]$ is $n \times n$ $\qquad\qquad\qquad (12\text{-}34g)$

$[Y_{AB}]$ is $n \times m$ $\qquad\qquad\qquad (12\text{-}34h)$

$[Y_{BA}]$ is $m \times n$ $\qquad\qquad\qquad (12\text{-}34i)$

$[Y_{BB}]$ is $m \times m$ $\qquad\qquad\qquad (12\text{-}34j)$

From equation (12-34a) we write

$$\tilde{0} = [Y_{BA}]\tilde{E} + [Y_{BB}]\tilde{V} \qquad (12\text{-}35a)$$

Solving for $\tilde{V}$:

$$\tilde{V} = -[Y_{BB}]^{-1}[Y_{BA}]\tilde{E} \qquad (12\text{-}35b)$$

Again, from (12-34a):

$$\tilde{I} = [Y_{AA}]\tilde{E} + [Y_{AB}]\tilde{V} \qquad (12\text{-}36a)$$

Substituting (12-35b) into (12-36a):

$$\tilde{I} = \{[Y_{AA}] - [Y_{AB}][Y_{BB}]^{-1}[Y_{BA}]\}\tilde{E} \tag{12-36b}$$

$$= [Y]\tilde{E} \tag{12-36c}$$

where

$$[Y] = [Y_{AA}] - [Y_{AB}][Y_{BB}]^{-1}[Y_{BA}] \tag{12-36d}$$

$= n \times n$ reduced admittance matrix with general entry y_{ij}.

The Y matrix defined in equation (12-36d) is similar to the Y bus matrix used in Chapter 7. The only difference is that the generator impedance values (X'_d) and loads are included. In the typical transient stability situation $[Y]$ may be thought of as having three states: prefault, faulted, and postfault. The prefault state is involved in the determination of initial conditions for the δ_i angles. The faulted state exists at $t = 0$, and persists until the fault is cleared, at $t = T_c$. For $t > T_c$, the postfault $[Y]$ is used. An example problem should clarify matters.

Example 12-6

For the system of example 12-1 compute the prefault, faulted, and postfault $[Y]$ matrices. Number the single generator source as one and the infinite bus as two.

Solution

An equivalent circuit corresponding to Figure 12-15 appears in Figure 12-16. The three states (prefault, faulted, and postfault) are created by proper positioning of the switches. Note the internal bus 3.

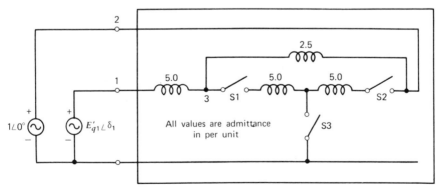

Figure 12-16 Circuit diagram for Example 12-6.

443

Prefault Condition: (S1, S2 closed; S3 open)

$$[Y'] = \begin{bmatrix} -j5 & 0 & +j5 \\ 0 & -j5 & +j5 \\ +j5 & +j5 & -j10 \end{bmatrix}$$

$$[Y] = \begin{bmatrix} -j5 & 0 \\ 0 & -j5 \end{bmatrix} + \frac{j}{10}\begin{bmatrix} 5 \\ 5 \end{bmatrix}[5 \quad 5]$$

$$= \begin{bmatrix} -j2.5 & +j2.5 \\ +j2.5 & -j2.5 \end{bmatrix}$$

Faulted Condition: (S1, S2, S3 closed)

$$[Y'] = \begin{bmatrix} -j5 & 0 & j5 \\ 0 & -j7.5 & j2.5 \\ j5 & j2.5 & -j12.5 \end{bmatrix}$$

$$[Y] = \begin{bmatrix} -j5 & 0 \\ 0 & -j7.5 \end{bmatrix} + \frac{j}{12.5}\begin{bmatrix} 5 \\ 2.5 \end{bmatrix}[5 \quad 2.5]$$

$$= \begin{bmatrix} -j3 & +j1 \\ +j1 & -j7 \end{bmatrix}$$

Postfault Condition: (S1, S2 open; S3 closed)

$$[Y'] = \begin{bmatrix} -j5 & 0 & +j5 \\ 0 & -j2.5 & +j2.5 \\ +j5 & +j2.5 & -j7.5 \end{bmatrix}$$

$$[Y] = \begin{bmatrix} -j5 & 0 \\ 0 & -j2.5 \end{bmatrix} + \frac{j}{7.5}\begin{bmatrix} 5 \\ 2.5 \end{bmatrix}[5 \quad 2.5]$$

$$= \begin{bmatrix} -j1.667 & +j1.667 \\ +j1.667 & -j1.667 \end{bmatrix}$$

The electrical power injected into the system from the ith machine is derived in the same manner as was equation (7-18a). We get:

$$P_{e_i} = \sum_{j=1}^{n} E_i E_j y_{ij} \cos(\delta_i - \delta_j - \gamma_{ij}) \tag{12-37}$$

We apply this to an example.

Example 12-7

Example 12-7

Continue example 12-6 by evaluating P_e for the generator in the prefault, faulted, and postfault conditions.

Solution

This is a two generator ($n = 2$) system, with bus 1 representing the generator under study and bus 2 representing the infinite bus. From example 12-1:

$$E_1 = E'_{q_1} = 1.2 \qquad \delta_1 = \delta$$

$$E_2 = 1.0 \qquad \delta_2 = 0$$

Prefault Condition
From example 12-6:

$$y_{11} = 2.5 \underline{/-90°}$$

$$y_{12} = 2.5 \underline{/+90°}$$

$$P_{e_1} = \sum_{j=1}^{2} E_1 E_j y_{1j} \cos(\delta_1 - \delta_j - \gamma_{1j})$$

$$= E_1^2 y_{11} \cos(-\gamma_{11}) + E_1 E_2 y_{12} \cos(\delta_1 - \gamma_{12})$$

$$= (1.2)^2 (2.5) \cos(+90°) + 1.2(1.0)(2.5) \cos(\delta - 90°)$$

$$= 0 + 3 \sin \delta$$

$$= 3 \sin \delta$$

which agrees with the result computed in example 12-1.

Faulted Condition
From example 12-6:

$$y_{11} = 3 \underline{/-90°}$$

$$y_{12} = 1 \underline{/90°}$$

$$P''_{e_1} = E_1^2 y_{11} \cos(+90°) + E_1 E_2 y_{12} \cos(\delta - 90°)$$

$$= 0 + (1.2)(1.0)(1.0) \sin \delta$$

$$= 1.2 \sin \delta$$

which agrees with the result computed in example 12-3.

445

Postfault Condition
From example 12-6:

$$y_{11} = 1.667\underline{/-90°}$$

$$y_{12} = 1.667\underline{/+90°}$$

$$P''_{e_1} = E_1^2 y_{11} \cos(+90°) + E_1 E_2 y_{12} \cos(\delta - 90°)$$

$$= 0 + (1.2)(1.0)(1.667) \sin \delta$$

$$= 2.0 \sin \delta$$

which agrees with the result calculated in example 12-3.

Each machine has its individual equation of motion. From equation (12-12) for the ith machine we write:

$$P_{m_i} - P_{e_i} = \frac{H_i}{\pi f} \left[\frac{d^2 \delta_i}{dt^2} + \frac{d\omega_r}{dt} \right] \tag{12-38}$$

Recall that ω_r is the angular velocity of an arbitrary rotating reference. We shall select the rotor of one of our machines as reference (r). Then by definition

$$\delta_i \big]_{i=r} = \delta_r = 0 \tag{12-39}$$

and

$$P_{m_r} - P_{e_r} = \frac{H_r}{\pi f} \frac{d\omega_r}{dt} \tag{12-40a}$$

$$\therefore \quad \frac{d\omega_r}{dt} = \frac{\pi f}{H_r} [P_{m_r} - P_{e_r}] \tag{12-40b}$$

Substituting (12-40b) into (12-38) and rearranging:

$$\frac{d^2 \delta_i}{dt^2} = \frac{\pi f}{H_i} \left[\left(P_{m_i} - \frac{H_i}{H_r} P_{m_r} \right) - \left(P_{e_i} - \frac{H_i}{H_r} P_{e_r} \right) \right] \tag{12-41}$$

A common power base must be selected for all machines. We solve the equation set by the Runge–Kutta method. Consider:

$$\frac{d\delta_i}{dt} = \alpha_i \tag{12-42a}$$

$$\frac{d\alpha_i}{dt} = \frac{\pi f}{H_i} \left[\left(P_{m_i} - \frac{H_i}{H_r} P_{m_r} \right) - \left(P_{e_i} - \frac{H_i}{H_r} P_{e_r} \right) \right] \tag{12-42b}$$

Example 12-7

A set of k's and l's (equation 12-24) is calculated for each machine. The time step Δt is chosen to be compatible with the rate of change of the swing curves, which in turn depend on H. The initial condition on $\tilde{\alpha}$ is zero. Initial values for E_i and δ_i are functions of system prefault loading and can be evaluated from the results of a power flow study. As in the single machine case, stability is determined by examining the δ, t plots as calculated for a particular clearing time. An example four bus system follows.

Example 12-8

For the system of Figure 12-17 investigate transient stability problems caused by a 3ϕ fault applied at bus 10. Specifically determine the critical clearing time. The fault is removed by opening breakers B1 and B2 simultaneously. Machine data is supplied in Table 12-2.

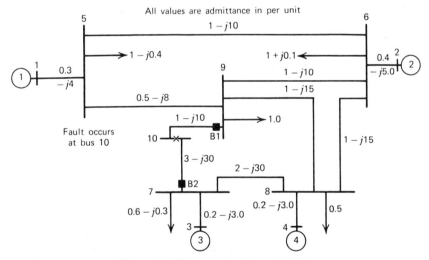

Figure 12-17 System for Example 12-6.

Table 12-2 Machine data for example 12-6.

	E_i PER UNIT	δ_i^0 RADIANS	P_{mi} PER UNIT	$H_i/\pi f$ SECONDS^2
Machine 1	1.30	0.0	0.185	0.05
Machine 2	1.40	0.2	1.556	0.04
Machine 3	1.35	0.6	2.357	0.03
Machine 4	1.20	0.5	1.793	0.03

Solution

The problem was solved by using a computer program that implemented the ideas presented in this section. The system admittance data is given in Figure 12-17. Examine the swing curves presented in Figure 12-18. The fault was cleared at 0.20 seconds. Observe that machine 3 is experiencing the most violent swing. This is predictable if we recognize that it is close to the fault, has the largest δ^0, and has a fairly low H_i. We would expect it to be the first to go unstable.

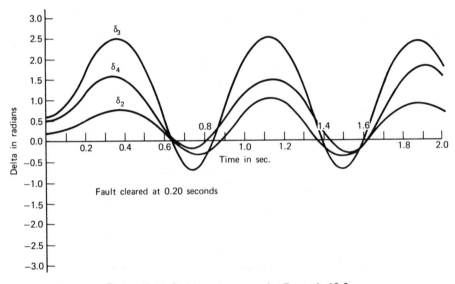

Figure 12-18 Stable swing curve for Example 12-6.

Now observe the situation illustrated in Figure 12-19, when the fault is cleared at 0.22 seconds. We observe an interesting phenomenon. If a system is going to be unstable, the machines usually will pull out of synchronism on the first swing. The reason for this is that the damping effects of the system usually insure that the first swing is the most severe. There is no guarantee, however, that the machine cannot drop out of synchronism during a later swing. As shown in Figure 12-19 the fault is cleared at $t = 0.22$ and the system survives the first swing, but synchronism is lost on the second swing. Clearing at $t = 0.24$ produces instability on the first swing, as shown in Figure 12-20. We would conclude $0.20 < T_{cc} < 0.22$ for this case.

We assume the system is unstable when and if $|\delta_i| > \pi$. We can claim no real accuracy for computation much past the first swing because many of the

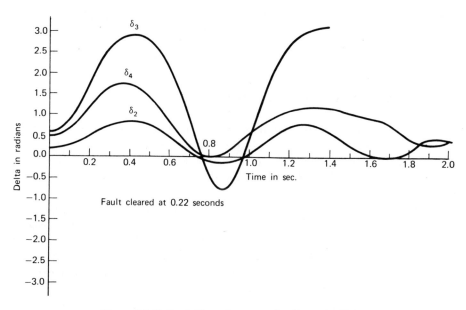

Figure 12-19 Unstable swing curve for Example 12-6.

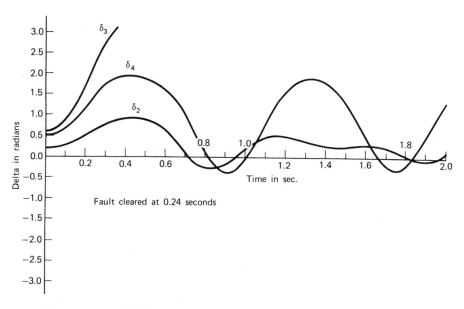

Figure 12-20 Unstable swing curve for Example 12-6.

simplifying assumptions are valid only for a short duration. One of the benefits of a computer approach to the solution is that more sophisticated models of system components can be used.

12-6 Generator Voltage Control

In the preceding sections we have held the generator internal voltage constant during transient stability calculations. If there is a need to make transient calculations over a long duration, the transient response of the voltage control loop should be considered. One second is a reasonable time limit to use for most systems.

Refer back to Figure 6-10 and concentrate on the voltage control loop. The source of generator field current is the exciter, whose output E_{fd} is applied to the generator rotor field windings. Physically, the exciter would be a dc generator or solid state polyphase rectifier. We will develop the equations for the dc generator. The exciter in this case is a dc self excited shunt generator. A control voltage V_a (the output of an amplifier) is serially inserted in the exciter field circuit for the purpose of controlling the exciter field current. Since it either adds or subtracts to E_{fd}, this arrangement is sometimes called a "buck or boost" system. An appropriate circuit model is shown in Figure 12-21a.

Observe that E_{fd} is a nonlinear function of the field current I_f. The functional relationship is the magnetization curve shown in Figure 12-21b. The air gap line is the linear extension of the magnetization curve with slope K_g. We divide the field current I_f into two components I_{fo} and ΔI_f, with the additional current ΔI_f required by saturation. Considering V_a to be zero, self excitation of the exciter is achieved when $K_g > R_f$, with R_f the field circuit resistance. Observe that if the exciter magnetization curve is linearized, the system is unstable; saturation is required to pull it into a stable mode.

In formulating the dynamic equations we have a notational problem. Recall in Chapter 6 we talked about "transient ac" signals, ac quantities with time varying magnitudes. The exciter variables are essentially dc, but do change under transient conditions. In the same sense think of E_{fd}, I_f, and so on, as "transient dc" values—that is, quantities that vary in time from normally fixed levels. We will use capital letters for these signals. It is also convenient to use operational calculus notation: the quantity p will represent time differentiation ($p = d/dt$).

We write

$$V_a - R_f I_f = V_l - E_{fd} \qquad\qquad (12\text{-}43\text{a})$$

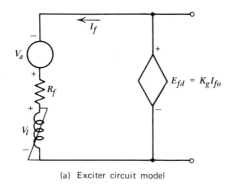

(a) Exciter circuit model

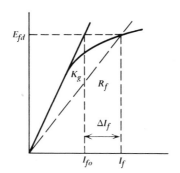

(b) Magnetization curve for the exciter

Figure 12-21 Exciter modeling.

where

$$I_f = I_{fo} + \Delta I_f \tag{12-43b}$$

$$V_l = pL_fI_{fo} \tag{12-43c}$$

$$E_{fd} = K_gI_{fo} \tag{12-43d}$$

$$p = \frac{d}{dt} \tag{12-43e}$$

It follows that

$$V_a - R_f\Delta I_f = (L_fp + R_f - K_g)I_{fo} \tag{12-44a}$$

$$= \frac{L_fp + R_f - K_g}{K_g}E_{fd} \tag{12-44b}$$

451

$$= \frac{\left(\dfrac{L_f}{K_g - R_f}\right)(p - 1)E_{fd}}{K_g/(K_g - R_f)} \tag{12-44c}$$

$$= \frac{T_e p - 1}{K_e} E_{fd} \tag{12-44d}$$

where

$$T_e = \frac{L_f}{K_g - R_f} = \text{Exciter time constant} \tag{12-44e}$$

$$K_e = \frac{K_g}{K_g - R_f} = \text{Exciter gain} \tag{12-44F}$$

The minus one is caused by the fact that $K_g > R_f$. We represent equation (12-44) in block diagram form as shown in Figure 12-22.

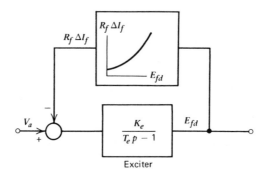

Figure 12-22 Exciter block diagram.

Operation of the exciter as viewed using the block diagram is interesting. Since the forward transfer function has a pole in the right half plane, the system appears to be unstable. For any nonzero error signal E_{fd} will increase. However, as E_{fd} increases, the feedback signal $R_f \Delta I_f$ increases at a greater rate. A stable equilibrium will be established when $R_f \Delta I_f$ matches the amplifier output V_a.

Observe that the nonlinear fedback function is due to saturation and derivable from Figure 12-21b. There are several possible mathematical functions that may be used to approximate saturation. One is the exponential expression

$$R_f \Delta I_f = A \, e^{BE_{fd}} \tag{12-45}$$

where A and B are constants determined from the device characteristics. The signal V_a is the output of an amplifier whose output is governed by the following relations

$$V_{\min} \le V_a \le V_{\max} \tag{12-46a}$$

$$pV_a = \frac{K_a}{T_a} V_1 - \frac{1}{T_a} V_a \tag{12-46b}$$

where

$V_1 =$ amplifier input voltage

Rearranging

$$V_a = \frac{K_a}{T_a p + 1} V_1 \tag{12-47}$$

The amplifier receives its input from the regulator, which in turn is driven by a difference signal $V_i - V$. Its transfer function is similar to the amplifier:

$$V_r = \frac{1}{T_r p + 1}(V_i - V) \tag{12-48}$$

where

$V_r =$ Regulator output voltage

$T_r =$ Regulator time constant

$V_i =$ Reference input voltage

$V =$ Generator output voltage

Analysis shows that the system as described has a dynamic response that is prone to excessive overshoot and stability problems. This can be corrected by the addition of a stabilizing transformer whose input is E_{fd} and whose output is V_f. The output V_f is subtracted from the regulator output V_r to provide the input to the amplifier V_1. An equivalent circuit for such a device is shown in Figure 12-23. We write

$$I_1 = \frac{E_{fd}}{R_1 + L_1 p} \tag{12-49}$$

$$V_f = MpI_1 \tag{12-50a}$$

$$= \frac{MpE_{fd}}{R_1 + L_1 p} \tag{12-50b}$$

$$= \frac{K_{st} p}{T_{st} p + 1} E_{fd} \tag{12-50c}$$

453

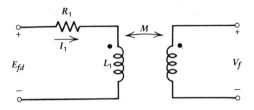

Figure 12-23 Circuit model for stabilizing transformer.

where

$$K_{st} = \frac{M}{R_1} = \text{Transformer gain} \qquad (12\text{-}50\text{d})$$

$$T_{st} = \frac{L_1}{R_1} = \text{Transformer time constant} \qquad (12\text{-}50\text{e})$$

The complete voltage control system is shown in Figure 12-24. Operation is as follows. Suppose the system is in equilibrium when the generator terminal voltage suddenly drops. The difference voltage $V_i - V$ increases, which after time delay causes an increase in V_a. The exciter will then adjust to a new and larger equilibrium value for E_{fd}. This in turn increases the generator field current that ultimately should raise V. The IEEE working group on exciters has categorized voltage control systems into several standard types, all of which are similar to the one discussed (see [7] in the end-of-chapter Bibliography).

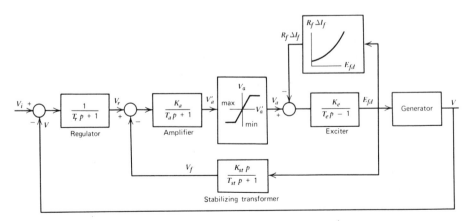

Figure 12-24 Block diagram for generator voltage control loop.

The last item that remains is to relate the effect of changes in E_{fd} to the generator terminal voltage. The equations interrelating V, E_f, and E'_q have already been discussed in section 12-2. The voltages E_f and E_{fd} are scaled in per unit in such a way that under steady state conditions $E_f = E_{fd}$. Under transient conditions any mismatch between E_f and E_{fd} will cause the voltage E'_q to change after some delay. Mathematically

$$pE'_q = \frac{1}{T'_{do}}(E_{fd} - E_f) \qquad (12\text{-}51)$$

where

T'_{do} = Open circuit generator direct axis transient time constant.

It is possible then to account for the transient response of the voltage control loop in the transient stability problem using the equations represented in the block diagram of Figure 12-24 and equations (12-13, 12-16, and 12-51). We shall not discuss the particulars; the reader is referred to [7] and [11] in the end-of-chapter Bibliography.

12-7 Turbine Speed Control

Our assumption that the generator input power P_m is constant was based on the contention that the turbine speed control loop response was too slow to be considered. If this is not the case, or if the transient calculation is to proceed over a prolonged period of time, then the turbine/governor dynamics should be considered.

A simplified governor/turbine simulation is presented in the block diagram shown in Figure 12-25. System parameters are defined as follows:

$$R = \frac{\Delta P_m}{\Delta \omega} = \text{System speed regulation}$$

T_c = Valve control time constant

T_s = Steam system time constant

The power supplied by the turbine is essentially controlled by the reference input value P_{m0}. Normally the speed error $\Delta \omega$ is zero. If it is desired to load the generator P_{m0} is increased. Since ΔP_m is zero, P'_{m1} equals P_{m0} and the signal is checked to see if it is within allowable limits. After a time delay introduced by the steam and control system dynamics P_m settles to P_{m0}.

Now consider a transient stability problem that causes ω to vary, for example, increase. The velocity error $\Delta \omega$ is now negative, causing ΔP_m to be

455

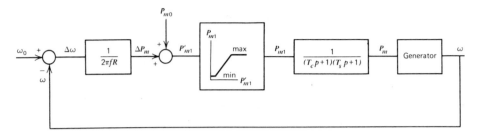

Figure 12-25 Turbine/governor speed control loop.

negative. P'_{m1} drops, as does P_{m1} assuming P'_{m1} is within limits. The output P_m will reduce, which in turn will tend to reduce ω, and therefore have an effect on transient stability.

12-8 Summary

We have discussed the transient stability problem for single and multiple machines. To avoid obscuring the stability problem, a particularly simple model for the machine was selected. In advanced work more refined models for the machine would be used. The student should postpone such study until he or she has acquired an in-depth understanding of synchronous machines.

When machine transient saliency $(X'_d \neq X_q)$ is considered, a complication arises. The equivalent circuit of Figure 12-4 is not applicable; in fact it is not possible to synthesize any conventional equivalent circuit. In such cases the expression for P_e is more involved:

$$P_e = \frac{E'_q V}{X'_d} \sin \delta + \frac{V^2}{2}\left(\frac{1}{X_q} - \frac{1}{X'_d}\right) \sin 2\delta \qquad (12\text{-}52)$$

See problem 12-4 for proof. Since the machine does not have an equivalent circuit, an arrangement as suggested in Figure 12-15 is not possible; the internal machine reactance X'_d cannot be included in $[Y']$. We then formulate $[Y']$, stopping at the generator terminals and use equations (12-16) and (12-52) to model the generator.

The direct approach to transient stability involves solution of the swing equation and study of the corresponding δ, t curves. The equal area method is an indirect approach, but unfortunately applicable only to the single machine problem. There are indirect methods applicable to the multi-machine problem, such as Liapunov's method, that are sometimes used.

It is probably surprising that ac methods may be used to analyze a transient situation. The time variations are of such rates that this approach is possible. It is important to realize that although machine speed (and therefore system frequency) changes, it doesn't change much relative to synchronous, and therefore we may use torque and power interchangeably and treat reactance constant with little error. Most of our approximations lose their validity for longer simulation times, and calculated results past about one second should be viewed with considerable skepticism.

The methods presented for solving the transient stability problem are amenable to refinement and modification to include exciter and governor effects, as well as better generator models. However, one should balance the benefits of higher accuracy against the difficulties of procuring accurate parameter data and increased complexity of calculation.

Large rotor swings will cause the voltage and speed control loops to respond. Tap changing transformers will react to large voltage and power fluctuations, as will protective relay schemes. It is possible that combinations of control actions taken will aggravate the system stability problem. This longer term problem, with time constants running into minutes, is referred to as the dynamic stability problem. This too is an important advanced study area within power system analysis.

Bibliography

[1] Brown, Homer E., *Solution of Large Networks by Matrix Methods*, Wiley, New York, 1975.

[2] Crary, Seldon B., *Power System Stability*, Vol. I, Wiley, New York, 1945.

[3] El-Abiad, A. H., and K. Hagappan, "Transient Stability Regions of Multimachine Power Systems,", *IEEE Transactions on PAS*, Feb. 1966, vol. PAS-85, No. 2, pp. 169–179.

[4] Elgerd, Olle I., *Electric Energy Systems Theory: An Introduction*, McGraw-Hill Inc., New York, 1971.

[5] Gless, G. E., "The Direct Method of Liapunov Applied to Transient Power System Stability," *IEEE Transactions on PAS*, Feb. 1966, vol. PAS-85, No. 2, pp. 159–168.

[6] Hahn, Wolfgang, *Theory and Application of Liapunov's Direct Method*. Prentice Hall, Inc., Englewood Cliffs, N.J., 1963.

[7] IEEE Committee Report, "Computer Representation of Excitation Systems," *IEEE Transactions on PAS*, vol. 87, June 1968, pp. 1460–1464.

[8] IEEE, *IEEE Standard Dictionary of Electrical and Electronics Terms*, Wiley-Interscience, New York, 1972.

[9] Kimbark, E. W., *Power System Stability*, Vol. 1, Wiley, New York, 1948.

[10] Neuenswander, John R., *Modern Power Systems*, International Textbook Co., Scranton, Pa., 1971.

[11] Stagg, Glenn W., and El-Abiad, Ahmed H., *Computer Methods in Power System Analysis*. McGraw-Hill, Inc., New York, 1968.

[12] Stevenson, Jr., William D., *Elements of Power Systems Analysis*, 3rd edition. McGraw-Hill, Inc., New York, 1975.

[13] Undrill, John M., "Dynamic Stability Calculations for an Arbitrary Number of Interconnected Synchronous Machines," *IEEE Transactions on PAS*, March 1968, vol. PAS-87, No. 3, pp. 835–844.

[14] Weedy, B. M., *Electric Power Systems*, 2nd edition. John Wiley and Sons Ltd., London, 1972.

[15] Yu, Yao-Nan and Khien Vongsuriya, "Non-Linear Power System Stability Study by Liapunov Function and Zubov's Method," *IEEE Transactions on PAS*, Dec. 1967, vol. PAS-86, No. 12, pp. 1480–1485.

[16] Zubov, V. I., *Methods of A. M. Lyapunov and Their Application*, Translated from the publishing house of Leningrad University, U.S. Atomic Energy Commission, Divisions of Technical Information, pp. 63–148, 1957.

Problems

12-1. The following equation is sometimes given for H:

$$H = \frac{2.31 \times 10^{-10} WR^2 (\text{rpm})^2}{S_{3\phi \text{ rating}}}$$

where

$W = $ Weight of rotating parts of generator in lbs.

$R = $ Radius of gyration in feet

rpm = synchronous speed in rpm

$S_{3\phi \text{ rating}} = $ Rated $S_{3\phi}$ in MVA

Derive the expression.

12-2. Given a synchronous machine of known H running unloaded at synchronous speed. Show that it will stop in $2H$ seconds if its rotor is subjected to a constant decelerating torque equal to T_{rated}.

12-3. A four pole 100 MVA 60 Hz generator has $H = 2.0$ seconds. Calculate J in kg-m^2.

12-4. Recall equations (12-15) and (12-16):

$$E'_q = V + jX_q I + j(X'_d - X_q)I_d$$

where for our purposes let

$$E'_q = E'_q \underline{/\delta}$$
$$V = V \underline{/0°}$$

(a) Prove that if $X'_d = X_q$

$$P_e = \frac{E'_q V}{X'_d} \sin \delta$$

(b) Prove that if $X'_d \neq X_q$

$$P_e = \frac{E'_q V}{X'_d} \sin \delta + \frac{V^2}{2}\left(\frac{1}{X_q} - \frac{1}{X'_d}\right) \sin 2\delta$$

12-5. Consider the system of Figure P12-5. If the infinite bus absorbs $S = 1.0 + j0.2$, prove that the corresponding values for E'_q and δ are 1.152 and 20.3°.

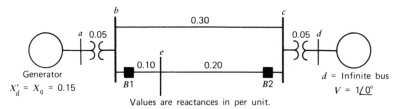

Figure **P12-5** System for Problem 12-5.

12-6. With the system of Figure P12-5 loaded as described in problem 12-5, breaker B1 inadvertently opens. Calculate the maximum rotor swing angle δ_3.

12-7. With the system of Figure P12-5 loaded as described in problem 12-5 calculate the critical clearing angle, given that a 3ϕ fault occurs at bus e, which is subsequently cleared by opening B1 and B2 simultaneously.

12-8. Continue problem 12-7 by writing a computer program to solve for the critical clearing time using the techniques discussed in section 12-4.

12-9. View the system of Figure P12-5 as a multimachine system. Treat the generator as machine 1 and the infinite bus as machine 2. Absorb the interior nodes a and e by series reduction, leaving nodes b and c ($m = 2$).

(a) Formulate prefault, faulted, and postfault $[Y']$ matrices, as understood from problem 12-7. $[Y']$ is 4×4.

(b) Calculate prefault, faulted, and postfault $[Y]$ matrices using equation (12-36d).

(c) For machine 1 determine prefault, faulted, and postfault expressions for P_e using equation (12-37). Use $E'_q = 1.152$ for E_1.

12-10. Determine values for A and B as used in equation (12-45) from the characteristics given in Figure P12-10.

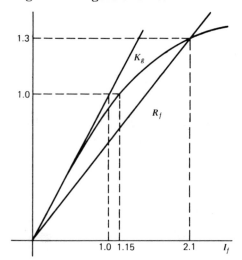

Figure P12-10 Exciter magnetization curves for Problem 12-10.

12-11. Suppose a generator has $X_d = 1.0$ pu and $X_q = 0.8$ pu. Calculate E_f from equation (12-13) if it is operating at rated terminal conditions and unity power factor. What is the corresponding E_{f_a} value?

12-12. Suppose the generator of Problem 12-11 has a voltage control system as diagrammed in Figure 12-24. Further suppose that its exciter constants are $K_g = 1.0$, $R_f = 0.619$, $A = 0.00035$, $B = 5.58$, and $T_e = 5$ seconds. For the unity pf case described in problem 12-11 determine the steady state amplifier output signal V_a.

12-13. Continue problem 12-12 and determine the input signal V_i if $K_a = 100$.

12-14. Write a computer program to simulate the transient response of the voltage control system of Figure 12-24. Use data from problems 12-11, 12-12, and 12-13 and additionally

$$V_{a_{max}} = +1.0 \qquad T_r = 0.03 \text{ seconds}$$

$$V_{a_{min}} = -1.0 \qquad T'_{do} = 0.50 \text{ seconds}$$

$$T_a = 0.10 \text{ seconds} \qquad X'_d = 0.30$$

$$K_{st} = 0.05$$

$$T_{st} = 0.50 \text{ seconds}$$

A specific problem to investigate would be to consider the loaded unity pf case described in problem 12-11 and subsequently the transient terminal voltage following a sudden disconnection from the system.

12-15. Consider the speed control loop shown in Figure 12-25. Assume that some external disturbance causes the speed ω to fluctuate as follows

$$\omega = 377 + 20(e^{-t} - e^{-2t})t \geq 0.$$

Assume the control system parameters are

$$R = 0.04 \qquad T_c = 0.0 \text{ seconds}$$

$$P_{m_1} = 1.0 \text{ (max)} \qquad T_s = 2.0 \text{ seconds}$$

$$= 0.0 \text{ (min)}$$

and the initial settings are:

$$\omega_0 = 377 \text{ rad/sec}$$

$$P_{m_0} = 0.5$$

Calculate and plot P_m and ω versus t.

APPENDIX TRANSMISSION LINE PARAMETERS

In Chapter 4 we were concerned with developing an equivalent circuit for the three phase power transmission line. We encountered certain complications that, had they been dealt with immediately, would have sidetracked us from our basic objective. We deal with these topics here to supplement Chapter 4 so that we might model practical transmission lines. This material should be considered an extension of that already presented and studied only after the reader has read and is familiar with Chapter 4.

A-1 Geometric Mean Radius and Equivalent Radius

A current carrying conductor will have current distributed throughout its cross-sectional area. Consequently, magnetic fields will exist internal to, as well as external to, the conductor. A correct consideration of this internal field requires a knowledge of the conductor cross-sectional geometry and the internal current distribution. A rigorous development for arbitrary conductor geometry and current distribution is beyond the scope of this book. The development for conductors of circular cross section and assumed uniform current distribution is reasonable and practical, and will give us insight into the basic problem. We write, from equation (4-2):

$$H_{\text{internal}} = \frac{i_{\text{enclosed}}}{2\pi\chi} \tag{A-1a}$$

$$= \frac{\pi\chi^2(i/\pi r^2)}{2\pi\chi} \tag{A-1b}$$

$$= \frac{i\chi}{2\pi r^2} \tag{A-1c}$$

where r = conductor radius.
The internal flux density is:

$$B_{\text{internal}} = \frac{\mu i \chi}{2\pi r^2} \tag{A-2}$$

where μ = permeability of the conductor material.

The flux linkage computation is somewhat tricky. A flux line positioned at χ, $0 \le \chi \le r$, links the fractional $(\pi\chi^2/\pi r^2)i$ current. Therefore

$$\lambda'_{\text{internal}} = \int_0^r \frac{\pi\chi^2}{\pi r^2} B_{\text{internal}} \, l \, d\chi \tag{A-3a}$$

$$= \int_0^r \frac{\mu i \chi^3}{2\pi r^4} l \, d\chi \tag{A-3b}$$

$$= \frac{\mu i \chi l^4}{8\pi r^4} \Big|_0^r \tag{A-3c}$$

$$= \frac{\mu i l}{8\pi} \tag{A-3d}$$

We recall for external flux linkage out to a remote point $\chi = D$:

$$\lambda'_{\text{external}} = \frac{\mu i l}{2\pi} \ln \frac{D}{r} \tag{4-5b}$$

The total flux linkage would be:

$$\lambda' = \lambda'_{\text{external}} + \lambda'_{\text{internal}} \tag{A-4a}$$

If the conductor is nonferromagnetic $\mu = \mu_0$, then:

$$\lambda' = \frac{\mu_0 i l}{2\pi} \left[\ln \frac{D}{r} + \frac{1}{4} \right] \tag{A-4b}$$

The flux linkage per unit length is

$$\lambda = \frac{\lambda'}{l} \tag{A-5a}$$

$$= \frac{\mu_0 i}{2\pi} \left[\ln \frac{D}{r} + \frac{1}{4} \right] \tag{A-5b}$$

We note the following arithmetic identity:

$$\frac{1}{4} = \ln e^{1/4} \tag{A-6}$$

Therefore (A-5b) simplifies to:

$$\lambda = \frac{\mu_0 i}{2\pi} \ln \frac{D}{re^{-1/4}} \tag{A-7}$$

Define

$$r' = re^{-1/4} \tag{A-8a}$$

$$= 0.7788r \tag{A-8b}$$

$$= \text{geometric mean radius}$$

forcing (A-7) to simplify to:

$$\lambda = \frac{\mu_0 i}{2\pi} \ln \frac{D}{r'} \tag{4-7b}$$

The quantity r' is therefore an equivalent radius of an infinitesimally thin walled conductor that could be used to account for the internal flux linkages of the original conductor. The idea of replacing the original conductor with an equivalent thin walled one is quite attractive since we can develop streamlined equations for the line inductance without bogging down in accounting for the internal field. *Gmr* values are tabulated for various power line conductors in many references (for examples, see [1] and [18] in the Bibliography at the end of Chapter 4). Example values are tabulated in Table 4-1. Note that a fair estimate for r' is about $0.8r$, if exact data is unavailable.

A common design practice for large power lines requires more than one conductor per phase, a practice referred to as "bundling." Again it is desirable to replace the "bundled" conductors with one equivalent conductor with an appropriate *gmr* and radius. To determine the equivalent *gmr* we assume equal current division among the conductors. Now we apply equations (4-9) and (4-10) to a bundle of n conductors, thinking point P sufficiently remote from the array to allow $D_{1p} \cong D_{2p} \cong \ldots \cong D_{np} = D$. We produce for flux linkages per unit length about the ith conductor.

$$\lambda_i = \frac{\mu}{2\pi} \left[\frac{i}{n} \ln \frac{D}{D_{i1}} + \frac{i}{n} \ln \frac{D}{D_{i2}} + \cdots + \frac{i}{n} \ln \frac{D}{D_{in}} \right] \tag{A-9a}$$

or

$$\lambda_i = \frac{\mu i}{2\pi} \ln \frac{D}{\sqrt[n]{D_{i1} D_{i2} \ldots D_{in}}} \tag{A-9b}$$

where $D_{ii} = r'_i = gmr$ of the ith conductor.
To obtain the average λ for the bundle:

$$\lambda = \frac{1}{n} \sum_{i=1}^{n} \lambda_i \tag{A-10a}$$

$$\lambda = \frac{1}{n} \sum_{i=1}^{n} \frac{\mu i}{2\pi} \ln \frac{D}{\sqrt[n]{D_{i1}D_{i2}\ldots D_{in}}} \tag{A-10b}$$

$$= \frac{\mu i}{2\pi} \ln \frac{D}{\sqrt[n^2]{(D_{11}D_{12}\ldots D_{1n})(D_{21}D_{22}\ldots D_{2n})\ldots(D_{n1}D_{n2}\ldots D_{nn})}} \tag{A-10c}$$

Comparing (A-10c) with (4-7b) we perceive the appropriate definition for the *gmr* of the bundle.

$$r' = \sqrt[n^2]{(D_{11}\ldots D_{1n})(D_{21}\ldots D_{2n})\ldots(D_{n1}\ldots D_{nn})} \tag{A-11}$$

where $D_{ii} = r'_i = gmr$ of the *i*th conductor.

In a similar manner an equivalent radius may be determined. Recall equation (4-31c):

$$v_{ij} = \frac{1}{2\pi\varepsilon} \sum_{k=1}^{n+1} \rho_k \ln \frac{D_{kj}}{D_{ki}} \tag{4-31c}$$

$D_{kk} = r_k = $ radius of the *k*th conductor

where we think of an array of $(n+1)$ conductors with the first n clustered together locally at the same potential and the $(n+1)$ conductor remote from the cluster. Then we conclude:

$$v_{ij} = 0 \text{ for } i, j = 1, 2, \ldots n \tag{A-12a}$$

$$v_{ij} = v_i \text{ for } i = 1, 2, \ldots n; j = 1+n \tag{A-12b}$$

Also assume

$$D_{k,n+1} \cong D \text{ for } k = 1, 2, \ldots n \tag{A-12c}$$

and

$$\rho_k = \frac{\rho}{n} \quad \text{for } k = 1, 2, \ldots n \tag{A-12d}$$

$$\rho_k = -\rho \quad \text{for } k = n+1 \tag{A-12e}$$

Equation (4-31c) becomes

$$v_i = \frac{\rho}{2\pi\varepsilon n} \left[\sum_{k=1}^{n} \ln \frac{D}{D_{ki}} - \ln \frac{r_{n+1}}{D} \right] \quad i = 1, 2, \ldots n \tag{A-13}$$

Recognizing that

$$-\rho \ln \frac{r_{n+1}}{D} = \frac{\rho}{n} \ln \frac{D^n}{r_{n+1}^n} \tag{A-14}$$

465

We write

$$v_i = \frac{\rho}{2\pi\varepsilon n}\left[\ln\frac{D^n}{D_{1i}D_{2i}\ldots D_{ni}}+\ln\left(\frac{D}{r_{n+1}}\right)^n\right] \tag{A-15}$$

To calculate $v_{av} = v$:

$$v = \frac{1}{n}\sum_{i=1}^{n}v_i \tag{A-16a}$$

$$= \frac{\rho}{2\pi\varepsilon n^2}\left[\ln\frac{D^{n2}}{(D_{11}\ldots D_{n1})(D_{12}\ldots D_{n2})\ldots(D_{n1}\ldots D_{nn})}+\ln\left(\frac{D}{r_{n+1}}\right)^{n2}\right]$$

$$\tag{A-16b}$$

or

$$v = \frac{\rho}{2\pi\varepsilon}\left[\ln\frac{D}{\sqrt[n^2]{(D_{11}\ldots D_{n1})\ldots(D_{n1}\ldots D_{nn})}}+\ln\frac{D}{r_{n+1}}\right] \tag{A-16c}$$

We recognize that the equivalent radius for the bundle must be

$$r = \sqrt[n^2]{(D_{11}\ldots D_{n1})(D_{21}\ldots D_{2n})\ldots(D_{n1}\ldots D_{nn})} \tag{A-17}$$

where $D_{ii} = r_i = $ radius of the ith conductor.

Example A-1

A phase bundle for one phase of a three phase line consists of 4 954 Mcm ACSR conductors arranged at the vertices of a square 25 cm on a side. Calculate the *gmr* and equivalent radius for the bundle.

Solution

First we will deal with the *gmr*. Look up the *gmr* for 954 Mcm ACSR conductor in Table 4-1.

From Table 4-1 we read $r_i' = 1.225$ cm. Then applying equation A-11:

$$r' = \sqrt[16]{[(1.225)(25)(25\sqrt{2})(25)]^4}$$

$$= \sqrt[4]{(1.225)(25)(25\sqrt{2})(25)}$$

$$= 12.83 \text{ cm}$$

Now find the equivalent radius:

From Table 4-1 read $r = 1.519$ cm.

Then $r = \sqrt[16]{[(1.519)(25)(25\sqrt{2})(25)]^4}$

$= \sqrt[4]{(1.519)(25)(25\sqrt{2})(25)}$

$= 13.53 \, \text{cm}$

Our results, equations (A-11) and (A-17), give us instruction as to how to calculate equivalent *gmr* and radii values for bundled phase conductors. Although not exact, this approach gives excellent results as long as the maximum spacing between members of the bundle is one tenth, or less, than the spacing between phases. For most line designs this condition is always met. The equivalent phase conductor is to be located at the geometric centroid of the bundle.

A-2 Consideration of Ground and an Arbitrary Number of Neutrals

Consider an array of parallel infinite conductors located above a semi-infinite ground plane of resistivity ρ. Of this array three conductors are considered phase conductors and are designated *a*, *b*, and *c* (actually the phase conductors may be bundled, but if so, have each been replaced by a single equivalent conductor by the method of section A-1.). The rest of the array is composed of conductors referred to as "neutral's" and are designated $n1, n2, \ldots nk$ (a total of k), numbered essentially arbitrarily. The distinguishing characteristic of the neutrals is that they are electrically connected together, and to the ground plane, at regular intervals along their length (for application to the power line, at each tower). The phase conductors are insulated from each other and ground. We shall model serial (resistive and inductive) and shunt (capacitive) effects differently and will separate our discussion into two parts.

A-2A SERIAL (INDUCTIVE AND RESISTIVE) EFFECTS

In a classical paper (see [5] in the Bibliography at the end of Chapter 4) Carson presented an approximate method for modeling the ground return path of a single infinite ground as a fictitious ground return conductor. Carson's equations, modified and applied to the 3 phase power transmission line problem by others (see [2] and [28] in the Bibliography at the end of Chapter 4), give us instruction as to where to place this equivalent ground conductor and what its resistance should be. From Chapter 4 equation (4-18), we write

$$\tilde{V}_{ii'} = l[Z]\tilde{I} \tag{A-18a}$$

where now $i = a, b, c, n1, n2, \ldots, nk, g$

and g = Carson's equivalent ground conductor.

$\tilde{V}_{ii'}$ is $(3 + k + 1) \times 1$

$\tilde{I}$ is $(3 + k + 1) \times 1$

and $[Z]$ is $(3 + k + 1) \times (3 + k + 1)$ with entries:

$$z_{ii} = R_i + j\omega l_{ii} \quad \text{ohm/m} \tag{A-18b}$$

$$z_{ij} = j\omega l_{ij}; \, i \neq j \quad \text{ohm/m} \tag{A-18c}$$

where

R_i = resistance per unit length for the ith conductor $(i \neq g)$ (A-18d)

$$R_g = 9.869 \times 10^{-7} f \text{ ohm/m} \tag{A-18e}$$

$$l_{ij} = \frac{\mu}{2\pi} \ln \frac{1}{D_{ij}} \tag{A-18f}$$

requiring that

D_{ij} = center-to-center distance in metres from the ith to the jth conductor, $i \neq j; i, j \neq g$. (A-18g)

D_{ii} = gmr in metres of the ith conductor, $i \neq g$. (A-18h)

$$D_{ig} = D_{gi} = 25.7 \sqrt[4]{\rho/f} \quad \text{metre}^{1/2} \tag{A-18i}$$

$$D_{gg} = 1 \tag{A-18j}$$

f = frequency in Hertz (A-18k)

Now subtract the last equation of the set, that is, the ground equation, from the top $3 + k$ equations, that is, the phase and neutral equations. Add to the set the constraint:

$$i_a + i_b + i_c + i_{n1} + \cdots + i_{nk} + i_g = 0 \tag{A-19a}$$

or

$$I_a + I_b + I_c + I_{n1} + \cdots + I_{nk} + I_g = 0 \tag{A-19b}$$

We get the equivalent of (4-22) for this more general situation:

$$
\begin{bmatrix} V_a \\ V_b \\ V_c \\ \hline 0 \\ \vdots \\ 0 \end{bmatrix} = \begin{bmatrix} z_{aa} & z_{ab} & z_{ac} & z_{a,n1} & \cdots & z_{ag} \\ z_{ba} & z_{bb} & z_{bc} & \vdots & & z_{bg} \\ z_{ca} & z_{cb} & z_{cc} & z_{c,n1} & \cdots & z_{cg} \\ \hline z_{n1,a} & \cdots & z_{n1,c} & z_{n1,n1} & \cdots & z_{n1,g} \\ & & & & & z_{n2,g} \\ \vdots & & \vdots & \vdots & & \vdots \\ 1 & \cdots & 1 & 1 & \cdots & 1 \end{bmatrix} \begin{bmatrix} I_a \\ I_b \\ I_c \\ \hline I_{n1} \\ \vdots \\ I_g \end{bmatrix}
$$

where

$$z_{ii} = R_i + j\frac{\omega\mu}{2\pi}\ln\frac{D_{ig}}{r_i'} \tag{A-20b}$$

$$z_{ij} = j\frac{\omega\mu}{2\pi}\ln\frac{D_{ig}}{D_{ij}} \tag{A-20c}$$

$$z_{ig} = -R_g + j\frac{\omega\mu}{2\pi}\ln\frac{1}{D_{ig}} \tag{A-20d}$$

for $i = a, b, c, n1, \ldots, nk; \ i \neq g$

We partition (A-20a) to produce:

$$\tilde{V}_{abc} = [Z_{\varnothing\varnothing}]\tilde{I}_{abc} + [Z_{\varnothing n}]\tilde{I}_n \tag{A-21a}$$

$$0 = [Z_{n\varnothing}]\tilde{I}_{abc} + [Z_{nn}]\tilde{I}_n \tag{A-21b}$$

where

$\tilde{V}_{abc}$ is 3×1

$\tilde{I}_{abc}$ is 3×1

$\tilde{I}_n$ is $(k+1) \times 1$

$[Z_{\varnothing\varnothing}]$ is 3×3

$[Z_{\varnothing n}]$ is $3 \times (k \times 1)$

$[Z_{n\varnothing}]$ is $(k+1) \times 3$

$[Z_{nn}]$ is $(k+1) \times (k+1)$

Precise identification of these matrices may be accomplished by comparing (A-20a) with (A-21a and b). We wish to eliminate $\tilde{I}_n$. From

(A-21b):

$$\tilde{I}_n = -[Z_{nn}]^{-1}[Z_{n\emptyset}]\tilde{I}_{abc}$$

Substituting into (A-21a):

$$\tilde{V}_{abc} = [Z_{\emptyset\emptyset}]\tilde{I}_{abc} - [Z_{\emptyset n}][Z_{nn}]^{-1}[Z_{n\emptyset}]\tilde{I}_{abc} \tag{A-22a}$$

$$= [[Z_{\emptyset\emptyset}] - [Z_{\emptyset n}][Z_{nn}]^{-1}[Z_{n\emptyset}]]\tilde{I}_{abc} \tag{A-22b}$$

$$= [Z_{abc}]\tilde{I}_{abc} \tag{A-22c}$$

where

$$[Z_{abc}] = [Z_{\emptyset\emptyset}] - [Z_{\emptyset n}][Z_{nn}]^{-1}[Z_{n\emptyset}] \tag{A-22d}$$

The sequence impedance matrix is then computed from equation (2-60):

$$[Z_{012}] = [T]^{-1}[Z_{abc}][T] \tag{2-60}$$

Computation of $[Z_{012}]$ is practical only by computer.

A-2B SHUNT (CAPACITIVE) EFFECTS

The effect of ground will be modeled by the image charge method. Although not exact, experience has shown that this method is a realistic and practical approximation. Imagine an image array of conductors placed below the ground plane (i.e., each image conductor is placed directly beneath its mate, equidistant from the ground plane). Each image conductor carries the negative charge of its mate. Denote the image conductors with primes (a', b', c', $n1', \ldots, nk'$; a total of $3+k$). We apply equation (4-31c) to the problem of computing the voltage between any conductor and its image:

$$v'_{ii} = \frac{1}{2\pi\varepsilon}\left[\sum_{j=a}^{nk} \rho_j \ln\frac{D_{ji'}}{D_{ji}} + \sum_{j'=a'}^{nk'} \rho'_j \ln\frac{D_{j'i'}}{D_{j'i}}\right] \tag{A-23a}$$

$$j = a, b, c, n1, n2, \ldots, nk \tag{A-23b}$$

$$j' = a', b', c', n1', n2', \ldots, nk' \tag{A-23c}$$

$$\rho'_j = -\rho_j \tag{A-23d}$$

Also

$$D_{ij'} = D_{i'j} \tag{A-24a}$$

and

$$D_{i'j'} = D_{ij} \tag{A-24b}$$

$$\therefore \quad v_{ii'} = \frac{1}{2\pi\varepsilon} \sum_{j=a}^{nk} \rho_j \ln \frac{D_{ji'}^2}{D_{ji}^2} \tag{A-25}$$

By symmetry the voltage from the ith conductor to the ground plane is one half $v_{ii'}$:

$$v_i = \tfrac{1}{2} v_{ii'} \tag{A-26a}$$

$$v_i = \frac{1}{2\pi\varepsilon} \sum_{j=a}^{nk} \rho_j \ln \frac{D_{ji'}}{D_{ji}} \tag{A-26b}$$

Since the neutrals are grounded

$$v_i = 0; \quad i = n1, \ n2, \ldots nk \tag{A-27}$$

For sinusoidal steady state both voltage and charge density may be represented as phasors. Writing equation (A-26b) in matrix form:

$$\begin{bmatrix} V_a \\ V_b \\ V_c \\ \hline 0 \\ \vdots \\ 0 \end{bmatrix} = \begin{bmatrix} f_{aa} & f_{ab} & f_{ac} & f_{a,n1} & \cdots & f_{a,nk} \\ f_{ba} & f_{bb} & f_{bc} & \vdots & & \vdots \\ f_{ca} & f_{cb} & f_{cc} & f_{c,n1} & \cdots & f_{c,nk} \\ \hline f_{n1,a} & \cdots & f_{n1,c} & f_{n1,n1} & \cdots & f_{n1,nk} \\ \vdots & & \vdots & \vdots & & \vdots \\ f_{nk,a} & \cdots & f_{nk,c} & f_{nk,n1} & \cdots & f_{nk,nk} \end{bmatrix} \begin{bmatrix} P_a \\ P_b \\ P_c \\ \hline P_{n1} \\ \vdots \\ P_{nk} \end{bmatrix}$$

where

$$f_{ij} = \frac{1}{2\pi\varepsilon} \ln \frac{D_{ji'}}{D_{ji}} \tag{A-28b}$$

$$i, j = a, b, c, n1, \ldots, nk \tag{A-28c}$$

$$i' = a', b', c', n1', \ldots, nk' \tag{A-28d}$$

$$D_{ii} = r_i \tag{A-28e}$$

We partition (A-28a) to produce:

$$\tilde{V}_{abc} = [F_{\emptyset\emptyset}]\tilde{P}_{abc} + [F_{\emptyset n}]\tilde{P}_n \tag{A-29a}$$

$$\check{O} = [F_{n\emptyset}]\tilde{P}_{abc} + [F_{nn}]\tilde{P}_n \tag{A-29b}$$

where

$$\tilde{V}_{abc} \text{ is } 3 \times 1$$

$$\tilde{P}_{abc} \text{ is } 3 \times 1$$

$$\tilde{P}_n \text{ is } k \times 1$$

$[F_{\emptyset\emptyset}]$ is 3×3

$[F_{\emptyset n}]$ is $3 \times k$

$[F_{n\emptyset}]$ is $k \times 3$

$[F_{nn}]$ is $k \times k$

Precise identification of these matrices may be accomplished by comparing (A-29a and b) with (A-28a). We wish to eliminate $\tilde{P}_n$. From (A-29b):

$$\tilde{P}_n = -[F_{nn}]^{-1}[F_{n\emptyset}]\tilde{P}_{abc} \tag{A-30a}$$

Substituting into (A-29a):

$$\tilde{V}_{abc} = [F_{\emptyset\emptyset}]\tilde{P}_{abc} - [F_{\emptyset n}][F_{nn}]^{-1}[F_{n\emptyset}]\tilde{P}_{abc} \tag{A-30b}$$

$$= [[F_{\emptyset\emptyset}] - [F_{\emptyset n}][F_{nn}]^{-1}[F_{n\emptyset}]]\tilde{P}_{abc} \tag{A-30c}$$

$$= [F_{abc}]\tilde{P}_{abc} \tag{A-30d}$$

where

$$[F_{abc}] = [F_{\emptyset\emptyset}] - [F_{\emptyset n}][F_{nn}]^{-1}[F_{n\emptyset}] \tag{A-30e}$$

We now have "absorbed" the neutrals into $[F_{abc}]$, which is 3×3. The development is now identical to that presented in Chapter 4, starting with equation (4-38) and finishing with (4-43). Again note that computation with these equations is practical only by computer except for special symmetrical cases.

A-3 Results for an Example Line

Study the line shown in Figure A-1. The following results were computer calculated.

ZABC Matrix (ohms per kilometer)

$(0.11956 + j0.63271)$	$(0.09752 + j0.30153)$	$(0.09547 + j0.25036)$
$(0.09752 + j0.30153)$	$(0.12266 + j0.63073)$	$(0.09752 + j0.30153)$
$(0.09547 + j0.25036)$	$(0.09752 + j0.30153)$	$(0.11956 + j0.63271)$

Z012 Matrix (ohms per kilometer)

$(0.31427 + j1.20099)$	$(0.01334 - j0.00968)$	$(-0.01506 - j0.00671)$
$(-0.01506 - j0.00671)$	$(0.02375 + j0.34758)$	$(-0.02995 + j0.01768)$
$(0.01334 - j0.00968)$	$(0.03028 + j0.01709)$	$(0.02375 + j0.34758)$

472

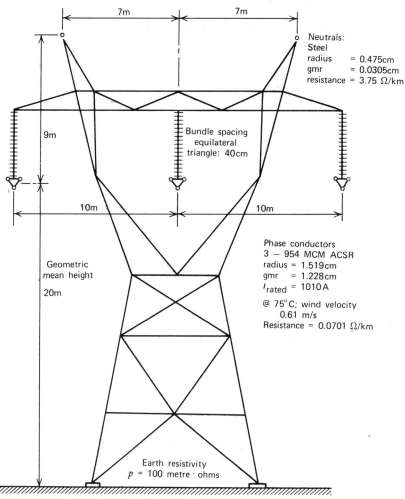

Figure A-1. Example 500 kV Power Transmission Line.

YABC Matrix (micromhos per kilometer)

$(0.00000 + j4.09490)$	$(0.00000 - j0.77167)$	$(0.00000 - j0.23867)$
$(0.00000 - j0.77167)$	$(0.00000 + j4.26525)$	$(0.00000 - j0.77167)$
$(0.00000 - j0.23867)$	$(0.00000 - j0.77167)$	$(0.00000 + j4.09490)$

Y012 Matrix (micromhos per kilometer)

$(0.00000 + j2.96367)$	$(-0.10469 + j0.06044)$	$(0.10469 + j0.06044)$
$(0.10469 + j0.06044)$	$(0.00000 + j4.74568)$	$(0.35690 - j0.20605)$
$(-0.10469 + j0.06044)$	$(-0.35690 - j0.20605)$	$(-0.00000 + j4.74568)$

473

Characteristic Impedances and Propagation Constants

$$Z_{c_0} = 642 - j82.6 \text{ ohms}$$

$$Z_{c_1} = 271 - j9.24 \text{ ohms}$$

$$Z_{c_2} = 271 - j9.24 \text{ ohms}$$

$$\gamma_0 = 0.245 + j1.902 \text{ per } 10^3 \text{ km}$$

$$\gamma_1 = 0.044 + j1.285 \text{ per } 10^3 \text{ km}$$

$$\gamma_2 = 0.044 + j1.285 \text{ per } 10^3 \text{ km}$$

For a line length of 300 km the following pi equivalent circuit parameters were calculated.

$$Z_0 = 84.4 + j343 \text{ ohms}$$

$$Z_1 = 6.78 + j102 \text{ ohms}$$

$$Z_2 = 6.78 + j102 \text{ ohms}$$

$$Y_0/2 = 3.31 + j457 \text{ microsiemens}$$

$$Y_1/2 = 0.620 + j721 \text{ microsiemens}$$

$$Y_2/2 = 0.620 + j721 \text{ microsiemens}$$

The line thermal limit is:

$$S_{3\phi\text{rated}} = V_{L\text{rated}} I_{L\text{rated}} \sqrt{3} \tag{4-77}$$

$$= (500)(3.03)\sqrt{3}$$

$$= 2624 \text{ MVA}$$

The steady state stability limit is calculated from equation (4-82b). Since A and B are functions of the line length, $P_{3\phi ss}$ is also. We compute the following values:

LINE LENGTH (KM)	$P_{3\phi ss}(MW)$
100	6706
200	3383
300	2289
400	1753
500	1441
600	1241
700	1106
800	1014
900	950
1000	908

INDEX

475